# Mechanical Engineering Series

Frederick F. Ling
*Series Editor*

**Springer**
*New York*
*Berlin*
*Heidelberg*
*Hong Kong*
*London*
*Milan*
*Paris*
*Tokyo*

# Mechanical Engineering Series

Frederick F. Ling
*Series Editor*

David G. Hull

# Optimal Control Theory
# for Applications

 Springer

David G. Hull
M.J. Thompson Regents Professor
Aerospace Engineering and
  Engineering Mechanics
The University of Texas at Austin
Austin, TX 78712-0235, USA
dghull@mail.utexas.edu

*Series Editor*
Frederick F. Ling
Ernest F. Gloyna Regents Chair in Engineering
Department of Mechanical Engineering
The University of Texas at Austin
Austin, TX 78712-1063, USA
  and
Distinguished William Howard Hart
  Professor Emeritus
Department of Mechanical Engineering,
  Aeronautical Engineering and Mechanics
Rensselaer Polytechnic Institute
Troy, NY 12180-3590, USA

Library of Congress Cataloging-in-Publication Data
Hull, David G.
    Optimal control theory for applications / David G. Hull.
      p. cm. — (Mechanical engineering series)
    Includes bibliographical references and index.
    ISBN 978-1-4419-2299-1
      1. Optimal control theory.   2. Calculus of variations.   I. Title.   II. Mechanical engineering
series (Berlin, Germany)
QA402.3.H85 2003
629.8′312—dc21                                                                    2003050495

Printed in the United States of America.

9 8 7 6 5 4 3 2 1

www.springer-ny.com

Springer-Verlag   New York  Berlin  Heidelberg
*A member of BertelsmannSpringer Science+Business Media GmbH*

To my teachers:

Angelo Miele
Jason Speyer

# Series Preface

Mechanical engineering, an engineering discipline born of the needs of the industrial revolution, is once again asked to do its substantial share in the call for industrial renewal. The general call is urgent as we face profound issues of productivity and competitiveness that require engineering solutions, among others. The Mechanical Engineering Series is a series featuring graduate texts and research monographs intended to address the need for information in contemporary areas of mechanical engineering.

The series is conceived as a comprehensive one that covers a broad range of concentrations important to mechanical engineering graduate education and research. We are fortunate to have a distinguished roster of consulting editors, each an expert in one of the areas of concentration. The names of the consulting editors are listed on page ii of this volume. The areas of concentration are applied mathematics, biomechanics, computational mechanics, dynamic systems and control, energetics, mechanics of materials, processing, thermal science, and tribology.

Austin, Texas                                                                      Frederick F. Ling

# Preface

Optimization is an area of mathematics that is concerned with finding the "best" points, curves, surfaces, and so on. "Best" is determined by minimizing some measure of performance subject to equality and inequality constraints. Points are constrained by algebraic equations; curves are constrained by ordinary differential equations and algebraic equations; surfaces are constrained by partial differential equations, ordinary differential equations, and algebraic equations.

The scope of this text is the analytical optimization of points and curves. Optimization of points is called parameter optimization here although the names function minimization and ordinary theory of maxima and minima are used elsewhere. Optimization of curves and points started in the 1600s as the calculus of variations. With the appearance of a wide variety of dynamical systems in the second half of the 1900s, optimization of the performance of dynamical systems began to be performed in a control format and was renamed optimal control theory. Calculus of variations problems are easily converted into optimal control problems. The name optimal control theory is used here, but it is emphasized that the subject matter is optimization, not control. It is also emphasized that this is a textbook on fundamentals.

This textbook is the outgrowth of teaching analytical optimization to aerospace engineering graduate students for many years. To make the material available to the widest audience, the prerequisites are limited to calculus and differential equations. Also, the example problems are not tied to any particular engineering discipline, and both geometric and dynamical system examples are used to exercise the theory.

This text is concerned with the derivation and application of the conditions for a relative minimum. The value of the performance index for an admissible comparison point or curve (defined by small changes from the minimum) is compared with that for the minimum. While Taylor series expansions are usually used to obtain the first- and second-order terms, it is shown here that the differential of calculus can be used to derive the Taylor series expansion one order at a time. Hence, the differential is used throughout this text to derive the first and second differentials. Conditions for an optimum are obtained from the vanishing of the first differential, and conditions for a minimum are obtained from the positiveness of the second differential.

My motive for writing this text was to try to create a "benchmark" text on analytical methods in parameter optimization and optimal control theory. To me, the term "benchmark" means that a uniformity exists in the terminology, the symbology, and the methodology used throughout the text. The principal unifying element of this text is the use of the differential to obtain the first- and second-order terms. This has long been done for parameter optimization. Here, differential calculus is applied to the elements of the optimal control problem: algebraic equations, ordinary differential equations, and integrals. The differential of an integral (Leibnitz' rule) contains differentials of the initial and final times as well as the integral of the differential of the integrand holding the time constant. In the calculus of variations, the differential holding the time constant is called a variation. The same terminology is employed here. Hence, the differential of an integral contains differentials and variations, but the overall operation is taking a differential. The use of the differential can be extended to optimization problems involving multiple integrals.

Before quitting the preface, several comments are made about the three parts of the text.

**Part I. Parameter Optimization.** Parameter optimization is the minimization of a scalar function of several variables subject to equality and inequality algebraic constraints. The value of the performance index at a neighboring point is compared with that at the minimum, leading to the first and second differentials and the conditions for an optimal point and the conditions for a minimal point. Unfortunately, these conditions were not named upon their discovery, but they have been known for several hundred years.

**Part II. Optimal Control Theory.** The optimal control problem is to minimize a scalar performance index subject to differential and algebraic constraints. In general, the first differential conditions that determine an optimal path are the Euler–Lagrange equations and the natural boundary conditions. If the problem does not contain the time explicitly, a first integral exists, and if the control has a discontinuity, the optimal jump is determined by the Erdmann–Weierstrass corner conditions. Before developing the second differential conditions, two tests are presented for a minimal path (Weierstrass condition and Legendre–Clebsch condition). The second differential yields the Jacobi conjugate point condition for a minimal path. When control inequality constraints are considered, the first differential conditions plus the Weierstrass condition constitute what is known as the Pontryagin minimum principle.

**Part III. Approximate Solutions.** This part contains two subjects that are related to optimization: perturbation problems and conversion of optimal control problems into parameter optimization problems. First, a perturbation problem is one in which the equations defining the problem (algebraic, differential, or optimal control) contain a small parameter. The objective is to show that the differential can be used to obtain the first-order equations, the second-order equations, and so on of the approximate analytical solution. Several methods exist for converting optimal control problems into parameter optimization problems. They depend on the use of explicit or implicit numerical integration. Differentials are used to derive the equations of condition of the integrators.

For a one semester course, the following chapters can be covered: Introduction: Chapter 1; Parameter optimization: Chapters 2 through 5. Optimal control theory: Chapters 6, 7, 9, 10, and 11; first differential parts of Chapters 13, 15, 16, and 17.

David G. Hull
Austin, Texas

# Contents

**Appendix**

# 1

# Introduction to Optimization

## 1.1   Introduction

This text is concerned with the derivation of conditions for finding the minimal performance of a nonlinear system and with the application of these conditions to the solution of selected minimization problems. In this chapter, optimization is divided into several parts depending on the complexity of the geometry involved. Then, example problems and general problems are presented for parameter optimization and optimal control theory. Next, the conversion of optimal control problems into parameter optimization problems is considered. Necessary conditions and sufficient conditions are discussed. Finally, a more detailed description of the text is presented.

## 1.2   Classification of Systems

In the order of increasing geometric complexity, systems can be classified as follows:

(a) Systems defined by algebraic equations where the unknowns are points;

(b) Systems defined by algebraic equations and ordinary differential equations where the unknowns are points and curves; and

(c) Systems defined by algebraic, ordinary differential, and partial differential equations where the unknowns are points, curves, surfaces, and so on.

Note that each time the defining system is made more difficult, all previous systems are included. While points, curves, surfaces, and so on are geometric quantities, the corresponding mathematical quantities are parameters, functions of one variable, and functions of multiple variables. These descriptions are used interchangeably.

Only the first two types of systems are considered in this text. Optimization of systems of type (a) is called parameter optimization and is covered in Chapters 2 through 5. Optimization of systems of type (b) is called optimal control theory and is discussed in Chapters 6 through 17. Optimization problems for these kinds of systems are described in the next two sections.

## 1.3   Parameter Optimization

The problem of finding the optimal performance of a system described by algebraic equations is called parameter optimization, function minimization, ordinary theory of maxima and minima, or nonlinear programming.

### 1.3.1   Distance Problem

An example of a parameter optimization problem is that of minimizing the distance from the origin to the parabola shown in Fig. 1.1. The quantity $a$ is fixed, but its value can be changed to move the parabola relative to the origin. Formally, the optimization problem is stated as follows: Find the point $x, y$ that minimizes the performance index

$$J = \sqrt{x^2 + y^2} \tag{1.1}$$

subject to the equality constraint that the point lie on the parabola

$$y - x^2 - a = 0. \tag{1.2}$$

Once the solution of the optimization problem has been found, the quantity $a$ can be changed to explore its effect on the solution. This problem has some interesting results.

This is a constrained optimization problem. However, in this particular problem, the constraint can actually be solved for $y$ which can then be

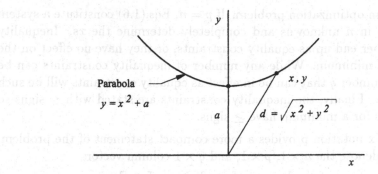

Figure 1.1: Minimum Distance to a Parabola

eliminated from the performance index to form the unconstrained problem of minimizing

$$J = \sqrt{x^2 + (x^2 + a)^2}. \tag{1.3}$$

Relative to the constrained problem, the restriction of the parabola to the first quadrant leads to the following inequality constraints:

$$x \geq 0, \quad y \geq 0. \tag{1.4}$$

## 1.3.2  General Parameter Optimization Problem

The general parameter optimization problem is to find the point $x_1, \ldots, x_n$ that minimizes the performance index

$$J = \phi(x_1, x_2, \ldots, x_n), \tag{1.5}$$

subject to the equality constraints

$$\psi_j(x_1, x_2, \ldots, x_n) = 0, \quad j = 1, \ldots, p, \tag{1.6}$$

and the inequality constraints

$$\theta_k(x_1, x_2, \ldots, x_n) \leq 0, \quad k = 1, \ldots, q. \tag{1.7}$$

Note that the performance index is a scalar function of the $n$ parameters. Furthermore, the number of equality constraints must be such that $p < n$ in order to have an optimization problem. If $p = n$, Eqs.(1.6) constitute a system of $n$ equations in $n$ unknowns and completely determine the $x$s. Inequality constraints either end up as equality constraints, or they have no effect on the location of the minimum. While any number of inequality constraints can be imposed, the number $\bar{q}$ that can be in effect as equality constraints will be such that $p + \bar{q} < n$. Finally, the inequality constraints are stated with $\leq$ signs so that conditions for a minimum have $\geq$ signs.

Matrix notation provides a more compact statement of the problem. If $x$, $\psi$, and $\theta$ denote the $n \times 1$, $p \times 1$, and $q \times 1$ column vectors

$$x = \begin{bmatrix} x_1 \\ \vdots \\ x_n \end{bmatrix}, \quad \psi = \begin{bmatrix} \psi_1 \\ \vdots \\ \psi_p \end{bmatrix}, \quad \theta = \begin{bmatrix} \theta_1 \\ \vdots \\ \theta_q \end{bmatrix}, \tag{1.8}$$

the parameter optimization problem becomes the following: Find the point $x$ which minimizes the performance index

$$J = \phi(x), \tag{1.9}$$

subject to the equality constraints

$$\psi(x) = 0, \tag{1.10}$$

and the inequality constraints

$$\theta(x) \leq 0. \tag{1.11}$$

It is possible to introduce a control-oriented notation for parameter optimization problems. Theoretically, the $p$ equality constraints (1.10) can be solved for $p$ $x$s in terms of the remaining $n - p$ $x$s. The former are dependent variables or, in optimal control terminology, state variables. The latter are independent variables, or control variables. Inequality constraints complicate the situation because once an inequality constraint becomes active (becomes an equality constraint), the number of independent controls is reduced. For parameter optimization to be presented in such a way that it can become an integral part of optimal control theory, it is not convenient to use an optimal control representation for the parameters.

## 1.4   Optimal Control Theory

The first path optimization problems were geometric in nature and were called calculus of variations problems. In time, the need to optimally control the behavior of dynamical systems has led to a reformulation of the problem. Both classes of problems can be solved within the framework of either class of problems.

Optimal control theory is concerned with finding curves and their end points which minimize a scalar performance index subject to point and path constraints. A natural division in the development of the theory is to consider fixed final time problems first and then consider free final time problems. These problems are so-named because the monotonically increasing time is chosen as the variable of integration. In practice, any variable can be used as the variable of integration, but if it is not monotonic, the problem is much more difficult to solve.

## 1.4.1  Distance Problem

The problem formulated in Section 1.3 can be posed as an optimal control problem by seeking the curve that minimizes the distance to the parabola. Mathematically, the distance between two points along an arbitrary curve $y(x)$ is given by

$$d = \int_{x_0}^{x_f} \sqrt{1 + y'^{\,2}} \, dx, \qquad (1.12)$$

where the prime denotes a derivative with respect to $x$. Then, the formal statement of the optimization problem is to find the curve $y(x)$ that minimizes the performance index

$$J = \int_{x_0}^{x_f} \sqrt{1 + y'^{\,2}} \, dx, \qquad (1.13)$$

subject to the prescribed initial conditions

$$x_0 = 0, \quad y_0 = 0, \qquad (1.14)$$

and the prescribed final condition

$$y_f - x_f^{\,2} - a = 0. \qquad (1.15)$$

Historically, this is known as a calculus of variations problem. However, it can be converted into an optimal control problem by defining the control to be the slope of the curve, that is, $u = y'$, and the state to be $y$. The optimal control problem is then to find the control history $u(x)$ that minimizes the performance index

$$J = \int_{x_0}^{x_f} \sqrt{1 + u^2} \, dx, \qquad (1.16)$$

subject to the differential constraint

$$y' = u \tag{1.17}$$

and the prescribed boundary conditions (1.14) and (1.15).

Because the final value of the variable of integration $x$ is not fixed, this problem is a "free final time" problem. An example of a "fixed final time" problem is to minimize the distance from the origin to a fixed final point.

If it is assumed that the optimal curve is a straight line, that is, $y' = $ Const, the optimal control problem reduces to the parameter optimization problem of Section 1.3.

## 1.4.2   Acceleration Problem

Suppose that it is possible to control the acceleration of a car moving in a straight line, and suppose that it is desired to move the car from one position and velocity to another position and velocity. If $x$ denotes the position, the motion of the car is governed by the differential equation $\ddot{x} = a$ where $a$ is the controlled acceleration. Since problems involving derivatives of all orders exist, it is customary to write the system equations in first order. Hence, if $v = \dot{x}$ denotes velocity, the car equation can be rewritten as a system of two first-order equations, that is, $\dot{x} = v$ and $\dot{v} = a$.

An example of a fixed final time problem is to make the transfer in a given time while minimizing the control energy. The formal statement of the problem is as follows: Find the acceleration history $a(t)$ that minimizes the performance index

$$J = \frac{1}{2} \int_{t_0}^{t_f} a^2 \, dt, \tag{1.18}$$

subject to the car equations of motion

$$\begin{aligned} \dot{x} &= v \\ \dot{v} &= a, \end{aligned} \tag{1.19}$$

the prescribed initial conditions (subscript $s$ denotes a specific value)

$$t_0 = t_{0_s}, \quad x_0 = x_{0_s}, \quad v_0 = v_{0_s}, \tag{1.20}$$

and the prescribed final conditions

$$t_f = t_{f_s}, \quad x_f = x_{f_s}, \quad v_f = v_{f_s}. \tag{1.21}$$

In the differential constraints (1.19), the differentiated variables $x$ and $v$ are called states, and the undifferentiated variable $a$ is called a control.

An example of a free final time problem is to minimize the transfer time. However, because the car equations of motion (1.19) are linear in the controlled acceleration, the optimal control is infinite acceleration at $t_0$ followed by zero acceleration followed by infinite negative acceleration (deceleration) at $t_f$. To prevent infinite accelerations that cannot be achieved physically, bounds must be imposed on the acceleration. Hence, the minimum final time problem is stated as follows: Find the control history $a(t)$ that minimizes the performance index

$$J = t_f \tag{1.22}$$

subject to the differential constraints (1.19), the prescribed initial conditions (1.20), the prescribed final conditions

$$x_f = x_{f_s}, \quad v_f = v_{f_s}, \tag{1.23}$$

and the acceleration limits

$$a \leq a_{\max}, \quad a_{\min} \leq a. \tag{1.24}$$

Equations (1.24) are control inequality constraints because they explicitly involve the control. An example of a state inequality constraint is to require the car not to exceed the speed limit, that is, $v \leq v_{\max}$. This constraint does not explicitly contain the control.

Since the states appear in differentiated form, it is possible to require them to be continuous (no jumps). In fact, in a physical problem such as this, position and velocity must be continuous. On the other hand, because the control is not differentiated, it cannot be required to be continuous, and jumps in the control can occur. For example, it is mathematically possible for the control to switch from $a_{\max}$ to $a_{\min}$ instantaneously.

Having to determine the times at which the control switches from one boundary to another may not be desirable. An approximate solution can be obtained by adding a term to the performance index that penalizes the final time if the acceleration becomes too large. The performance index can be written as

$$J = t_f + k \int_{t_0}^{t_f} a^2 \, dt, \tag{1.25}$$

where $k$ is a positive constant, and the control no longer appears linearly. The optimal control problem is now to minimize the performance index (1.25) subject to the system dynamics (1.19), the prescribed initial conditions (1.20), and the prescribed final conditions (1.23). After the optimization problem is solved, the constant $k$ is chosen so that the control inequality constraints (1.24) are satisfied.

## 1.4.3  Navigation Problem

As another example of an optimal control problem, consider the problem of flying a constant speed aircraft in a constant speed crosswind from one point to another in minimum time. Figure 1.2 is a picture of this simplified form of Zermelo's problem. Here, $x$ and $y$ are Cartesian coordinates; $V$ is the constant speed of the aircraft relative to the air; $\theta$ is the controllable orientation of the aircraft velocity vector relative to the $x$-axis; $w$ is the speed of the air relative to the ground. Hence, the equations of motion of the aircraft relative to the ground are $\dot{x} = V \cos \theta$ and $\dot{y} = V \sin \theta + w$, where the dot denotes a derivative with respect to time. Furthermore, the aircraft is to be flown from the initial point $t_0 = 0$, $x_0 = 0$, $y_0 = 0$ to the final point $x_f = 1$, $y_f = 0$. Note that $t_f$ is free, as it must be to have a minimum time problem.

The optimal control problem is stated as follows: Find the control history $\theta(t)$ that minimizes the final time

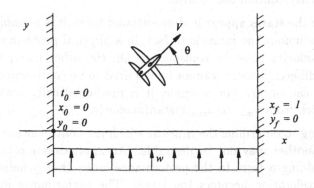

Figure 1.2: Navigation Problem

$$J = t_f, \tag{1.26}$$

subject to the differential constraints

$$\dot{x} = V \cos \theta \tag{1.27}$$
$$\dot{y} = V \sin \theta + w, \tag{1.28}$$

the prescribed initial conditions

$$t_0 = 0, \quad x_0 = 0, \quad y_0 = 0, \tag{1.29}$$

and the prescribed final conditions

$$x_f = 1, \quad y_f = 0. \tag{1.30}$$

Here, $t$ is the variable of integration; the differentiated variables $x(t)$ and $y(t)$ are the states; and the undifferentiated variable $\theta(t)$ is the control.

If there is no condition imposed on $y_f$, the differential equation (1.28) has no effect on the problem and can be ignored. It can be used after the optimal control problem is solved to find the value of $y_f$ where the aircraft crosses the line $x_f = 1$.

Note that this is a free final time problem because the final value of the variable of integration is not prescribed. However, it can be converted into a "fixed final time" problem by introducing the time transformation

$$\tau = \frac{t}{t_f}. \tag{1.31}$$

The optimal control problem is now stated as follows: Find the control history $\theta(\tau)$ and the parameter $t_f$ that minimize the performance index

$$J = t_f, \tag{1.32}$$

subject to the differential constraints

$$\frac{dx}{d\tau} = t_f V \cos \theta \tag{1.33}$$

$$\frac{dy}{d\tau} = t_f (V \sin \theta + w), \tag{1.34}$$

the prescribed initial conditions

$$\tau_0 = 0, \quad x_0 = 0, \quad y_0 = 0, \tag{1.35}$$

and the prescribed final conditions

$$\tau_f = 1, \quad x_f = 1, \quad y_f = 0. \tag{1.36}$$

Here, the final value of the variable of integration $\tau$ is fixed. The variables $x(\tau)$ and $y(\tau)$ are the states; $\theta(\tau)$ is the control. Unfortunately, this problem now has the complexity that $t_f$ is an unknown parameter which can be converted into a state by adding the differential constraint $dt_f/d\tau = 0$, but the problem then has a free initial state $(t_f)$.

The navigation problem can also be converted into a "fixed final time" problem by choosing $x$ to be the variable of integration. Because the aircraft is expected to move directly toward the final point, $x$ is expected to be monotonic. The optimal control problem is now stated as follows: Find the control history $\theta(x)$ that minimizes the performance index

$$J = t_f, \tag{1.37}$$

subject to the differential constraints

$$\frac{dt}{dx} = \frac{1}{V \cos \theta} \tag{1.38}$$

$$\frac{dy}{dx} = \frac{V \sin \theta + w}{V \cos \theta}, \tag{1.39}$$

the prescribed initial conditions

$$x_0 = 0, \quad t_0 = 0, \quad y_0 = 0, \tag{1.40}$$

and the prescribed final conditions

$$x_f = 1, \quad y_f = 0. \tag{1.41}$$

Here, the final value of the variable of integration $x$ is fixed. The variables $t(x)$ and $y(x)$ are the states; $\theta(x)$ is the control.

This problem can also be converted into one that has no states. If the integration of the state equations (1.38) and (1.39) is carried out subject to the boundary conditions (1.40) and (1.41), the navigation problem can be stated as follows: Find the control history $\theta(x)$ that minimizes the performance index

$$J = \int_{x_0}^{x_f} \frac{1}{V \cos \theta} \, dx, \tag{1.42}$$

subject to the prescribed boundary conditions

$$x_0 = 0, \quad x_f = 1 \tag{1.43}$$

and the integral constraint

$$\int_{x_0}^{x_f} \frac{V \sin \theta + w}{V \cos \theta} \, dx = 0. \tag{1.44}$$

If $y_f$ is free, the integral constraint can be disregarded.

## 1.4.4   General Optimal Control Problem

To derive necessary conditions and sufficient conditions for a minimum, it is necessary to pose a somewhat general optimal control problem. In matrix format, the optimal control problem to which most of this text is addressed is the following: Find the control history $u(t)$ that minimizes the performance index

$$J = \phi(t_f, x_f) + \int_{t_0}^{t_f} L(t, x, u) \, dt, \tag{1.45}$$

subject to the system dynamics

$$\dot{x} = f(t, x, u), \tag{1.46}$$

the prescribed initial conditions

$$t_0 = t_{0_s}, \quad x_0 = x_{0_s}, \tag{1.47}$$

and the prescribed final conditions

$$\psi(t_f, x_f) = 0. \tag{1.48}$$

Here, $x$ denotes the $n \times 1$ state vector; $u$ denotes the $m \times 1$ control vector; $\phi$ and $L$ are scalars; $f$ is an $n \times 1$ vector; and $\psi$ is a $p \times 1$ vector, where $p \leq n+1$. Historically, this problem is known as the Bolza problem. If $\phi = 0$, it is called a Lagrange problem, and if $L = 0$, a Mayer problem.

Other point and path constraints can be included. For example, a final point inequality constraint

$$\theta(t_f, x_f) \leq 0 \tag{1.49}$$

can be imposed, as can initial or internal point constraints

$$\alpha(t_c, x_c) = 0, \quad \beta(t_c, x_c) \leq 0, \tag{1.50}$$

where $t_0 \leq t_c < t_f$. Additional path constraints include integral constraints,

$$\int_{t_0}^{t_f} M(t, x, u)\, dt = K, \quad \int_{t_0}^{t_f} \bar{M}(t, x, u)\, dt \leq \bar{K}, \tag{1.51}$$

where $K$ and $\bar{K}$ are constants, control constraints

$$C(t, x, u) = 0, \quad \bar{C}(t, x, u) \leq 0, \tag{1.52}$$

and state constraints

$$S(t, x) = 0, \quad \bar{S}(t, x) \leq 0. \tag{1.53}$$

All of the constraints are in matrix form. Note that the inequalities are stated with $\leq$ signs. This is done to make all conditions for a minimum have a $\geq$ sign.

The minimum time acceleration problem can be put in the general form by making the following definitions:

$$
\begin{aligned}
x_1 &= x, \quad x_2 = v, \quad u = a \\
\phi &= t_f, \quad L = 0, \quad f_1 = x_2, \quad f_2 = u \\
\psi_1 &= x_{1_f} - x_{1_{f_s}}, \quad \psi_2 = x_{2_f} - x_{2_{f_s}} \\
\bar{C}_1 &= u - u_{\max} \leq 0, \quad \bar{C}_2 = u_{\min} - u \leq 0.
\end{aligned}
\tag{1.54}
$$

It is interesting to note that the two control inequality constraints can be rewritten as one, that is,

$$\bar{C} = -(u - u_{\min})(u_{\max} - u) \leq 0. \tag{1.55}$$

The navigation problem can be put in the form of the general optimal control problem by defining the following matrix elements:

$$
\begin{aligned}
x_1 &= x, \quad x_2 = y, \quad u = \theta \\
\phi &= t_f, \quad L = 0, \quad f_1 = V \cos\theta, \quad f_2 = V \sin\theta + w \\
\psi_1 &= x_f - 1, \quad \psi_2 = y_f.
\end{aligned}
\tag{1.56}
$$

### 1.4.5 Conversion of an Optimal Control Problem into a Parameter Optimization Problem

An optimal control problem can be approximated by a parameter optimization problem by requiring the control to be piecewise linear (or piecewise constant, piecewise cubic, etc.). For the navigation problem in the form of Eqs.(1.32) through (1.36), the steering angle $\theta(\tau)$ can be replaced by the piecewise linear function $\theta(\tau)$ shown in Fig. 1.3. Then, the unknowns become the nodes $\theta_1, \ldots, \theta_5$ of the piecewise linear control and the final time $t_f$. In matrix notation, the parameter vector

$$X = [t_f \ \ \theta_1 \ \ \theta_2 \ \ \theta_3 \ \ \theta_4 \ \ \theta_5]^T \tag{1.57}$$

contains all of the unknown parameters.

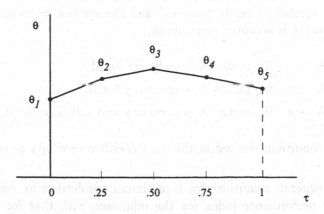

Figure 1.3: Piecewise Linear Control

Given values for the elements of $X$, the equations of motion (1.33) and (1.34) can be integrated to give $x_f = x_f(X)$, $y_f = y_f(X)$. The minimum time problem is then restated as follows: Find the parameter vector $X$ that minimizes the performance index

$$J = X_1, \tag{1.58}$$

subject to the equality constraints

$$x_f(X) - 1 = 0 \tag{1.59}$$

$$y_f(X) = 0. \tag{1.60}$$

This is now a parameter optimization problem with the unknowns being the elements of the parameter vector $X$.

The general optimal control problem defined in Section 1.4.4 can be converted into a parameter optimization problem. Solution of these problems is numerical in nature, and the solution process is called nonlinear programming.

## 1.5   Necessary Conditions and Sufficient Conditions

One objective of this text is to derive conditions for minimal points and curves. These conditions fall in the categories of necessary conditions and sufficient conditions. As a review, assume that A and B are mathematical statements. Then, if the symbol $\Rightarrow$ reads "implies" and the symbol $\Leftarrow$ reads "is implied by," the following is accepted terminology:

> A  $\Rightarrow$  B   means A is sufficient for B.
>
> A  $\Leftarrow$  B   means A is necessary for B.
>
> A  $\Leftrightarrow$  B   means A is necessary and sufficient for B.

Each of the conditions derived in this text is either necessary or sufficient but not both.

In general, conditions for a minimum are derived by comparing the value of the performance index for the minimum with that for an admissible comparison point or path. Necessary conditions are derived by considering particular admissible comparison points or paths. They must be satisfied by the minimum and are used to identify candidates for the minimum. On the other hand, sufficient conditions are derived for arbitrary admissible comparison points or paths. If satisfied, a candidate for a minimum becomes a minimum.

## 1.6   Text Organization

This text is divided into three parts. Part I, Chapters 2 through 5, contains parameter optimization. Part II, Chapters 6 through 17, covers optimal control theory. Part III, Chapters 18 through 21, discusses approximate solutions. At the beginning of each part, a detailed description is presented.

An underlying theme in this text is the use of differentials to make Taylor series expansions. In Part I, differentials are reviewed, and it is shown that Taylor expansions can be obtained one term at a time by taking differentials. Then, the first and second differentials are used to obtain the optimality conditions for parameter optimization. Differentials are extended in Part II to differential equations and integrals. The results are applied to the derivation of the first and second differentials for optimal control problems. In Part III, differentials are used to derive the various-order equations for perturbation problems involving algebraic, differential, and optimal control equations. Also, it is used to derive the equations of condition for the creation of numerical integration formulas which are used to convert optimal control problems into parameter optimization problems.

# Part I. Parameter Optimization

Parameter optimization is the minimization of a scalar function of several variables subject to a number of equality constraints and inequality constraints. The unknown quantities are points. Optimal points can be classified as interior optimal points or boundary optimal points. The latter are those optimal points that satisfy inequality constraints as equality constraints. Differential calculus is used to derive the first and second differentials. In general, the first differential must vanish, and if the second differential is positive, the optimal point is a minimal point. From these equations, the necessary conditions and sufficient conditions for a minimal point are derived.

In Chapter 2, it is first shown that a Taylor series can be obtained on a term-by-term basis by taking differentials. Then, the minimization of a function of unconstrained (independent) variables is investigated. The first differential conditions for a minimal point are that the derivative of the performance index with respect to each variable must vanish. The second differential condition for a minimum is the following: If the second derivative matrix of the performance index is positive definite, the optimal point is a minimum. It is also shown that if the second derivative matrix is positive semidefinite, several orders of differentials may be involved in determining whether or not an optimal point is a minimum.

In Chapter 3, equality constraints are added to the problem. Two approaches are used to derive the first and second differentials: the direct approach and the Lagrange multiplier approach. The direct approach is to derive the first and second differentials of the performance index and the constraints. Then, the dependent differentials are eliminated to form the first differential and the second differential in terms of the independent differentials only. In the multiplier approach, the constraints are multiplied by undetermined constants (multipliers) and added to the performance index to form the augmented performance index. Then, the first and second differentials are obtained by taking differentials of the augmented performance index. The multipliers are chosen to remove the dependent differentials from the first differential. In the second differential, the relationships between dependent differentials and independent differentials must still be determined.

Inequality constraints are added in Chapter 4. They are converted into equality constraints through the use of slack variables. The general parameter optimization problem is defined at the end of this chapter.

Up to this point, indicial notation is used in order to concentrate on optimization. Matrix notation, algebra, and calculus are introduced in Chapter 5 and used from then on. The optimality conditions for both the unconstrained and constrained minimization problems are rederived in matrix notation.

The minimization of a scalar performance index subject to equality constraints using Lagrange multipliers has been known since the time of Euler and Lagrange. Inequality constraints have been used for some time, but converting an inequality constraint into an equality constraint by setting it equal to the square of a new real variable was reported in Ref. Ha, p. 150, without a reference to previous work. While these new variables are being called slack variables, perhaps they should be called Hancock variables.

# 2

# Unconstrained Minimization

## 2.1  Introduction

This chapter is begun by discussing differentials and how they can be used
to derive Taylor series expansions one term at a time. Parameter optimization begins by considering the minimization of a function of unconstrained
(independent) variables. First, the conditions for minimizing a function of one
variable are derived. It is shown that necessary conditions for a minimum are
that the first differential of the performance index must vanish and that the
second differential must be nonnegative. Given a point that satisfies the necessary conditions, the sufficient condition for a minimum is that the second
differential be positive. A function of two independent variables is considered
next, and then the generalization to $n$ variables follows.

## 2.2  Taylor Series and Differentials

In this section, it is shown that the Taylor series can be developed on a term-
by-term basis by taking differentials. A Taylor series is a power series with the
coefficients given by derivatives at a point. For the function (Fig. 2.1)

$$y = y(x) \tag{2.1}$$

the power series at $x_1$ is given by

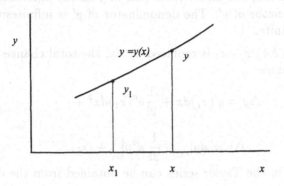

Figure 2.1: Neighboring Points

$$y = y_1 + a_1(x - x_1) + a_2(x - x_1)^2 + \cdots. \tag{2.2}$$

Then,

$$
\begin{aligned}
a_1 &= \left.\frac{dy}{dx}\right|_{x_1} = y'(x_1) \\[2mm]
a_2 &= \left.\frac{1}{2!}\frac{d^2y}{dx^2}\right|_{x_1} = \frac{1}{2!}y''(x_1)
\end{aligned}
\tag{2.3}
$$

$$\vdots$$

so that

$$y = y(x_1) + y'(x_1)(x - x_1) + \frac{1}{2!}y''(x_1)(x - x_1)^2 + \cdots. \tag{2.4}$$

This is the Taylor series expansion of the function $y = y(x)$ about the point $x_1$.

In calculus (Refs. Co and Ta), the differential of the function $y = y(x)$ satisfies the relation

$$dy = y'(x)dx, \tag{2.5}$$

where $dx$ is an independent differential and $dy$ is a dependent differential. There are higher-order differentials defined by

$$d^2y = y''(x)dx^2 \tag{2.6}$$

and so on. For $d$ to behave as a differential operator and for Eq. (2.6) to be the differential of Eq.(2.5), differentials of independent differentials must be zero, that is, $d(dx) = 0$.

In Eq.(2.5) there is no limit on the size of $dx$, but in optimization it is assumed to be small, but finite. Note that in $y'dx$ the differential $dx$ cannot cancel the denominator of $y'$. The denominator of $y'$ is infinitesimal, and the differential $dx$ is finite.

If in Eq.(2.4) $x - x_1$ is replaced by $dx$, the total change in $y$, that is, $\Delta y = y - y_1$, becomes

$$\Delta y = y'(x_1)dx + \frac{1}{2!}y''(x_1)dx^2 + \cdots \tag{2.7}$$

or

$$\Delta y = dy\big|_{x_1} + \frac{1}{2!}d^2y\big|_{x_1} + \cdots. \tag{2.8}$$

Hence, each term in the Taylor series can be obtained from the differential of the previous term.

Two changes are now made in the nomenclature for future derivations. First, since matrices are used for multiple variables, the derivative $y'$ is replaced by $y_x$ which can be used for $x$ being a scalar or a vector. Second, the point $x_1$ at which the expansion is made is replaced by $x$. With these changes, Eqs. (2.7) and (2.8) become

$$\Delta y = y_x(x)dx + \frac{1}{2!}y_{xx}(x)dx^2 + \cdots \tag{2.9}$$

and

$$\Delta y = dy + \frac{1}{2!}d^2y + \cdots. \tag{2.10}$$

Now, in $y_x(x)$, the subscript $x$ denotes the general $x$, and the argument $x$ denotes the point about which the expansion is made. In this way, each term of the Taylor series expansion can be obtained in one step rather than two. This nomenclature is no different than that used in dynamics where the general coordinates (two dimensions) are $x, y$ and the position of the center of gravity is called $x, y$.

In general, the differential satisfies all the formulas associated with taking a derivative. In particular, the differential of a function of multiple variables, that is, $f(x, y)$, is given by

$$df = f_x(x, y)dx + f_y(x, y)dy. \tag{2.11}$$

Note that the ordinary derivative is obtained from the limiting process

$$\lim_{dx \to 0} \frac{\Delta y}{dx} = \frac{dy}{dx}. \tag{2.12}$$

This means that for small $dx$ (but finite) the series (2.10) is a descending series. Hence, we can define differentials $dx, dy, d^2y, \ldots$ in terms of stopping short of the limit, but making $dx$ sufficiently small that higher-order terms are negligible with respect to a previous term. In this way all of the derivative formulas hold, but $dx$ is not zero.

Consider the implicit representation of the curve $y = y(x)$, that is,

$$f(x, y) = 0, \tag{2.13}$$

where $x$ is assumed to be the independent variable and $y$ is the dependent variable. The first differential of Eq.(2.13) is given by

$$df = f_x dx + f_y dy = 0, \tag{2.14}$$

where $dx$ is the independent differential. Since differentials of independent differentials are zero, the differential of the first differential becomes

$$d^2 f = f_{xx} dx^2 + 2 f_{xy} dx dy + f_{yy} dy^2 + f_y d^2 y = 0. \tag{2.15}$$

Higher-order differentials are obtained in the same way. Note that all the derivatives are evaluated at $x, y$.

Equation (2.14) can be solved for $dy$ as

$$dy = -\frac{f_x}{f_y} dx. \tag{2.16}$$

Next, Eq.(2.15) can be solved for $d^2y$ and combined with Eq. (2.14) to obtain

$$d^2 y = -\frac{1}{f_y} \left[ f_{xx} - 2 f_{xy} \frac{f_x}{f_y} + f_{yy} \left( \frac{f_x}{f_y} \right)^2 \right] dx^2. \tag{2.17}$$

Hence, while Eq.(2.13) may not be solvable for $y$ in terms of $x$, $\Delta y$ can be solved in terms of $dx$ for neighboring points.

To check the validity of these results, assume that

$$f = y - y(x) \tag{2.18}$$

which makes the implicit form of a curve equal to the explicit form (2.1). Then,

$$f_x = -y_x, \quad f_y = 1, \quad f_{xx} = -y_{xx}, \quad f_{xy} = 0, \quad f_{yy} = 0, \tag{2.19}$$

and Eqs.(2.16) and (2.17) become Eqs.(2.5) and (2.6), as they should.

In summary, the differential of a variable is a small but finite quantity; the differential satisfies all the formulas for taking a derivative; the differential of an independent differential is zero; and Taylor series can be developed one term at a time by taking differentials.

## 2.3   Function of One Variable

As an introduction to the terminology of parameter optimization, consider the function $\phi$ of a single variable $x$ shown in Fig. 2.2.

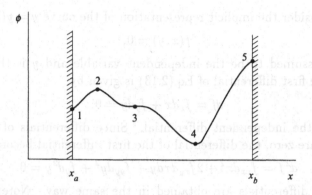

Figure 2.2:  Examples of Optimal Points

The optimal points of $\phi$, labeled 1 through 5, can be divided into the following two groups:

Interior optimal points

(2)  relative maximum

(3)  saddle point, neither a maximum nor a minimum

(4)  relative minimum, absolute minimum in $[x_a,\ x_b]$

Boundary optimal points

(1)  relative minimum

(5)  relative maximum, absolute maximum $[x_a,\ x_b]$.

It should be mentioned that there exist other types of optimal points such as those illustrated in Fig. 2.3. The distinguishing characteristic of these points is that the derivative of $\phi$ is not defined at the optimal point. Such points cannot be determined analytically and are not considered any further. In this text, it is assumed that all derivatives needed to develop the theory exist.

In the development of conditions for determining an optimal point, it is observed that, if $\phi$ has a maximal point, $-\phi$ has a minimal point (see

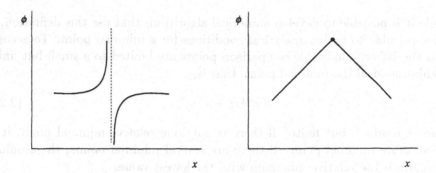

Figure 2.3: Nonanalytic Optimal Points

Fig. 2.4). Hence, it is only necessary to derive conditions for a minimal point. Conditions for a maximal point are obtained by replacing $\phi$ by $-\phi$.

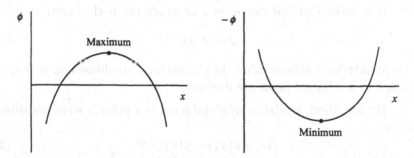

Figure 2.4: Converting a Maximum into a Minimum

**Note.** In writing $\phi = \phi(x)$, $x$ denotes the arbitrary independent variable. However, in what follows, the symbol $x$ is also used to denote the minimal point. This is done to avoid having to use a lot of subscripted variables in the development and application of optimality conditions.

Let $x$ denote a minimal point and $x_*$ denote an admissible comparison point, which by definition satisfies all of the constraints. The mathematical definition of an absolute minimum $x$ in the interval $[x_a, x_b]$ is given by

$$\Delta\phi = \phi(x_*) - \phi(x) > 0, \tag{2.20}$$

where an admissible comparison point $x_*$ is defined as

$$x_a \leq x_* \leq x_b, \quad x_* \neq x. \tag{2.21}$$

While it is possible to develop numerical algorithms that use this definition, it is not possible to derive analytical conditions for a minimal point. To accomplish the latter, admissible comparison points are limited to a small but finite neighborhood of the minimal point, that is,

$$0 < |x_* - x| < \varepsilon, \qquad (2.22)$$

where $\varepsilon$ is small, but finite. If there is only one relative minimal point, it is the absolute minimal point. If there are several minimal points, the absolute minimum is the relative minimum with the lowest value.

In the light of the above paragraph, interior minimal points can be imbedded in a neighborhood whereas boundary minimal points must be on either side of the neighborhood. Because of this difference, interior minimal points are discussed here, and boundary minimal points are discussed in the chapter on inequality constraints (Chapter 4).

It is desired to find the point $x$ at which the performance index

$$J = \phi(x) \qquad (2.23)$$

takes on an interior minimal value. In this section, conditions on $\phi$ that must be satisfied at a minimal point are derived.

By definition, a relative minimal point is a point $x$ which satisfies the condition

$$\Delta J = \phi(x_*) - \phi(x) > 0 \qquad (2.24)$$

for all admissible comparison points $x_*$ in the small but finite neighborhood of $x$, that is,

$$0 < |x_* - x| < \varepsilon. \qquad (2.25)$$

This equation can be rewritten as (see Fig. 2.5)

$$x_* = x + dx, \quad 0 < |dx| < \varepsilon, \qquad (2.26)$$

where $dx$ is a small but finite displacement that can be positive or negative, that is, $x$ is an interior optimal point so that admissible comparison points exist to the right and to the left of $x$. At this point, the condition for a minimum becomes

$$\Delta J = \phi(x + dx) - \phi(x) > 0 \qquad (2.27)$$

regardless of the choice of the admissible comparison point, that is, regardless of the choice of $dx$.

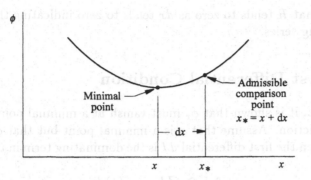

Figure 2.5: Definition of $dx$

By expanding Eq.(2.27) in a Taylor series, the following relation is obtained:

$$\Delta J = dJ + \frac{1}{2!}d^2 J + \ldots > 0, \tag{2.28}$$

where

$$dJ = \phi_x(x)dx \tag{2.29}$$

$$d^2 J = \phi_{xx}(x)dx^2. \tag{2.30}$$

The first and second differentials can be obtained by taking differentials and recalling that the differential of an independent differential ($dx$) is zero.

If $\phi(x)$ and all of its derivatives $\phi^{(k)}(x)$ are continuous, the series is a descending series. In other words, as long as $\phi$ and its derivatives are finite, a $dx$ can be found such that each term in the series is smaller, even negligibly smaller, than the previous term. This can be demonstrated by forming the ratio $R$ of two consecutive terms in the series, that is,

$$R = \frac{\frac{1}{(k+1)!}\phi^{(k+1)}(x)dx^{k+1}}{\frac{1}{k!}\phi^{(k)}(x)dx^k} \tag{2.31}$$

or

$$R = \frac{1}{k+1}\frac{\phi^{(k+1)}(x)}{\phi^{(k)}(x)}dx. \tag{2.32}$$

Hence, if all the derivatives of $\phi$ are continuous, they are finite, and

$$R \sim dx. \tag{2.33}$$

This means that $R$ tends to zero as $dx$ tends to zero indicating that the series is a descending series.

## 2.3.1   First Differential Condition

In this section, it is shown that $\phi_x$ must vanish at a minimal point. The proof is by contradiction. Assume that $x$ is a minimal point but that $\phi_x(x) \neq 0$. If $\phi_x(x) \neq 0$, then the first differential $dJ$ is the dominating term in the expansion (2.28) so that

$$\Delta J \cong dJ = \phi_x(x)dx \tag{2.34}$$

and must be positive regardless of the choice of $dx \neq 0$. At $x$, $\phi_x(x)$ has a specific value — either positive or negative but not zero. Assume that $\phi_x(x) > 0$. If $dx > 0$, then $\Delta J > 0$. On the other hand, if $dx < 0$, then $\Delta J < 0$. Since $\Delta J$ must be positive at a minimal point for all $dx \neq 0$, $x$ cannot be a minimal point. This is a contradiction. Hence,

$$\phi_x(x) = 0 \tag{2.35}$$

must hold at a minimal point. Eq.(2.35) constitutes one equation in one unknown, but it may admit several solutions.

The first differential condition for a maximum ($\Delta J < 0$) is obtained by replacing $\phi$ in Eq.(2.35) by $-\phi$ and, hence, is also $\phi_x(x) = 0$. Since

$$\phi_x(x) = 0 \quad \Leftarrow \quad x \text{ is a minimal point}, \tag{2.36}$$

it is a necessary condition for a minimum (see Section 1.5). The points that satisfy the equation $\phi_x(x) = 0$ are called optimal points, and they can be minima, maxima, or even saddle points which are neither minima nor maxima (look at a sketch of $\phi = x^3$).

## 2.3.2   Second Differential Conditions

Assume that $x$ is a solution of the optimality condition (2.35). Since the first differential vanishes, the value of $\Delta J$ can be approximated by the second differential, that is,

$$\Delta J \cong \frac{1}{2!}d^2J = \frac{1}{2!}\phi_{xx}(x)dx^2. \tag{2.37}$$

From this relation, it is apparent that if

$$\phi_{xx}(x) > 0, \tag{2.38}$$

$\Delta J$ is positive for all $dx \neq 0$ and that $x$ is a minimal point. Hence, Eq.(2.38) is a sufficient condition for a minimum because

$$\phi_{xx}(x) > 0 \quad \Rightarrow \quad x \text{ is a minimum}. \tag{2.39}$$

However, it is possible for $\phi_{xx}$ to vanish (look at $\phi = x^4$) and a minimal point still exist. Hence, the condition

$$\phi_{xx}(x) \geq 0 \tag{2.40}$$

is a necessary condition for a minimum because

$$\phi_{xx}(x) \geq 0 \quad \Leftarrow \quad x \text{ is a minimum}. \tag{2.41}$$

### 2.3.3  Higher-Order Differentials

If the second derivative vanishes, the nature of the optimal point is determined by examining higher-order differentials. With the first and second differentials being zero,

$$\Delta J = \frac{1}{3!}d^3 J + \frac{1}{4!}d^4 J + \cdots, \tag{2.42}$$

where

$$d^3 J = \phi^{(3)}(x)dx^3, \quad d^4 J = \phi^{(4)}(x)dx^4. \tag{2.43}$$

Hence, if $\phi^{(3)}(x) \neq 0$, the optimal point is a saddle point because $\Delta J$ is proportional to $dx^3$ and $dx$ can be positive or negative. If $\phi^{(3)}(x) = 0$ and $\phi^{(4)}(x) > 0$, the optimal point is a minimum. Examples of such functions are $\phi = x^3$ and $\phi = x^4$, respectively.

### 2.3.4  Summary

In summary, optimal points $x$ are located by applying the first differential necessary condition

$$\phi_x(x) = 0. \tag{2.44}$$

Then, potential minimal points are identified by applying the second differential necessary condition

$$\phi_{xx}(x) \geq 0. \tag{2.45}$$

If an $x$ obtained from Eq.(2.44) satisfies the sufficient condition

$$\phi_{xx}(x) > 0, \tag{2.46}$$

it is a minimum. However, should the second derivative vanish, higher- order differentials must be investigated to determine the nature of the optimal point.

In terms of just the first and second differentials, there are no conditions that are both necessary and sufficient. The condition $\phi_x = 0$ is necessary for a minimum but not sufficient; a maximum satisfies the same condition. Given a solution of $\phi_x = 0$, $\phi_{xx} \geq 0$ is necessary for a minimum but not sufficient; the second derivative can vanish at a maximum. The condition $\phi_{xx} > 0$ is sufficient for a minimum but not necessary; the second derivative can vanish and a minimum still exist.

As stated previously, conditions for a maximum can be obtained from those for a minimum by replacing $\phi$ by $-\phi$. Hence,

$$\phi_x = 0, \quad \phi_{xx} \leq 0, \quad \phi_{xx} < 0 \tag{2.47}$$

are the corresponding conditions for a maximum.

The conditions for a minimum can be stated in terms of the first and second differentials. The necessary conditions are that the first differential must vanish at an optimal point,

$$dJ = 0, \tag{2.48}$$

and that the second differential must be nonnegative,

$$d^2J \geq 0. \tag{2.49}$$

An optimal point is a minimum if the sufficient condition

$$d^2J > 0 \tag{2.50}$$

is satisfied.

## 2.4 Distance Problem

The problem is to minimize the distance from the origin $d = \sqrt{x^2 + y^2}$ to the parabola $y = x^2 + a$, where $a$ is a known constant (see Fig. 2.6). In this case, minimizing $d^2$ is equivalent to minimizing $d$ so the performance index can be written as

$$J = x^2 + (x^2 + a)^2 \triangleq \phi. \tag{2.51}$$

Find and identify its optimal points for all possible values of the constant $a$.

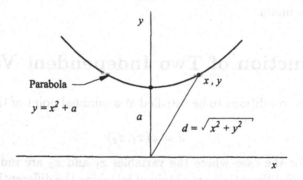

Figure 2.6: Minimum Distance to a Parabola

The first differential condition (2.44) can be written as

$$\phi_x = 2x[1 + 2(x^2 + a)] = 0 \tag{2.52}$$

and yields the following optimal points:

$$
\begin{aligned}
a > -1/2, \quad & x = 0 \\
a = -1/2, \quad & x = 0 \\
a < -1/2, \quad & x = 0 \quad \text{and} \quad x = \pm\sqrt{-a - 1/2}.
\end{aligned}
\tag{2.53}
$$

The nature (minimum or maximum) of an optimal point is determined by examining the sign of the second derivative

$$\phi_{xx} = 2(6x^2 + 1 + 2a).\qquad(2.54)$$

For the optimal points defined by Eq.(2.53), the following results are obtained:

$$
\begin{array}{llll}
a > -1/2, & x = 0, & \phi_{xx} > 0, & \text{minimum} \\
a = -1/2, & x = 0, & \phi_{xx} = 0, & \text{unknown} \\
a < -1/2, & x = 0, & \phi_{xx} < 0, & \text{maximum} \\
& x = \pm\sqrt{-a - 1/2}, & \phi_{xx} > 0, & \text{minimum.}
\end{array}
\qquad(2.55)
$$

For the case where $a = -1/2$, it is necessary to analyze higher-order differentials. Since $\phi^{(3)} = 24x$ and $\phi^{(4)} = 24$, it is concluded that the optimal point $x = 0$ is a minimum.

## 2.5 Function of Two Independent Variables

In this section, conditions to be satisfied at a minimal point of the performance index

$$J = \phi(x_1, x_2)\qquad(2.56)$$

are derived for the case where the variables $x_1$ and $x_2$ are independent. The first and second differentials are obtained by taking the differentials of Eq.(2.56) and evaluating them on the minimal path (see Section 2.1).

### 2.5.1 First Differential Conditions

The first differential of $J$ is obtained by taking the differential of Eq.(2.56), that is,

$$dJ = \phi_{x_1}(x_1, x_2)dx_1 + \phi_{x_2}(x_1, x_2)dx_2,\qquad(2.57)$$

and applying it at the minimal point $x_1$, $x_2$. The differentials $dx_1$ and $dx_2$ that locate an admissible comparison point are independent (arbitrary) since an admissible comparison point can be located anywhere in the neigborhood of the minimal point (see Fig. 2.7). Hence, $dJ$ must vanish regardless of the choice of $dx_1$ and $dx_2$.

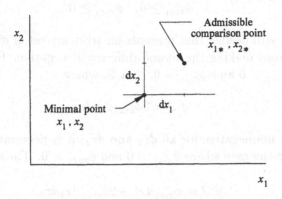

Figure 2.7: Definition of $dx_1$ and $dx_2$

Since the differentials $dx_1$ and $dx_2$ are independent, the choice $dx_1 \neq 0$ and $dx_2 = 0$ reduces Eq.(2.56) to the form of Eq.(2.15). Then, an argument similar to that of Sec. 2.3.1 leads to $\phi_{x_1} = 0$. Similarly, the choice $dx_1 = 0$ and $dx_2 \neq 0$ leads to $\phi_{x_2} = 0$. Finally, the conditions

$$\phi_{x_1} = 0, \quad \phi_{x_2} = 0 \tag{2.58}$$

cause the first differential to vanish for an arbitrary comparison point $dx_1 \neq 0$, $dx_2 \neq 0$. Eqs.(2.58) are necessary conditions for a minimum.

Equations (2.58) constitute two equations in two unknowns $x_1$ and $x_2$. Since the same conditions must be satisfied by a maximal point, the solutions of Eq.(2.58) are only optimal points.

## 2.5.2  Second Differential Conditions

The second differential is obtained by taking the differential of the first differential (2.57) and is given by

$$d^2J = \phi_{x_1 x_1} dx_1^2 + 2\phi_{x_1 x_2} dx_1 dx_2 + \phi_{x_2 x_2} dx_2^2, \tag{2.59}$$

where the derivatives are evaluated at the optimal point. For a solution of Eqs.(2.58) to be a minimum, the second differential must be nonnegative regardless of the choice of $dx_1$ and $dx_2$.

Since $dx_1$ and $dx_2$ are independent, the particular differentials $dx_1 \neq 0$, $dx_2 = 0$ and $dx_1 = 0, dx_2 \neq 0$ lead to the following necessary conditions:

$$\phi_{x_1 x_1} \geq 0, \quad \phi_{x_2 x_2} \geq 0. \tag{2.60}$$

A condition is still needed that prevents an arbitrary set of differentials ($dx_1 \neq 0$, $dx_2 \neq 0$) from making the second differential negative. First, consider the case where $\phi_{x_1 x_1} = 0$ and $\phi_{x_2 x_2} = 0$, that is, where

$$d^2 J = 2\phi_{x_1 x_2} dx_1 dx_2. \tag{2.61}$$

For $d^2 J$ to be nonnegative for all $dx_1$ and $dx_2$, it is necessary that $\phi_{x_1 x_2} = 0$. Next, consider the case where $\phi_{x_1 x_1} > 0$ and $\phi_{x_2 x_2} = 0$. The second differential is given by

$$d^2 J = \phi_{x_1 x_1} dx_1^2 + 2\phi_{x_1 x_2} dx_1 dx_2. \tag{2.62}$$

It is zero along the lines

$$dx_1 = 0, \quad dx_1 = -(2\phi_{x_1 x_2}/\phi_{x_1 x_1})dx_2, \tag{2.63}$$

and negative between these lines. Hence, to eliminate the region where $d^2 J$ is negative, it is necessary that $\phi_{x_1 x_2} = 0$. The same result is obtained when $\phi_{x_1 x_1} = 0$ and $\phi_{x_2 x_2} > 0$. Finally, for the case where $\phi_{x_1 x_1} > 0$ and $\phi_{x_2 x_2} > 0$, the second differential (2.59) is negative between the lines

$$dx_1 = \{[-\phi_{x_1 x_2} \pm (\phi_{x_1 x_2}^2 - \phi_{x_1 x_1}\phi_{x_2 x_2})^{1/2}]/\phi_{x_1 x_1}\}dx_2. \tag{2.64}$$

Hence, if it is required that

$$\phi_{x_1 x_1}\phi_{x_2 x_2} - \phi_{x_1 x_2}^2 \geq 0, \tag{2.65}$$

the two lines degenerate into one or disappear entirely (become imaginary), and the region in which the second differential is negative no longer exists.

Note that Eq.(2.65) contains the previous particular cases so that the second differential necessary conditions for a minimum are given by

$$\phi_{x_1 x_1} \geq 0, \quad \phi_{x_2 x_2} \geq 0, \quad \phi_{x_1 x_1}\phi_{x_2 x_2} - \phi_{x_1 x_2}^2 \geq 0. \tag{2.66}$$

If any one of these quantities is zero, it is necessary to investigate higher-order differentials to completely determine the nature of an optimal point.

Sufficient conditions are obtained by removing the equalities in Eq. (2.66) and observing that the first and third conditions imply the second. Hence, sufficient conditions for a minimum are given by

$$\phi_{x_1 x_1} > 0, \quad \phi_{x_1 x_1}\phi_{x_2 x_2} - \phi_{x_1 x_2}^2 > 0. \tag{2.67}$$

It should be noted that the second differential is a quadratic form which can be rewritten in matrix notation as

$$d^2 J = [dx_1 \ dx_2] \begin{bmatrix} \phi_{x_1 x_1} & \phi_{x_1 x_2} \\ \phi_{x_2 x_1} & \phi_{x_2 x_2} \end{bmatrix} \begin{bmatrix} dx_1 \\ dx_2 \end{bmatrix}. \tag{2.68}$$

Note that the second derivative matrix is symmetric and that the $dx$s are independent. The second differential is nonnegative if the second derivative matrix is positive semidefinite. Given a square matrix, a *minor* is defined as the determinant of a square submatrix obtained by deleting rows and columns. A *principal minor* is the determinant of a square submatrix whose diagonal terms are diagonal terms in the original matrix. With these definitions, the necessary and sufficient conditions for the second derivative matrix to be positive semidefinite are that *all* principal minors be nonnegative (Ref. Ga, p. 307). Hence, for the second derivative matrix to be positive semidefinite, the following conditions must be satisified:

$$|\phi_{x_1 x_1}| \geq 0, \quad |\phi_{x_2 x_2}| \geq 0, \quad \begin{vmatrix} \phi_{x_1 x_1} & \phi_{x_1 x_2} \\ \phi_{x_2 x_1} & \phi_{x_2 x_2} \end{vmatrix} \geq 0, \tag{2.69}$$

which are identical with Eqs.(2.66). On the other hand, the second differential is positive if the second derivative matrix is positive definite. Necessary and sufficient conditions for the second derivative matrix to be positive definite are that the *leading* principal minors be positive (Ref. Ga), that is,

$$|\phi_{x_1 x_1}| > 0, \quad \begin{vmatrix} \phi_{x_1 x_1} & \phi_{x_1 x_2} \\ \phi_{x_2 x_1} & \phi_{x_2 x_2} \end{vmatrix} > 0. \tag{2.70}$$

These conditions are the same as those given in Eq.(2.48).

If most of the leading principal minors are positive, but one or more is zero, the necessary conditions should be checked to be sure that all principal minors are $\geq 0$. If they are not, the optimal point is not a minimal point.

Another approach to the conditions for definiteness is to examine the eigenvalues of the second derivative matrix that are the values of $\lambda$ satisfying the equation

$$\begin{vmatrix} \phi_{x_1 x_1} - \lambda & \phi_{x_1 x_2} \\ \phi_{x_2 x_1} & \phi_{x_2 x_2} - \lambda \end{vmatrix} = 0. \tag{2.71}$$

Expansion of the determinant and solution of the resulting quadratic equation lead to

$$2\lambda = (\phi_{x_1 x_1} + \phi_{x_2 x_2}) \pm [(\phi_{x_1 x_1} + \phi_{x_2 x_2})^2 - 4(\phi_{x_1 x_1} \phi_{x_2 x_2} - \phi_{x_1 x_2}^2)]^{1/2}. \tag{2.72}$$

Since the term under the square root can be rewritten as

$$(\phi_{x_1 x_1} - \phi_{x_2 x_2})^2 + 4\phi_{x_1 x_2}^2, \tag{2.73}$$

the eigenvalues are real. Then, in view of Eq.(2.66), the second differential necessary conditions for a minimum are equivalent to the eigenvalues being nonnegative. Finally, the sufficient conditions are equivalent to the eigenvalues being positive.

## 2.5.3   Summary

In summary, the necessary conditions to be satisifed by an interior minimal point of the performance index $J = \phi(x_1, x_2)$ are that the first differential must vanish ($dJ = 0$) and that the second differential must be nonnegative ($d^2 J \geq 0$). If the second differential is positive, the optimal point is a minimum. The corresponding conditions on $\phi$ are given by

$$\phi_{x_1} = 0, \quad \phi_{x_2} = 0 \tag{2.74}$$

and

$$\phi_{x_1 x_1} \geq 0, \quad \phi_{x_2 x_2} \geq 0, \quad \phi_{x_1 x_1} \phi_{x_2 x_2} - \phi_{x_1 x_2}^2 \geq 0. \tag{2.75}$$

Given a point that satisfies the necessary conditions (2.74), sufficient conditions for a minimum are

$$\phi_{x_1 x_1} > 0, \quad \phi_{x_1 x_1} \phi_{x_2 x_2} - \phi_{x_1 x_2}^2 > 0. \tag{2.76}$$

If any of the second derivative terms in Eq.(2.75) vanishes, higher-order differentials must be examined to completely determine the nature of the optimal point. In this case, however, $\Delta J$ may contain several orders of differentials, making it more difficult to verify minimality (see Section 2.7).

It is interesting to note that in generating the equivalent conditions for a maximum ($\phi \rightarrow -\phi$) not all of the inequalities change direction.

In trying to find the optimal point(s) for a particular problem, it is often convenient to derive the first and second differentials by taking differentials and deriving the above results directly. This is demonstrated in the next section.

## 2.6  Examples

Consider the minimization of the performance index

$$J = x_1^2 + x_2^2 \triangleq \phi(x_1, x_2). \tag{2.77}$$

Application of the conditions (2.74) leads to the optimal point $x_1 = 0$, $x_2 = 0$ so that the second derivative terms become

$$\phi_{x_1 x_1} = 2, \quad \phi_{x_1 x_1} \phi_{x_2 x_2} - \phi_{x_1 x_2}^2 = 4. \tag{2.78}$$

Since the sufficient conditions (2.76) are satisfied, the optimal point is a minimum. If the sign of either term in the performance index is made negative, the optimal point becomes a saddle point.

As a second example, find the point $x, y$ that minimizes the performance index

$$J = (y - x)(y - 2x). \tag{2.79}$$

The first differential is given by

$$dJ = (-3y + 4x)dx + (2y - 3x)dy. \tag{2.80}$$

The vanishing of the first differential leads to

$$-3y + 4x = 0, \quad 2y - 3x = 0 \tag{2.81}$$

whose solution (the optimal point) is

$$x = y = 0. \tag{2.82}$$

Next, the second differential is obtained by taking the differential of the first differential (2.80) and is given by

$$d^2 J = 4dx^2 - 6dxdy + 2dy^2. \tag{2.83}$$

It must be nonnegative regardless of the choice of the admissible comparison point $dx, dy$. The second differential is now rewritten in the form (2.68), that is,

$$d^2J = [dy\ dx] \begin{bmatrix} 4 & -3 \\ -3 & 2 \end{bmatrix} \begin{bmatrix} dy \\ dx \end{bmatrix} \geq 0. \tag{2.84}$$

By looking at the leading principal minors, that is,

$$|4| > 0, \quad \begin{vmatrix} 4 & -3 \\ -3 & 2 \end{vmatrix} = -1 < 0, \tag{2.85}$$

it is seen the matrix is not positive definite. Hence the optimal point is not a minimum. Incidentally, the optimal point is not a maximum either. Hence, it is a saddle point.

## 2.7  Historical Example

In the early days of parameter optimization, the problem of identifying the optimal point of the performance index

$$J = (y - ax^2)(y - bx^2) \tag{2.86}$$

caused some concern (Ref. Ha, p. 34). Here, $a$ and $b$ are positive constants, and $b > a$.

The first differential of $J$ is given by

$$dJ = [-2(a + b)xy + 4abx^3]dx + [2y - (a + b)x^2]dy \tag{2.87}$$

and for $dJ = 0$ with independent $dx$ and $dy$ leads to

$$-2(a + b)xy + 4abx^3 = 0, \quad 2y - (a + b)x^2 = 0. \tag{2.88}$$

The solution of the first differential conditions is

$$x = y = 0. \tag{2.89}$$

Next, from $d(dJ)$ the second differential

$$d^2J = [-2(a + b)y + 12abx^2]dx^2 - 4(a + b)xdxdy + 2dy^2 \tag{2.90}$$

has the following value at the optimal point:

$$d^2 J = 2dy^2. \tag{2.91}$$

In general, the second differential has the form

$$d^2 J = [dx \ dy] \begin{bmatrix} 0 & 0 \\ 0 & 2 \end{bmatrix} \begin{bmatrix} dx \\ dy \end{bmatrix}, \tag{2.92}$$

but the second derivative matrix is neither positive definite nor negative definite. Higher-order differentials at the optimal point are given by

$$d^3 J = -6(a+b)dx^2 dy \tag{2.93}$$

$$d^4 J = 24abdx^4. \tag{2.94}$$

Determining the nature of the optimal point (0,0) involves several differentials.

The total change in the performance index at the optimal point is given by

$$\Delta J = dJ + \frac{1}{2!}d^2 J + \frac{1}{3!}d^3 J + \frac{1}{4!}d^4 J \tag{2.95}$$

or

$$\Delta J = dy^2 - (a+b)dx^2 dy + abdx^4. \tag{2.96}$$

This equation can be rewritten as

$$\Delta J = (dy - adx^2)(dy - bdx^2), \tag{2.97}$$

which is plotted in Fig. 2.8. For $dy$s such that

$$dy < adx^2, \quad dy > bdx^2, \tag{2.98}$$

the total change in the performance index is positive. However, for

$$adx^2 < dy < bdx^2, \tag{2.99}$$

$\Delta J$ is negative. Hence, since $\Delta J$ is not greater than zero or less than zero for all admissible comparison points ($dx \neq 0, dy \neq 0$), the optimal point is neither a minimum nor a maximum. Therefore, it is a saddle point.

This is an example where the nature of the optimal point is not determined by the sign of a single differential.

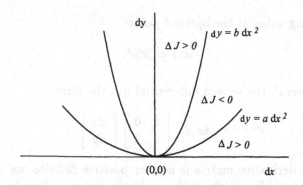

Figure 2.8: Historical Example

## 2.8   Function of $n$ Independent Variables

The most general problem of unconstrained optimization is to minimize a function of $n$ independent variables, that is,

$$J = \phi(x_1, x_2, \ldots, x_n). \qquad (2.100)$$

The first differential is given by

$$dJ = \sum_{k=1}^{n} \phi_{x_k} dx_k \qquad (2.101)$$

and must vanish regardless of the choice of $dx_k$. Since the variables are independent (each $dx_k$ can be chosen arbitrarily), the first differential leads to

$$\phi_{x_k} = 0, \quad k = 1, \ldots, n. \qquad (2.102)$$

Next, the second differential is given by the quadratic form

$$d^2 J = \sum_{j=1}^{n} \sum_{k=1}^{n} \phi_{x_j x_k} \, dx_j \, dx_k, \qquad (2.103)$$

which must be nonnegative for a minimum. Since the second derivative matrix is symmetric and the $dx$s are independent, the second differential is nonnegative if the second derivative matrix is positive semidefinite, that is,

$$[\phi_{x_j x_k}] \geq 0, \quad j, k = 1, \ldots, n. \qquad (2.104)$$

This is so if all principal minors are nonnegative. However, if any principal minor vanishes, it is necessary to investigate higher-order differentials to verify minimality. Finally, the second differential is positive if the leading principal minors are positive, that is,

$$|\phi_{x_1 x_1}| > 0, \quad \begin{vmatrix} \phi_{x_1 x_1} & \phi_{x_1 x_2} \\ \phi_{x_2 x_1} & \phi_{x_2 x_2} \end{vmatrix} > 0, \ldots, \quad |\phi_{x_j x_k}| > 0 \qquad (2.105)$$

where $j, k = 1, \ldots, n$.

An equivalent set of necessary conditions and sufficient conditions for the second derivative matrix to be positive semidefinite (definite) is that its eigenvalues be nonnegative (positive). The eigenvalues $\lambda$ of the second derivative matrix are the solutions of the equation

$$|\phi_{x_j x_k} - \lambda d_{jk}| = 0, \quad j, k = 1, \ldots, n, \qquad (2.106)$$

where $d_{jk}$ is the Kronecker delta ($= 1$ if $j = k$, $= 0$ if $j \neq k$).

If the second derivative matrix is positive definite or if the eigenvalues are positive, the nature of the optimal point is determined by the second differential. If the second derivative matrix is positive semidefinite or if one or more eigenvalues are zero, the nature of the optimal point is determined by looking at the second and higher-order differentials. If the second differential is zero, the nature of the optimal point is determined by higher-order differentials.

# Problems

2.1 Find and identify the optimal points of the performance index

$$J = x(x - 2)^2.$$

Show that a relative maximum occurs at $x = 2/3$, and a relative minimum, at $x = 2$.

2.2 Find and identify the optimal points of the function

$$J = \frac{x^2}{(x - 2)}.$$

Show that a relative maximum occurs at $x = 0$, and a relative minimum at $x = 4$. Show that the performance index at the maximum has a lower value than it has at the minimum. How can this be?

2.3  Find the point $A$ that maximizes the performance index

$$J = \sin A \sin(\pi - A).$$

Limit the investigation to values of $A$ between 0 and $\pi$. Show that the maximum is at $A = \pi/2$.

2.4  Find and identify the optimal points of the performance index

$$J = \sin A \cos(\pi - A).$$

Only look at points where $0 \le A \le \pi$. Show that a minimum is at $A = \pi/4$, and a maximum is at $A = 3\pi/4$.

2.5  Maximize the performance index

$$J = [4 - 3\cos^2 2\theta]^{1/2}.$$

Show that the optimal points occur at $\theta = \pm n\pi/4$ and that minima occur for $n$ even and maxima, for $n$ odd.

2.6  Show that the second variation is negative between the lines (2.53) and between the lines (2.54).

2.7  Show that the function

$$J = x^2 + y^2 + (1 - x + 2y)^2$$

has a minimum at $x = 1/6$, $y = -1/3$.

2.8  Show that the performance index

$$J = \frac{1}{2}x_1^2 + x_1 x_2$$

has an optimal point at $x_1 = 0, x_2 = 0$, but that it is neither a minimum nor a maximum.

2.9  Find and identify the point $x, y$ that optimizes the performance index

$$J = x^2 - y^4.$$

The optimal point is $(0,0)$, but it is a saddle point.

2.10 Consider the performance index

$$J = \frac{1}{2}(x_1 + x_2)^2 + (x_1 + x_2)x_3,$$

which has the second derivative matrix

$$\phi_{xx} = \begin{bmatrix} 1 & 1 & 1 \\ 1 & 1 & 1 \\ 1 & 1 & 0 \end{bmatrix}.$$

(a) Using principal minors, show that the sufficient conditions for a minimum are not satisfied. Then, show that the necessary conditions are not satisfied.

(b) Repeat (a) using eigenvalues .

(c) Show that the optimal point is not a maximum either.

(d) Finally, looking at the first variation conditions, show that a unique optimal point does not exist.

# 3

# Constrained Minimization: Equality Constraints

## 3.1  Introduction

To have an optimization problem, the number of variables must exceed the number of constraints. Thus, the subject of constrained optimization is introduced by minimizing a function of two variables where the variables are related through one algebraic constraint. First, the function is minimized using the direct approach, Here, differentials of the performance index and the constraint are taken and the dependent differential is eliminated to form the first and second differentials in terms of the independent differential. Next, the Lagrange multiplier approach is introduced. Here, the constraint is multiplied by an undetermined constant (Lagrange multiplier) and is added to the performance index. The multiplier is used later to eliminate the dependent differential. The results are compared with those of the direct approach to confirm the validity of the multiplier method. This part of the chapter is concluded by a discussion of higher-order differentials. Next, the problem of $n$ constrained variables is analyzed using an indicial notation; the same problem is investigated in Chapter 5 using matrix notation.

## 3.2   Function of Two Constrained Variables

Consider the problem of minimizing the function or performance index

$$J = \phi(x_1, x_2), \tag{3.1}$$

where the variables $x_1$ and $x_2$ are not independent but must satisfy the constraint

$$\psi(x_1, x_2) = 0. \tag{3.2}$$

Since the choice is arbitrary, $x_1$ is identified as the independent variable, and hence, $x_2$ becomes the dependent variable.

The constraint (3.2) represents a curve in the $x_1$, $x_2$ plane as shown in Fig. 3.1 along with contours of the performance index.

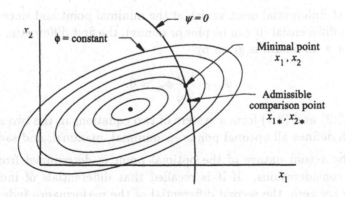

Figure 3.1: Problem Definition: Equality Constraint

If the coordinates $x_1, x_2$ define the minimal point, then admissible comparison points $x_{1_*}$, $x_{2_*}$ lie on the curve in the neighborhood of the minimal point.

The conditions for a minimum are first derived using the direct approach. Then, the Lagrange multiplier method is introduced and shown to be equivalent to the direct approach.

### 3.2.1  Direct Approach

The first differential of the performance index (3.1) is given by

$$dJ = \phi_{x_1} dx_1 + \phi_{x_2} dx_2. \tag{3.3}$$

Then, since $\psi = 0$,

$$d\psi = \psi_{x_1} dx_1 + \psi_{x_2} dx_2 = 0. \tag{3.4}$$

The next step is to express $dJ$ in terms of the independent differential $dx_1$ so that the condition for a minimum can be obtained. Solving Eq.(3.4) for the dependent differential $dx_2$ leads to

$$dx_2 = -\frac{\psi_{x_1}}{\psi_{x_2}} dx_1, \tag{3.5}$$

and subsequently to

$$dJ = \left[ \phi_{x_1} - \frac{\psi_{x_1}}{\psi_{x_2}} \phi_{x_2} \right] dx_1. \tag{3.6}$$

Since the first differential must vanish at the minimal point and since $dx_1$ is an independent differential (it can be plus or minus), the first differential necessary condition for a minimum is given by

$$\phi_{x_1} - \frac{\psi_{x_1}}{\psi_{x_2}} \phi_{x_2} = 0. \tag{3.7}$$

Equations (3.2) and (3.7) form a system of two equations in the two unknowns $x_1$, $x_2$ which defines all optimal points — minimal, maximal, and saddle.

The actual nature of the optimal point is determined from second differential considerations. If it is recalled that differentials of independent differentials are zero, the second differential of the performance index is given by

$$d^2 J = \phi_{x_1 x_1} dx_1^2 + 2\phi_{x_1 x_2} dx_1 dx_2 + \phi_{x_2 x_2} dx_2^2 + \phi_{x_2} d^2 x_2, \tag{3.8}$$

and the second differential of the constraint becomes

$$d^2 \psi = \psi_{x_1 x_1} dx_1^2 + 2\psi_{x_1 x_2} dx_1 dx_2 + \psi_{x_2 x_2} dx_2^2 + \psi_{x_2} d^2 x_2 = 0. \tag{3.9}$$

Solving Eq.(3.9) for $d^2 x_2$ and using Eq.(3.5) to eliminate $dx_2$ leads to

$$d^2 x_2 = -\frac{1}{\psi_{x_2}} \left[ \psi_{x_1 x_1} - 2\psi_{x_1 x_2} \frac{\psi_{x_1}}{\psi_{x_2}} + \psi_{x_2 x_2} \left( \frac{\psi_{x_1}}{\psi_{x_2}} \right)^2 \right] dx_1^2. \tag{3.10}$$

Finally, $d^2J$ is expressed in terms of the independent differential $dx_1$ by combining Eqs.(3.5), (3.8), and (3.10), that is,

$$d^2J = A \, dx_1^2,  \tag{3.11}$$

where

$$A = \left[ \phi_{x_1x_1} - \frac{\phi_{x_2}}{\psi_{x_2}} \psi_{x_1x_1} \right] - 2 \left[ \phi_{x_1x_2} - \frac{\phi_{x_2}}{\psi_{x_2}} \psi_{x_1x_2} \right] \frac{\psi_{x_1}}{\psi_{x_2}}$$
$$+ \left[ \phi_{x_2x_2} - \frac{\phi_{x_2}}{\psi_{x_2}} \psi_{x_2x_2} \right] \left( \frac{\psi_{x_1}}{\psi_{x_2}} \right)^2.  \tag{3.12}$$

It is seen from this relation that the necessary condition for a minimum is given by

$$A \geq 0.  \tag{3.13}$$

The sufficient condition is obtained by removing the equal sign.

The direct approach can involve a considerable amount of manipulation for more general problems. Hence, a simpler approach is needed.

## 3.2.2 Lagrange Multiplier Approach

It is observed that the problem

$$\text{Minimize } J = \phi(x_1, x_2)$$
$$\text{subject to } \psi(x_1, x_2) = 0  \tag{3.14}$$

is identical with the problem

$$\text{Minimize } J' = \phi(x_1, x_2) + \nu\psi(x_1, x_2) \triangleq G(x_1, x_2, \nu)$$
$$\text{subject to } \psi(x_1, x_2) = 0,  \tag{3.15}$$

regardless of the choice of the undetermined constant $\nu$ which is called a Lagrange multiplier. In other words, zero is being added to the performance index. The new performance index is called the augmented performance index. Again, $x_1$ is chosen to be the independent variable, and $x_2$ the dependent variable.

At this point, it is not known how to treat $\nu$ during the differential process. The most general case is to treat it as a dependent variable and take its differential. The first differential becomes

$$dJ' = G_{x_1}dx_1 + G_{x_2}dx_2 + G_\nu d\nu = 0 \tag{3.16}$$

or

$$dJ' = (\phi_{x_1} + \nu\psi_{x_1})dx_1 + (\phi_{x_2} + \nu\psi_{x_2})dx_2 + \psi d\nu = 0. \tag{3.17}$$

The last term vanishes because $\psi = 0$. Since $\nu$ is arbitrary, choose its value such that the coefficient of the dependent differential $dx_2$ vanishes. Then, $dx_1$ is an independent differential and its coefficient must vanish. Hence, the following equations must be satisfied at an optimal point:

$$G_{x_1} = \phi_{x_1} + \nu\psi_{x_1} = 0, \quad G_{x_2} = \phi_{x_2} + \nu\psi_{x_2} = 0, \quad G_\nu = \psi = 0 \tag{3.18}$$

and constitute a system of three equations in the three unknowns $x_1$, $x_2$, and $\nu$. Either of the first two equations can be solved for $\nu$, which can then be substituted into the other equation to obtain a condition for $x_1$ and $x_2$. Doing so gives

$$\nu = -\frac{\phi_{x_1}}{\psi_{x_1}} = -\frac{\phi_{x_2}}{\psi_{x_2}} \tag{3.19}$$

or

$$\frac{\phi_{x_1}}{\psi_{x_1}} = \frac{\phi_{x_2}}{\psi_{x_2}}. \tag{3.20}$$

This is the same result obtained by the direct approach; see Eq. (3.7).

The second differential is obtained by taking the differential of the first differential (3.16) and is given by

$$
\begin{aligned}
d^2J' = {} & G_{x_1x_1}dx_1^2 + G_{x_1x_2}dx_1dx_2 + G_{x_1\nu}dx_1d\nu \\
& + G_{x_2x_1}dx_1dx_2 + G_{x_2x_2}dx_2^2 + G_{x_2\nu}dx_2d\nu \\
& + G_{\nu x_1}d\nu dx_1 + G_{\nu x_2}d\nu dx_2 + G_{\nu\nu}d\nu^2 \\
& + G_{x_2}d^2x_2 + G_\nu d^2\nu.
\end{aligned}
\tag{3.21}
$$

The last two terms vanish because of the first differential conditions (3.18). Next, $G_{\nu\nu} = 0$ because $G$ is linear in $\nu$. Then, with $G_{x_1\nu} = G_{\nu x_1} = \psi_{x_1}$ and $G_{x_2\nu} = G_{\nu x_2} = \psi_{x_2}$,

$$
\begin{aligned}
G_{x_1\nu}dx_1d\nu + G_{x_2\nu}dx_2d\nu = d\psi d\nu = 0 \\
G_{\nu x_1}d\nu dx_1 + G_{\nu x_2}d\nu dx_2 = d\psi d\nu = 0
\end{aligned}
\tag{3.22}
$$

since $d\psi = 0$. At this point, the second differential can be rewritten as

$$d^2 J' = G_{x_1 x_1} dx_1^2 + 2 G_{x_1 x_2} dx_1 dx_2 + G_{x_2 x_2} dx_2^2, \qquad (3.23)$$

where the differentials $dx_1$ and $dx_2$ are not independent but must be consistent with $d\psi = 0$. If the dependent differential is eliminated through the use of Eq.(3.5), the second differential becomes

$$d^2 J' = \left[ G_{x_1 x_1} - 2 G_{x_1 x_2} \frac{\psi_{x_1}}{\psi_{x_2}} + G_{x_2 x_2} \left( \frac{\psi_{x_1}}{\psi_{x_2}} \right)^2 \right] dx_1^2. \qquad (3.24)$$

Since the second differential must be nonnegative at a minimum and since $dx_1$ is independent, a necessary condition for a minimum is that

$$G_{x_1 x_1} - 2 G_{x_1 x_2} \frac{\psi_{x_1}}{\psi_{x_2}} + G_{x_2 x_2} \left( \frac{\psi_{x_1}}{\psi_{x_2}} \right)^2 \geq 0. \qquad (3.25)$$

It is easily verified that this relation is identical with Eq.(3.13) by taking the partial derivatives of $G$ defined in Eq.(3.15) and using the relation (3.19) for $\nu$. The sufficient condition comes from replacing $\geq 0$ by $> 0$.

In summary, to minimize the performance index $J = \phi(x_1, x_2)$ subject to the constraint $\psi(x_1, x_2) = 0$, form the augmented function $J' = \phi + \nu \psi \triangleq G$ and apply the conditions

$$G_{x_1} = 0, \quad G_{x_2} = 0, \quad \psi = 0 \qquad (3.26)$$

and

$$G_{x_1 x_1} - 2 G_{x_1 x_2} \frac{\psi_{x_1}}{\psi_{x_2}} + G_{x_2 x_2} \left( \frac{\psi_{x_1}}{\psi_{x_2}} \right)^2 \geq 0. \qquad (3.27)$$

The second differential (3.27) is only valid if $dx_1$ can be chosen to be the independent differential. Since this is not always the case, it is probably better to derive the second differential condition for each problem.

## 3.2.3   Remarks

The use of a Lagrange multiplier has the effect of making all differentials independent in the first differential. However, this is not true in the second differential, and dependent differentials must be eliminated. The advantage of using a Lagrange multiplier is that it is only necessary to work with $d\psi = 0$ and not $d\psi = 0$ and $d^2\psi = 0$ as in the direct approach.

In parameter optimization, the Lagrange multipliers do not need to be varied. The coefficients of $d\nu$, $d^2\nu$, ... are zero.

In terms of the first and second differentials, it is not necessary to take the differential of dependent differentials. This is not true if higher-order differentials need to be derived.

The last two remarks are true because differentials are being taken from the minimal point. In numerical methods, differentials are taken relative to an initial guess or nominal point where $\psi \neq 0$, $d\psi \neq 0$, and $G_{x_2} \neq 0$.

The optimality condition (3.20) has an interesting geometric interpretation. A set of contours for a function $\psi = \psi(x_1, x_2)$, that is, $\psi =$ const, is shown in Fig. 3.2. At a point $P$ on the contour $\psi = 0$, there exists a unit vector normal to the contour and in the direction of increasing $\psi$. A linear element $ds$ of $\psi = 0$ at $P$ is shown in Fig. 3.3 where the orientation of $ds$ is determined by $dx_1$ and $dx_2$. Since $\psi = 0$, then $d\psi = \psi_{x_1} dx_1 + \psi_{x_2} dx_2 = 0$ and $dx_2 = -(\psi_{x_1}/\psi_{x_2})dx_1$. If $i$ and $j$ are unit vectors in the $x_1$ and $x_2$ directions, the unit normal vector is given by

$$n = -\sin\beta i + \cos\beta j. \tag{3.28}$$

From Fig. 3.3, it is seen that

$$\sin\beta = \frac{dx_2}{(dx_1^2 + dx_2^2)^{1/2}}, \quad \cos\beta = \frac{dx_1}{(dx_1^2 + dx_2^2)^{1/2}} \tag{3.29}$$

or, in terms of the derivatives of $\psi$, that

$$\sin\beta = -\frac{\psi_{x_1}/\psi_{x_2}}{[1 + (\psi_{x_1}/\psi_{x_2})^2]^{1/2}}, \quad \cos\beta = \frac{1}{[1 + (\psi_{x_1}/\psi_{x_2})^2]^{1/2}}. \tag{3.30}$$

Hence, the unit normal vector becomes

$$n_\psi = \frac{(\psi_{x_1}/\psi_{x_2})i + j}{[1 + (\psi_{x_1}/\psi_{x_2})^2]^{1/2}}. \tag{3.31}$$

A contour of the performance index ($\phi =$ Const) also passes through $P$ and similar reasoning leads to the following expression for its normal:

$$n_\phi = \frac{(\phi_{x_1}/\phi_{x_2})i + j}{[1 + (\phi_{x_1}/\phi_{x_2})^2]^{1/2}}. \tag{3.32}$$

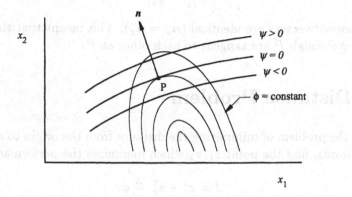

Figure 3.2: Unit Normal Vector

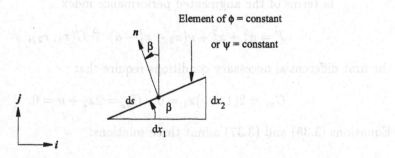

Figure 3.3: Orientation of Unit Normal Vector

Because of the optimality conditions (3.20), that is, because

$$\frac{\psi_{x_1}}{\psi_{x_2}} = \frac{\phi_{x_1}}{\phi_{x_2}}, \tag{3.33}$$

the two normal vectors are identical ($n_\psi = n_\phi$). This means that the contours of $\psi$ and $\phi$ through $P$ are tangent to each other at $P$.

## 3.3   Distance Problem

Consider the problem of minimizing the distance from the origin to a parabola. In other words, find the point $x_1, x_2$ which minimizes the performance index

$$J = x_1^2 + x_2^2 \triangleq \phi, \tag{3.34}$$

subject to the constraint

$$\psi \triangleq x_2 - x_1^2 - a = 0, \tag{3.35}$$

where $a$ is a constant. For this particular problem, the constraint can be solved for $x_2$, and $x_2$ can be eliminated from the performance index (see Section 2.3). The intent here, however, is to solve the problem using the Lagrange multiplier approach.

In terms of the augmented performance index

$$J' = x_1^2 + x_2^2 + \nu(x_2 - x_1^2 - a) \triangleq G(x_1, x_2), \tag{3.36}$$

the first differential necessary conditions require that

$$G_{x_1} = 2(1 - \nu)x_1 = 0, \quad G_{x_2} = 2x_2 + \nu = 0. \tag{3.37}$$

Equations (3.35) and (3.37) admit three solutions:

$$x_1 = 0, \quad x_2 = a, \quad \nu = -2a \tag{3.38}$$

and

$$x_1 = \pm\sqrt{-a - 1/2}, \quad x_2 = -1/2, \quad \nu = 1. \tag{3.39}$$

Note that for $a > -1/2$, the last two solutions become imaginary.

The next step is to determine the nature (minimum, maximum, saddle) of the optimal points by looking at the sign of the second differential (3.25), that is,

$$d^2 J' = 2(1 - \nu + 4x_1^2)dx_1^2. \tag{3.40}$$

The results for the solution (3.38) are given by

$$a > -1/2, \quad d^2 J' > 0, \quad \text{minimum}$$
$$a = -1/2, \quad d^2 J' = 0, \quad \text{unknown} \tag{3.41}$$
$$a < -1/2, \quad d^2 J' < 0, \quad \text{maximum},$$

and those for the other two solutions (3.39) are

$$a > -1/2, \quad \text{solutions are imaginary}$$
$$a = -1/2, \quad d^2 J' = 0, \quad \text{unknown} \tag{3.42}$$
$$a < -1/2, \quad d^2 J' > 0, \quad \text{minimum}.$$

The nature of the solution $a = -1/2$, $x_1 = 0$ is still unknown so higher-order differentials must be examined. The starting point is the general form of the second differential (3.21) without the $d\nu$ terms, that is,

$$d^2 J' = 2(1 - \nu)dx_1^2 + 2dx_2^2 + (2x_2 + \nu)d^2 x_2. \tag{3.43}$$

The third differential is the differential of the second differential and is given by

$$d^3 J' = 6dx_2 d^2 x_2 + (2x_2 + \nu)d^3 x_2 = 24x_1 dx_1^3. \tag{3.44}$$

Taking differentials of the contraint (3.35) leads to

$$dx_2 = 2x_1 dx_1$$
$$d^2 x_2 = 2dx_1^2 \tag{3.45}$$
$$d^3 x_2 = 0.$$

Hence, at $x_1 = 0$, the third differential vanishes. The fourth differential is given by

$$d^4 J' = 6d^2 x_2^2 + 8dx_2 d^3 x_2 + (2x_2 + \nu)d^4 x_2 = 24dx_1^4, \tag{3.46}$$

since $d^4 x_2 = 0$, and is positive. Therefore, the solution $a = -1/2$, $x_1 = 0$ is a minimum because the third differential is zero and the fourth differential is positive. Note that it has been necessary to include the term $G_{x_2}d^2 x_2$ in the second differential in order to obtain the correct higher-order differentials.

## 3.4   Function of $n$ Constrained Variables

The general problem of constrained minimization is to find the point $x_1, x_2, \ldots, x_n$ that minimizes the performance index

$$J = \phi(x_1, x_2, \ldots, x_n), \tag{3.47}$$

subject to the constraints

$$\begin{aligned}
\psi_1\,(x_1, x_2, \ldots, x_n) &= 0 \\
\psi_2\,(x_1, x_2, \ldots, x_n) &= 0 \\
&\vdots \\
\psi_p\,(x_1, x_2, \ldots, x_n) &= 0.
\end{aligned} \tag{3.48}$$

Each constraint is multiplied by a Lagrange multiplier and added to the performance index to form the augmented performance

$$J' = \phi(x_1, \ldots, x_n) + \sum_{j=1}^{p} \nu_j \psi_j(x_1, \ldots, x_n) \triangleq G(x_1, \ldots, x_n). \tag{3.49}$$

The first differential is given by

$$dJ' = \sum_{k=1}^{n} G_{x_k} dx_k, \tag{3.50}$$

where, however, the differentials $dx_k$ are not independent. If the $p$ Lagrange multipliers are chosen such that the coefficients of the $p$ dependent differentials (which can be arbitrarily specified) vanish, then the remaining $n-p$ coefficients must be zero because the corresponding differentials are independent. Hence, the first differential necessary conditions are given by

$$G_{x_k} = 0, \quad k = 1, \ldots, n \tag{3.51}$$

and are solved in conjunction with the constraints (3.48) for the $x$s and $\nu$s.

Next, the second differential of the augmented function is derived from Eq.(3.50) to be

$$d^2 J' = \sum_{k=1}^{n} \sum_{l=1}^{n} G_{x_k x_l} dx_k dx_l \geq 0, \tag{3.52}$$

where the term involving $G_{x_k}$ has been omitted because of Eq.(3.51). The relationships between the differentials are obtained from $d\psi_j = 0$ and are given by

$$\sum_{k=1}^{n} \psi_{j_{x_k}} dx_k = 0 , \quad j = 1, \ldots, p. \tag{3.53}$$

These $p$ equations are solved for the $p$ dependent differentials that are eliminated from Eq.(3.52) to form a new quadratic form in terms of the independent differentials. This matrix must be symmetric. The conditions for a minimum are then found by the procedure stated in Section 2.7. It is possible to solve for the reduced quadratic form and its associated second derivative matrix using matrix manipulation. This is done in Chapter 5.

For multiple constraints, it is not always possible to solve Eqs.(3.51) for the Lagrange multipliers $\nu$. Consider the case of three variables $x, y$, and $z$ and two constraints for which Eqs.(3.51) become

$$\begin{aligned} G_x &= \phi_x + \nu_1 \psi_{1_x} + \nu_2 \psi_{2_x} = 0 \\ G_y &= \phi_y + \nu_1 \psi_{1_y} + \nu_2 \psi_{2_y} = 0 \\ G_z &= \phi_z + \nu_1 \psi_{1_z} + \nu_2 \psi_{2_z} = 0. \end{aligned} \tag{3.54}$$

To solve any two of these equations by Cramer's rule for the multipliers $\nu_1$ and $\nu_2$, the determinant of the coefficient matrix must be nonzero. This is the same as saying that the normals to the constraints at the optimal point must be linearly independent. Two constraints can be linearly dependent if they are tangent at the optimal point or if they form a cusp at the optimal point.

## 3.5  Example

Find the point $x_1, x_2, x_3$ that minimizes the performance index

$$J = x_1^2 - x_2 + 2x_3, \tag{3.55}$$

subject to the constraint

$$x_1^2 + x_2^2 + x_3^2 = 2. \tag{3.56}$$

For this problem,

$$J' = (x_1^2 - x_2 + 2x_3) + \nu(x_1^2 + x_2^2 + x_3^2 - 2) \triangleq G, \tag{3.57}$$

so that the first differential conditions become

$$G_{x_1} = 2(1+\nu)x_1 = 0$$
$$G_{x_2} = -1 + 2\nu x_2 = 0$$
$$G_{x_3} = 2(1+\nu x_3) = 0 \tag{3.58}$$
$$\psi = x_1^2 + x_2^2 + x_3^2 - 2 = 0.$$

These equations admit four solutions:

(a)  $x_1 = 0$,           $x_2 = 1/2\nu$,     $x_3 = -1/\nu$,   $\nu = (5/8)^{1/2}$

(b)  $x_1 = 0$,           $x_2 = 1/2\nu$,     $x_3 = -1/\nu$,   $\nu = -(5/8)^{1/2}$

(c)  $x_1 = (3/4)^{1/2}$,  $x_2 = -(1/2)$,    $x_3 = 1$,        $\nu = -1$

(d)  $x_1 = -(3/4)^{1/2}$, $x_2 = -(1/2)$,    $x_3 = 1$,        $\nu = -1$.

$$\tag{3.59}$$

The nature of these optimal points is determined from the second differential

$$d^2 J' = 2(1+\nu)dx_1^2 + 2\nu dx_2^2 + 2\nu dx_3^2 \tag{3.60}$$

and $d\psi = 0$, that is,

$$x_1 dx_1 + x_2 dx_2 + x_3 dx_3 = 0. \tag{3.61}$$

For the first two solutions, the second differential reduces to

$$d^2 J' = [dx_1 \ dx_3] \begin{bmatrix} 2(1+\nu) & 0 \\ 0 & 10\nu \end{bmatrix} \begin{bmatrix} dx_1 \\ dx_3 \end{bmatrix} \tag{3.62}$$

and the principal minors of the second derivative matrix are given by

$$P_1 = 2(1+\nu), \quad P_2 = 10\nu, \quad P_3 = 20(1+\nu)\nu. \tag{3.63}$$

If $\nu = (5/8)^{1/2}$, the principal minors become

$$P_1 > 0, \quad P_2 > 0, \quad P_3 > 0, \tag{3.64}$$

which means solution (a) is a minimum. On the other hand, if $\nu = -(5/8)^{1/2}$,

$$P_1 > 0, \quad P_2 < 0, \quad P_3 < 0, \tag{3.65}$$

meaning that solution (b) is neither a minimum or maximum. For the last two solutions ($\nu = -1$), the second differential (3.60) becomes

$$d^2 J' = -2(dx_2^2 + dx_3^2) \tag{3.66}$$

regardless of the relationship between dependent and independent differentials. Hence, since $d^2 J' < 0$, the last two solutions are maxima.

# Problems

3.1  A tin can manufacturer wants to find the dimensions of a cylindrical can (closed top and bottom) such that, for a given amount of tin, the volume of the can is a maximum. If the thickness of the tin stock is constant, a given amount of tin implies a given surface area of the can. Use height and radius as variables, and use a Lagrange multiplier. Show that the optimal height is twice the radius.

3.2  Determine the point $x_1, x_2$ at which the function

$$J = x_1 + x_2$$

is a minimum subject to the constraint

$$x_1^2 + x_1 x_2 + x_2^2 = 1 .$$

Show that $x_1 = x_2 = -1/\sqrt{3}$.

3.3  Show that the results of Section 3.2.2 do not change if the Lagrange multiplier $\nu$ is treated (a) as an independent variable, (b) as a dependent variable, or (c) a constant.

3.4  Minimize the performance index

$$J = \frac{1}{2}(x^2 + y^2 + z^2)$$

subject to the constraints

$$x + 2y - z = 3$$

$$x - y + 2z = 12.$$

Show that

$$x = 5, \quad y = 1, \quad z = 4, \quad \nu_1 = -2, \quad \nu_2 = -3.$$

3.5  Find the point $x, y, z$ that minimizes the performance index

$$J = \frac{1}{2}(x^2 + y^2 + z^2)$$

subject to the constraint

$$x + 2y - 3z - 7 = 0.$$

Show that $x = 1/2$, $y = 1$, $z = -3/2$ and that the reduced second differential (in terms of independent variables only) is positive definite.

3.6  Minimize the performance index

$$J = x - y + 2z$$

subject to the constraint

$$x^2 + y^2 + z^2 = 2.$$

Show that $x = -1/\sqrt{3}$, $y = 1/\sqrt{3}$, $z = -2/\sqrt{3}$ and that the reduced second differential matrix is positive definite.

3.7  Maximize the performance index

$$J = x_1 x_2$$

subject to the constraint

$$x_1 + x_2 - 1 = 0.$$

Show that $x_1 = x_2 = 1/2$ and that the sufficient conditions for a maximum are satisfied.

3.8  Minimize the performance index

$$J = -x_1 x_2 + x_2 x_3 + x_3 x_1$$

subject to the constraint

$$x_1 + x_2 - x_3 = 1.$$

Show that the optimal point is $x_1 = x_2 = -x_3 = 1/3$ and that it is a minimum.

3.9 Find the point $x, y$ that minimizes the performance index

$$J = \sqrt{x^2 + y^2}$$

subject to the constraint
$$y = 1 - x.$$

Show that the minimum occurs at $x = 1/2, y = 1/2$. The problem is to minimize the distance from the origin to the given line but to use the square root performance index.

3.10 Find the point $A, B$ that maximizes the performance index

$$J = \sin A \sin B$$

subject to the constraint
$$A + B = \pi.$$

Use a Lagrange multiplier. Consider only values of $A$ in the range $0 \leq A \leq \pi/2$, but do not adjoin it as a constraint.

Ans. $A = B = \pi/2$

3.11 Find and identify the optimal points of the performance index

$$J = \sin A \cos B$$

subject to the constraint
$$A + B = \pi.$$

Use a Lagrange multiplier. Consider only the solutions for which $0 \leq A \leq \pi$, but do not adjoin it as a constraint.

Ans. $A = \pi/4$, minimum; $A = 3\pi/4$, maximum

3.12 With reference to Figs. 3.3 and 3.4, show that the displacement of fixed magnitude which maximizes the change in $\psi$ is in the direction of the normal vector. In other words, find the displacement $dx_1$, $dx_2$ that maximizes

$$d\psi = \psi_{x_1} dx_1 + \psi_{x_2} dx_2$$

subject to the constraint

$$dx_1^2 + dx_2^2 = dr^2 = \text{Const.}$$

3.13 Find the points on the ellipse

$$(x/a)^2 + (y/b)^2 = 1,$$

where $a$ and $b$ are positive constants and $a > b$, that optimize the distance from the origin to the ellipse, that is,

$$d^2 = x^2 + y^2.$$

Using the second differential, show that the minima occur at $x = 0$, and the maxima, at $y = 0$. What happens to the first and second differential conditions for the case where $a = b$?

3.14 Find the rectangle of maximum area that can be inscribed in an ellipse, that is, maximize the performance index

$$J = 4xy$$

subject to the constraint

$$(x/a)^2 + (y/b)^2 = 1,$$

where $a$ and $b$ are positive constants and $a > b$. Show that the maximum occurs when $x = a/\sqrt{2}$ and $y = b/\sqrt{2}$. Does the problem have a solution when $a = b$?

3.15 Minimize the distance from the origin to the line $y = mx + b$, where $m$ and $b$ are known constants. Show that, for the optimum, the line from the origin is orthogonal to the given line.

3.16 Minimize the distance from the point $(x_0, 0)$ to the circle $x^2 + y^2 = r^2$, where $r > x_0$ is a known constant and $x_0 > 0$. What happens when $x_0 = 0$?

3.17 Minimize the distance from the point $(x_0, 0)$ to the ellipse $(x/a)^2 + (y/b)^2 = 1$, where $x_0$, $a$, and $b$ are known constants where $a > x_0$ and $a > b$.

# 4

# Constrained Minimization: Inequality Constraints

## 4.1  Introduction

The minimization of a function whose variables must satisfy inequality constraints is considered here. Because of their nature, any number of inequality constraints can be imposed. To introduce the subject, conditions for locating a boundary minimal point of a function of one independent variable (see Section 2.2) are determined using the direct approach. Next, a device for converting an inequality constraint into an equality constraint, called a slack variable, is used to solve the same problem. Then, inequality constraints are treated more formally by minimizing a function of two variables subject to one inequality constraint. For a minimal point on the boundary, two second variation conditions must be satisfied. The optimal point must be a minimum with respect to admissible comparison points on the boundary as if the boundary were an equality constraint, and it must be a minimum with respect to admissible comparison points off the boundary. The latter requires that the Lagrange multiplier associated with the inequality constraint ($\leq 0$) must be nonnegative at a minimal point on the boundary. Next, eliminating bounded variables in favor of unbounded variables is discussed. This approach is demonstrated by solving some example problems from linear programming. Finally, some comments are made about the general parameter optimization problem.

## 4.2   Boundary Minimal Points

The development in Section 2.2 for a function of one independent variable has been for interior minimal points. Boundary minimal points differ from interior minimal points in that admissible comparison points must lie to the right or to the left of the minimal point. To develop the necessary conditions for a boundary minimum, the procedure of Section 2.2 is reconsidered. It is assumed that the first differential does not vanish so that it dominates the Taylor series expansion, that is,

$$\Delta J \cong dJ = \phi_x(x)dx. \tag{4.1}$$

Then, $\Delta J$ must be positive regardless of the choice of the admissible comparison point or, equivalently, regardless of the choice of $dx \neq 0$.

For a left boundary minimal point (see Fig. 4.1), $x = x_a$, and $dx$ must be positive to ensure that $x_* > x_a$. Hence, Eq. (4.1) with $dx > 0$ leads to

$$\phi_x(x_a) > 0. \tag{4.2}$$

This means that a left boundary point is a minimum if the slope is positive. However, it is possible for $\phi_x(x_a)$ to be zero and a minimum still exist (see Fig. 4.1). If this happens, higher-order differentials must be examined. As a consequence,

$$\phi_x(x_a) \geq 0 \tag{4.3}$$

is a necessary condition for a left boundary minimal point, and Eq. (4.2) is a sufficient condition.

For a right boundary minimum, $x = x_b$ and $dx < 0$, so that

$$\phi_x(x_b) \leq 0 \tag{4.4}$$

is the corresponding necessary condition.

The conditions developed in this section are actually those for minimizing the performance index

$$J = \phi(x) \tag{4.5}$$

subject to the inequality constraint

$$x \geq x_a \quad \text{or} \quad x \leq x_b. \tag{4.6}$$

The displacement procedure, however, can become tedious for more complex problems, and it is desired to have a more general approach. Such an approach is offered by introducing a new variable called a slack variable and is discussed in the next section.

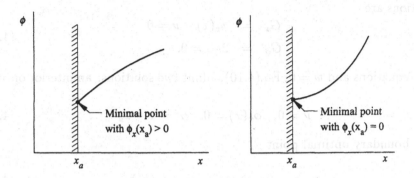

Figure 4.1: Boundary Minimal Points

# 4.3   Introduction to Slack Variables

To introduce slack variables, consider the problem of minimizing the performance index

$$J = \phi(x) \tag{4.7}$$

subject to the inequality constraint

$$x \geq x_a \tag{4.8}$$

or, in the standard form,

$$\theta \triangleq x_a - x \leq 0. \tag{4.9}$$

This inequality constraint can be converted into an equality constraint (Ref. Ha, p. 150) by introducing a real variable $\alpha$, called a slack variable, defined by the equation $\theta = -\alpha^2$. Hence,

$$\psi \triangleq \theta + \alpha^2 = x_a - x + \alpha^2 = 0. \tag{4.10}$$

The variable $\alpha$ is unbounded, and its optimal value is to be determined. While the problem can now be solved with known methods, the slack variable approach increases the number of variables.

The augmented performance index for this problem is given by

$$J' = \phi(x) + \nu(x_a - x + \alpha^2) \triangleq G(x, \alpha), \tag{4.11}$$

where $\nu$ is the Lagrange multiplier. In terms of $G$, the first differential necessary conditions are

$$
\begin{aligned}
G_x &= \phi_x(x) - \nu = 0 \\
G_\alpha &= 2\nu\alpha = 0.
\end{aligned}
\tag{4.12}
$$

These equations and $\psi = 0$, Eq.(4.10), admit two solutions: an interior optimal point

$$
\nu = 0, \quad \phi_x(x) = 0, \quad \alpha^2 = x - x_a
\tag{4.13}
$$

and a boundary optimal point

$$
\alpha = 0, \quad x = x_a, \quad \nu = \phi_x(x_a).
\tag{4.14}
$$

Next, the second differential (3.23), where $x_1 = x$ and $x_2 = \alpha$, is given by

$$
d^2 J' = \phi_{xx}(x)dx^2 + 2\nu d\alpha^2,
\tag{4.15}
$$

where $dx$ and $d\alpha$ are not independent but must satisfy the relation $d\psi = 0$. Hence, choosing $d\alpha$ to be the independent differential,

$$
dx = 2\alpha d\alpha.
\tag{4.16}
$$

Since $d\alpha^2 \geq 0$, the second differential (4.15) leads to the following necessary condition:

$$
2\alpha^2 \phi_{xx}(x) + \nu \geq 0.
\tag{4.17}
$$

Hence, at an interior minimal point ($\nu = 0$), the necessary condition for a minimum is

$$
\phi_{xx}(x) \geq 0
\tag{4.18}
$$

and that for a boundary minimal point ($\alpha = 0$) is

$$
\nu \geq 0.
\tag{4.19}
$$

The latter leads to

$$
\phi_x(x_a) \geq 0
\tag{4.20}
$$

from Eq.(4.14).

## 4.4 Function of Two Variables

In order to make the study of inequality constraints more formal, consider the problem of minimizing the performance index

$$J = \phi(x_1, x_2) \tag{4.21}$$

subject to the inequality constraint

$$\theta(x_1, x_2) \leq 0 \tag{4.22}$$

as illustrated in Fig. 4.2. The inequality constraint can be converted into an equality constraint by introducing a slack variable $\alpha$ such that

$$\psi \triangleq \theta(x_1, x_2) + \alpha^2 = 0. \tag{4.23}$$

Hence, for any real value of $\alpha$, positive or negative, the inequality constraint is satisfied. The problem is to find the values of $x_1, x_2$, and $\alpha$ that minimize the performance index (4.21) subject to the equality constraint (4.23). The augmented performance index is given by

$$J' = \phi(x_1, x_2) + \nu[\theta(x_1, x_2) + \alpha^2] \triangleq G(x_1, x_2, \alpha) \tag{4.24}$$

and the first differential becomes

$$dJ' = G_{x_1}dx_1 + G_{x_2}dx_2 + G_\alpha d\alpha. \tag{4.25}$$

The vanishing of $dJ'$ leads to

$$
\begin{aligned}
G_{x_1} &= \phi_{x_1} + \nu\theta_{x_1} = 0 \\
G_{x_2} &= \phi_{x_2} + \nu\theta_{x_2} = 0 \\
G_\alpha &= 2\nu\alpha = 0.
\end{aligned}
\tag{4.26}
$$

The equation for determining the optimal value of $\alpha$ implies two possible solutions: a solution on the boundary ($\alpha = 0$) and a solution away from the boundary ($\nu = 0$).

For the solution on the boundary, the optimality conditions lead to the following value of $\nu$:

$$\nu = -\frac{\phi_{x_2}}{\theta_{x_2}} = -\frac{\phi_{x_1}}{\theta_{x_1}}, \tag{4.27}$$

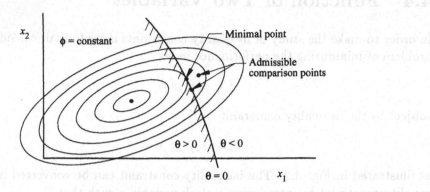

Figure 4.2: Problem Description: Inequality Constraint

and the following equations for determining the values of $x_1$ and $x_2$:

$$\phi_{x_2}\theta_{x_1} - \phi_{x_1}\theta_{x_2} = 0$$
$$\theta = 0. \tag{4.28}$$

To be a minimum, the optimal point obtained from those equations must satisfy the second differential necessary condition

$$d^2J' = G_{x_1x_1}dx_1^2 + 2G_{x_1x_2}dx_1dx_2 + G_{x_2x_2}dx_2^2$$
$$+ 2G_{x_1\alpha}dx_1d\alpha + 2G_{x_2\alpha}dx_2d\alpha + G_{\alpha\alpha}d\alpha^2 \geq 0, \tag{4.29}$$

where the differentials are not independent but must satisfy the relation

$$d\psi = \theta_{x_1}dx_1 + \theta_{x_2}dx_2 + 2\alpha d\alpha = 0. \tag{4.30}$$

Choose $dx_1$ and $d\alpha$ to be the independent differentials. Then, since

$$\alpha = 0, \quad G_{x_1\alpha} = 0, \quad G_{x_2\alpha} = 0, \quad G_{\alpha\alpha} = 2\nu, \tag{4.31}$$

the second differential becomes

$$d^2J' = \left[ G_{x_1x_1} - 2G_{x_1x_2}\left(\frac{\theta_{x_1}}{\theta_{x_2}}\right) + G_{x_2x_2}\left(\frac{\theta_{x_1}}{\theta_{x_2}}\right)^2 \right]dx_1^2$$
$$+ 2\nu d\alpha^2 \geq 0. \tag{4.32}$$

Finally, the conditions to be satisfied by a minimum on the boundary are given by

$$G_{x_1 x_1} - 2 G_{x_1 x_2} \frac{\theta_{x_1}}{\theta_{x_2}} + G_{x_2 x_2} \left( \frac{\theta_{x_1}}{\theta_{x_2}} \right)^2 \geq 0 \qquad (4.33)$$

$$\nu \geq 0. \qquad (4.34)$$

The first condition indicates a minimum with respect to differentials along the boundary, and the second condition indicates a minimum on the boundary with respect to differentials away from the boundary into the admissible region. That the differentials are into the admissible region ($\theta < 0$) is seen from the relation

$$\Delta \theta = d\theta + \frac{1}{2!} d^2 \theta = -2\alpha d\alpha - d\alpha^2 \qquad (4.35)$$

which for $\alpha = 0$ becomes

$$\Delta \theta = -d\alpha^2. \qquad (4.36)$$

Hence, $\Delta \theta$ is negative for all differentials $d\alpha$.

That $\nu$ should be nonnegative is apparent from Eqs.(4.27). In Fig. 4.2, it is seen that at the minimum $\phi_{x_1} > 0$ and $\theta_{x_1} < 0$ so that $\nu > 0$. Had the constraint passed through the unconstrained minimum, then $\phi_{x_1} = 0$ and $\nu = 0$. Hence, geometrically, $\nu \geq 0$ at a boundary minimum.

For the solution off the boundary ($\nu = 0$), the first and second differentials must give back the conditions for a minimum of a function of two independent differentials. In this connection, the first differential conditions (4.26) lead to

$$\phi_{x_1} = 0, \quad \phi_{x_2} = 0. \qquad (4.37)$$

Next, since

$$G_{x_1 x_1} = \phi_{x_1 x_1}, \quad G_{x_1 x_2} = \phi_{x_1 x_2}, \quad G_{x_2 x_2} = \phi_{x_2 x_2}, \quad \nu = 0, \qquad (4.38)$$

the second differential (4.29) becomes

$$d^2 J' = \phi_{x_1 x_1} dx_1^2 + 2\phi_{x_1 x_2} dx_1 dx_2 + \phi_{x_2 x_2} dx_2^2 \geq 0, \qquad (4.39)$$

where $dx_1$ and $dx_2$ can be identified as independent differentials. Hence, the conditions for a minimum are given by (see Section 2.5)

$$\phi_{x_1 x_1} \geq 0, \quad \phi_{x_2 x_2} \geq 0, \quad \phi_{x_1 x_1} \phi_{x_2 x_2} - \phi_{x_1 x_2}^2 \geq 0. \qquad (4.40)$$

## 4.5   Example

As an example, find the point $x_1, x_2$ that minimizes the performance index

$$J = -x_1 x_2, \tag{4.41}$$

subject to the inequality constraint

$$-x_1 - x_2 + 1 \leq 0. \tag{4.42}$$

From the augmented performance index

$$J' = -x_1 x_2 + \nu[-x_1 - x_2 + 1 + \alpha^2] \triangleq G(x_1, x_2, \alpha), \tag{4.43}$$

it is seen that the first differential conditions lead to

$$\begin{aligned}
G_{x_1} &= -x_2 - \nu = 0 \\
G_{x_2} &= -x_1 - \nu = 0 \\
G_{\alpha} &= 2\nu\alpha = 0.
\end{aligned} \tag{4.44}$$

The solution on the boundary ($\alpha = 0$) is given by

$$x_1 = x_2 = \frac{1}{2}, \quad \nu = -\frac{1}{2}. \tag{4.45}$$

However, since

$$G_{x_1 x_1} = 0, \quad G_{x_1 x_2} = -1, \quad G_{x_2 x_2} = 0, \quad \theta_{x_1} = 1, \theta_{x_2} = 1, \tag{4.46}$$

the second-differential conditions (4.33) and (4.34) become

$$2 \geq 0, \quad -\frac{1}{2} \geq 0 \tag{4.47}$$

and are not satisfied for a minimum.

The solution off the boundary ($\nu = 0$) is given by

$$x_1 = x_2 = 0, \quad \alpha^2 = -1. \tag{4.48}$$

Since $\alpha$ is imaginary, there is no solution off the boundary.

The conclusion reached from this analysis is that there exists no analytical minimum for this problem. Indeed, inspection of Eqs. (4.41) and (4.42) shows that the minimum occurs at $x_1 = x_2 = \infty$, and it is not an analytical minimum.

## 4.6 Eliminating Bounded Variables

Consider the problem of minimizing the performance index

$$J = \phi(x), \tag{4.49}$$

subject to the inequality constraint

$$x \geq x_a. \tag{4.50}$$

Introduce a new variable $\alpha$ such that

$$x = x_a + \alpha^2. \tag{4.51}$$

Instead of adjoining this equation as an inequality constraint, the bounded variable $x$ is simply eliminated from the problem, that is, the performance index is expressed as

$$J = \phi[x(\alpha)]. \tag{4.52}$$

The first differential necessary condition for a minimum $d\phi/d\alpha = 0$ leads to

$$\frac{d\phi}{d\alpha} = \frac{d\phi}{dx}\frac{dx}{d\alpha} = 0 \tag{4.53}$$

which implies that

$$\phi_x(x) = 0 \tag{4.54}$$

or

$$\frac{dx}{d\alpha} = 2\alpha = 0. \tag{4.55}$$

The former is the interior optimal, and the latter is the boundary optimal.

The second differential necessary condition requires that

$$\frac{d^2\phi}{d\alpha^2} = \frac{d^2\phi}{dx^2}\left(\frac{dx}{d\alpha}\right)^2 + \frac{d\phi}{dx}\frac{d^2x}{d\alpha^2} \geq 0 \tag{4.56}$$

which can be rewritten as

$$\frac{d^2\phi}{d\alpha^2} = 4\phi_{xx}(x)\alpha^2 + 2\phi_x(x) \geq 0. \tag{4.57}$$

Hence, for the interior minimal point ($\phi_x = 0$), this condition leads to $\phi_{xx}(x) \geq 0$, and for the boundary minimal point ($\alpha = 0$), to $\phi_x(x) \geq 0$.

The general idea is to introduce an unbounded variable to replace the bounded variable. For a constraint of the form

$$x_a \leq x \leq x_b, \qquad (4.58)$$

the approach of Eq.(4.50) is not valid because $\alpha$ is bounded by $x \leq x_b$. Here, it is observed that

$$0 \leq \frac{x - x_a}{x_b - x_a} \leq 1. \qquad (4.59)$$

A function of an unbounded variable which varies between 0 and 1 is $\sin^2 \alpha$. Hence, set

$$\frac{x - x_a}{x_b - x_a} = \sin^2 \alpha, \qquad (4.60)$$

so that

$$x = x_a + (x_b - x_a) \sin^2 \alpha. \qquad (4.61)$$

The constrained variable $x$ can be eliminated from the performance index in favor of the unconstrained variable $\alpha$.

This approach is not general in nature in that it may not always be possible to solve for the bounded variable in terms of the unbounded variable. However, devices such as (4.51) or (4.60) are often useful in numerical work.

## 4.7  Linear Programming Examples

The linear programming problem involves the minimization of a linear performance index subject to linear constraints. Because the performance index is linear, the minimum occurs on the boundary of the admissible region (where the constraints are satisfied). A simple example is to minimize the performance index

$$J = -x, \qquad (4.62)$$

subject to the inequality constraints

$$y + x - 1 \leq 0, \quad -x \leq 0, \quad -y \leq 0. \qquad (4.63)$$

The constraints form a closed region in the $xy$-plane as shown in Fig. 4.3. The inequality constraints can be converted into equality constraints by introducing slack variables $\alpha, \beta, \gamma$ as

$$y + x - 1 + \alpha^2 = 0, \quad -x + \beta^2 = 0, \quad -y + \gamma^2 = 0. \qquad (4.64)$$

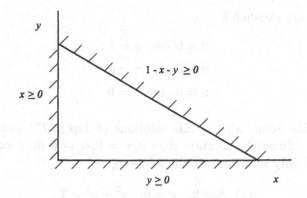

Figure 4.3: Linear Programming Problem

Because the constraints are linear, $x$ and $y$ can be eliminated from the problem to form the new problem of minimizing

$$J = -\beta^2,\tag{4.65}$$

subject to the equality constraint

$$\gamma^2 + \beta^2 - 1 + \alpha^2 = 0.\tag{4.66}$$

Note that there are three variables and one constraint so that there are two independent variables. If $\beta$ is eliminated between the performance index and the constraint, the ability to optimize with respect to $\beta$ is lost.

In terms of the augmented performance index

$$J' = -\beta^2 + \nu(\gamma^2 + \beta^2 - 1 + \alpha^2),\tag{4.67}$$

the first differential is given by

$$dJ' = 2\beta(-1 + \nu)d\beta + 2\nu\gamma d\gamma + 2\nu\alpha d\alpha.\tag{4.68}$$

If the independent differentials are chosen to be $d\gamma$ and $d\alpha$, the multiplier is chosen so that the coefficient of the dependent differential $d\beta$ vanishes. Then, the coefficients of the independent differentials must vanish. The first differential conditions for the optimal points are given by

$$2\beta(\nu - 1) = 0$$
$$2\nu\gamma = 0\tag{4.69}$$
$$2\nu\alpha = 0.$$

In order, they are satisfied if

$$\beta = 0 \quad \text{or} \quad \nu = 1$$
$$\nu = 0 \quad \text{or} \quad \gamma = 0 \tag{4.70}$$
$$\nu = 0 \quad \text{or} \quad \alpha = 0.$$

At this point, all possible solutions of Eqs.(4.70) and (4.66) must be considered. Since the solution $\beta = 0, \gamma = 0, \alpha = 0$ does not satisfy the constraint, the only solutions are the following:

$$\text{(a)} \quad \beta = 0, \quad \nu = 0, \quad \gamma^2 + \alpha^2 = 1 \tag{4.71}$$
$$\text{(b)} \quad \nu = 1, \quad \gamma = 0, \quad \alpha = 0, \quad \beta^2 = 1. \tag{4.72}$$

Solution (a) is all the points between $(0,0)$ and $(0,1)$, and solution (b) is the point $(1,0)$.

Next, the second differential is given by

$$d^2 J' = 2(\nu - 1)d\beta^2 + 2\nu d\gamma^2 + 2\nu d\alpha^2 \geq 0 \tag{4.73}$$

and must be nonnegative for a minimum. The differentials are not independent. Taking the differential of the constraint (4.66) leads to

$$\gamma d\gamma + \beta d\beta + \alpha d\alpha = 0. \tag{4.74}$$

For solution (a), these equations become

$$d^2 J' = -2d\beta^2 \tag{4.75}$$
$$\gamma d\gamma + \alpha d\alpha = 0. \tag{4.76}$$

Since Eq.(4.76) does not impose any conditions on $d\beta$, it is arbitrary, and Eq.(4.75) indicates that solution (a) is not a minimum. For solution (b), Eqs.(4.73) and (4.66) become

$$d^2 J' = 2d\gamma^2 + 2d\alpha^2 \geq 0 \tag{4.77}$$
$$d\beta = 0. \tag{4.78}$$

Hence, with $d\gamma$ and $d\alpha$ as independent differentials, the second differential is positive. Hence, solution (b) is the minimum.

Is solution (b) a maximum? Since there are no restrictions on $d\beta$, it can be zero. Therefore, there is not a unique maximum. The performance index has the highest value all along the line $x = 0$. This problem does not have a maximal point.

There are constraint configurations for which it is not possible to solve for the Lagrange multipliers. Consider the problem (Ref. Wa) of minimizing the performance index

$$J = -x, \tag{4.79}$$

subject to the constraints

$$y - (x - 1)^3 \le 0, \quad -x \le 0, \quad -y \le 0. \tag{4.80}$$

The $xy$-plane is shown in Fig. 4.4. It is seen that if the minimal point is at $(1,0)$, it is not possible to solve for the multipliers because the normals to the two constraints at this point are colinear. This point has been discussed in Section 3.4.

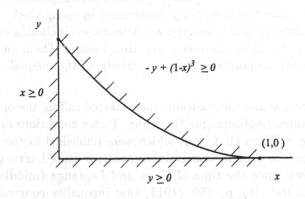

Figure 4.4: Problematic Constraints

## 4.8  General Problem

The general parameter optimization problem is to find the point $x_1, x_2, \ldots, x_n$ that minimizes the performance index

$$J = \phi(x_1, x_2, \ldots, x_n), \tag{4.81}$$

subject to the equality constraints

$$\psi_1(x_1, x_2, \ldots, x_n) = 0$$
$$\psi_2(x_1, x_2, \ldots, x_n) = 0$$
$$\vdots$$
$$\psi_p(x_1, x_2, \ldots, x_n) = 0,$$

(4.82)

and the inequality constraints

$$\theta_1(x_1, x_2, \ldots, x_n) \leq 0$$
$$\theta_2(x_1, x_2, \ldots, x_n) \leq 0$$
$$\vdots$$
$$\theta_q(x_1, x_2, \ldots, x_n) \leq 0.$$

(4.83)

Analytically, the solution of the general problem is straightforward. Inequality constraints are converted to equality constraints by using slack variables. All optimal points, interior and boundary, are determined, including optimal points that do not satisfy all of the inequality constraints. Then, in the solution process, those optimal points that do not satisfy all the inequality constraints are discarded.

In recent years, some authors have started calling the optimality conditions for parameter optimization the Kuhn–Tucker conditions or the Karush–Kuhn–Tucker conditions (Ref. Ku) which were published in the 1940s. Optimality conditions for equality constrained problems using Lagrange multipliers have been known since the time of Euler and Lagrange (middle 1700s), and it was stated in Ref. Ha, p. 150, (1917) that inequality constraints could be converted into equality constraints by using slack variables, although they were not called that at the time. Unfortunately, these conditions were not named at the time they were created.

# Problems

4.1 Find and identify the optimal points of the performance index

$$J = (x - 1)^2.$$

subject to the inequality constraint

$$x \geq 2.$$

Show that the only optimal point is on the boundary ($x = 2$) and that it is a minimum.

4.2 Maximize the performance index

$$J = (x - 1)^2$$

subject to the inequality constraint

$$x \leq 2.$$

Show that there are two optimal points. The optimal point off the boundary ($x = 1$) is a minimum, and the optimal point on the boundary ($x = 2$) is the desired maximum.

4.3 Maximize the performance index

$$J = xu,$$

subject to the inequality constraint

$$x + u \leq 1.$$

Show that the maximum occurs on the boundary at $x = u = 1/2$.

4.4 Consider the problem of maximizing the performance index

$$J = x_1 x_2,$$

subject to the constraint

$$x_1 + x_2 \geq 1.$$

Show that an optimal point occurs on the boundary but that it is not a maximum with respect to admissible comparison points off the boundary.

4.5 Find the point $x, u$ that optimizes the performance index

$$J = x^2 + u^2,$$

subject to the inequality constraint

$$x - 2u \geq 1.$$

Determine whether it is a minimum or a maximum. Show that a minimum occurs at $x = 1/5, u = -2/5$.

4.6 Find the point $x, u$ that minimizes the performance index

$$J = x^2 + y^2,$$

subject to the inequality constraint

$$y \geq 1 + x^2.$$

Show that the optimal point occurs at $x = 0$, $y = 1$ and that it is a minimum.

4.7 Minimize the peformance index

$$J = (4 - 3x^2)^{1/2},$$

subject to the constraint

$$0 \leq x \leq 1.$$

Show that the minimum occurs at $x = 1$. What happens at $x = 0$?

4.8 Maximize the performance index

$$J = 4 - 3x^2,$$

subject to the inequality constraint

$$x \geq 0.$$

Use a slack variable and a Lagrange multiplier. For this problem, the second differential vanishes, and it is necessary to use higher-order differentials. The maximum occurs at $x = 0$, and the results can be verified by eliminating the bounded variable $x$ in terms of the unbounded slack variable.

4.9 Find and identify the optimal points of the performance index

$$J = x_1^2 - x_2^2,$$

subject to the constraint

$$x_1 \geq 0.$$

Show that the only optimal point is $x_1 = x_2 = 0$ and that it is neither a minimum nor a maximum.

4.10 Find the minimal points of the function

$$J = (x_1 - x_2)^2,$$

where the variables $x_1$ and $x_2$ must satisfy the inequality constraint

$$x_1 + x_2 \geq 1.$$

Use a Lagrange multiplier and apply both the first differential and second differential necessary conditions. To explain the results, show that a unique optimal point does not exist; rather, there is a line of optimal points.

4.11 Find the point $x, y$ that minimizes the performance index

$$J = (x - 1)^2 + y^2,$$

subject to the constraint

$$y \geq 0.$$

Show that the optimal point is $x = 1$, $y = 0$ but that higher-order differentials are needed to prove it is a minimum.

4.12 Find the point $x, y, z$ that minimizes the performance index

$$J = (1/2)(x^2 + y^2 + z^2),$$

subject to the constraints

$$x + 2y - z = 3$$

$$x - y + 2z \leq 12.$$

Show that there are two optimal points. The off-boundary point ($x = 1/2, y = 1, z = -1/2$) is a minimum, and the on-boundary point ($x = 5, y = 1, z = 4$) is a saddle point.

4.13 Find and identify the optimal points of the performance index

$$J = \sin^2 A$$

that satisfy the inequality constraint

$$\frac{\pi}{4} \leq A \leq \frac{3\pi}{4}.$$

Show that a maximum is located at $A = \pi/2$ and that there are minima at $a = \pi/4$ and $a = 3\pi/4$.

# 5

# Minimization Using Matrix Notation

## 5.1 Introduction

After a brief review of matrix algebra and the development of matrix calculus, the conditions to be satisfied by a minimum of a function of $n$ independent variables and a function of $n$ constrained variables are derived in matrix notation.

## 5.2 Matrix Algebra

A matrix is an ordered array of elements for which the following terminology is used:

$$\text{scalar} \qquad a \qquad 1 \times 1 \text{ matrix} \qquad (5.1)$$

$$\text{column vector} \qquad \begin{bmatrix} a_1 \\ \vdots \\ a_m \end{bmatrix} \qquad m \times 1 \text{ matrix} \qquad (5.2)$$

$$\text{row vector} \qquad \begin{bmatrix} a_1 & \cdots & a_n \end{bmatrix} \qquad 1 \times n \text{ matrix} \qquad (5.3)$$

$$\text{matrix} \quad \begin{bmatrix} a_{11} & \cdots & a_{1n} \\ \vdots & & \vdots \\ a_{m1} & \cdots & a_{mn} \end{bmatrix} \quad m \times n \text{ matrix} \qquad (5.4)$$

where the first subscript denotes the row, and the second, the column.

## 5.2.1 Addition

To add or subtract two matrices, the matrices must be of the same size. The sum

$$c = a + b \qquad (5.5)$$

means that each element in $a$ is added to the corresponding element in $b$ to obtain $c$, that is,

$$c_{ij} = a_{ij} + b_{ij}. \qquad (5.6)$$

Addition is both commutative and associative:

$$a + b = b + a, \quad a + (b + c) = (a + b) + c. \qquad (5.7)$$

## 5.2.2 Multiplication

A matrix can be multiplied by a constant or by another matrix. If $k$ denotes a constant, the operation

$$b = ka \qquad (5.8)$$

means that

$$b_{ij} = ka_{ij}. \qquad (5.9)$$

To multiply one matrix by another, the number of columns of the first must equal the number of rows of the second. The operation

$$c = ab \qquad (5.10)$$

means that

$$c_{ij} = \sum_{k=1}^{n} a_{ik} b_{kj}. \qquad (5.11)$$

Matrix multiplication is associative, distributive, but not commutative:

$$a(bc) = (ab)c, \quad a(b + c) = ab + bc, \quad ab \neq ba. \qquad (5.12)$$

## 5.2.3   Special Matrices

The special matrices considered here are the null, the identity, the transpose, and the inverse matrices. The null or zero matrix is the matrix whose elements are all zero. The identity or unit matrix is the square matrix whose diagonal elements are ones and whose off-diagonal elements are zeroes. The transpose matrix is obtained from another matrix $a = [a_{ij}]$ by interchanging its rows and columns, that is,

$$a^T = [a_{ji}].\tag{5.13}$$

For a product of two matrices, the transpose satisfies the relation

$$(ab)^T = b^T a^T.\tag{5.14}$$

The inverse of a square matrix is defined by the relations

$$a^{-1} = [\alpha_{ij}], \quad \alpha_{ij} = \frac{A_{ji}}{|a|}, \quad A_{ij} = (-1)^{i+j} M_{ij},\tag{5.15}$$

where $|a|$ is the determinant of $a$ and where $M_{ij}$ is the determinant of the array that remains after the $i$th row and $j$th column have been deleted. The product of a matrix and its inverse results in the identity matrix $(aa^{-1} = I)$. Finally, for the product of two square matrices, the inverse is given by

$$(ab)^{-1} = b^{-1} a^{-1}.\tag{5.16}$$

# 5.3   Matrix Calculus

The operations discussed in this section are differential of a scalar function of a scalar, differential of a scalar function of a vector, differential of a vector function of a vector, and integration.

## 5.3.1   Differential

By definition, the differential of a matrix $a = [a_{ij}]$ is the matrix of differentials of its elements, that is,

$$da = [da_{ij}].\tag{5.17}$$

It is easily shown that

$$d(a + b) = da + db$$
$$d(ab) = (da)b + a(db)$$

(5.18)

by working in terms of elements. Note that the order of the matrices must be maintained.

Consider a scalar function $a$ of the scalar variable $x$ or $a = a(x)$. From Section 1.6, it is known that

$$da = a_x dx,$$

(5.19)

where $a_x$ denotes the derivative $da/dx$. In the following derivations, the form (5.19) is maintained.

Next, consider a scalar function $a$ of the column vector $x = [x_1 \cdots x_n]^T$, that is, $a = a(x_1, \ldots, x_n)$. From the chain rule,

$$da = a_{x_1} dx_1 + \cdots + a_{x_n} dx_n.$$

(5.20)

Now, define the matrices

$$a_x = [a_{x_1} \cdots a_{x_n}], \quad dx = \begin{bmatrix} dx_1 \\ \vdots \\ dx_n \end{bmatrix},$$

(5.21)

where $a_x$ is a row vector and, since $x$ is a column vector, $dx$ is a column vector. Then, from the definition of matrix multiplication (5.11), the differential (5.20) becomes

$$da = a_x dx.$$

(5.22)

The next case is where $a$ is a column vector whose elements are functions of the column vector $x$. Since $a = [a_1 \cdots a_m]^T$,

$$da = \begin{bmatrix} da_1 \\ \vdots \\ da_m \end{bmatrix} = \begin{bmatrix} a_{1x_1} dx_1 + \cdots + a_{1x_n} dx_n \\ \vdots \\ a_{mx_1} dx_1 + \cdots + a_{mx_n} dx_n \end{bmatrix}$$

(5.23)

or

$$da = \begin{bmatrix} a_{1x_1} & \cdots & a_{1x_n} \\ \vdots & & \vdots \\ a_{mx_1} & \cdots & a_{mx_n} \end{bmatrix} \begin{bmatrix} dx_1 \\ \vdots \\ dx_n \end{bmatrix}.$$

(5.24)

Then, if $a_x$ is defined as

$$a_x = \begin{bmatrix} a_{1x_1} & \cdots & a_{1x_n} \\ \vdots & & \vdots \\ a_{mx_1} & \cdots & a_{mx_n} \end{bmatrix}, \tag{5.25}$$

Eq. (5.24) can be rewritten as

$$da = a_x dx. \tag{5.26}$$

Hence, by the proper definition of $a_x$, the simple form (5.19) can be maintained as the dimensions of $a$ and $x$ change.

As an application, consider the differential of the column vector $x$ with respect to itself. Taking the differential of $a = x$ leads to

$$da = dx. \tag{5.27}$$

Comparison with Eq.(5.26) yields

$$a_x = I, \tag{5.28}$$

or $x_x = I$. Next, assume that

$$a = Ax, \tag{5.29}$$

where $A$ is an $n \times n$ constant matrix. By taking the differential directly,

$$da = Adx \tag{5.30}$$

so that

$$a_x = A. \tag{5.31}$$

The differential of a row vector $b = [b_1 \cdots b_n]$ can be taken by transposing it into a column vector and applying Eq. (5.26). Hence,

$$d(b^T) = (b^T)_x\, dx, \tag{5.32}$$

and transposing the result leads to

$$db = dx^T b^T_x{}^T. \tag{5.33}$$

As an application of this formula, consider the differential of the product of a row vector and a column vector, which is a scalar. Here, if $c = ba$,

$$dc = (db)a + b(da) = dx^T b^T_x{}^T a + ba_x dx = (a^T b^T_x + ba_x)dx. \tag{5.34}$$

If the differential is written as $dc = c_x dx$, the expression for $c_x$ is given by

$$c_x = a^T b^T{}_x + ba_x. \tag{5.35}$$

A frequent use of this result is for the case where

$$c = \frac{1}{2} x^T A x, \tag{5.36}$$

where $A$ is an $n \times n$ constant matrix. If $b = (1/2)x^T$ and $a = Ax$, Eq.(5.35) gives

$$c_x = \frac{1}{2} x^T (A^T + A), \tag{5.37}$$

or, if $A$ is symmetric,

$$c_x = x^T A. \tag{5.38}$$

It is interesting to verify Eq.(5.37) by taking the differential of Eq.(5.36). This process leads to

$$dc = \frac{1}{2}[d(x^T)Ax + x^T A dx]. \tag{5.39}$$

Then, since

$$d(x^T) = (dx)^T, \tag{5.40}$$

transposing the first term yields

$$dc = \frac{1}{2} x^T (A^T + A) dx \tag{5.41}$$

which leads to Eq.(5.37).

## 5.3.2  Integration

By definition, the integral of a matrix equals the matrix of the integrals of the elements, that is,

$$\int a \, dt = \left[\int a_{ij} \, dt\right]. \tag{5.42}$$

Integration by parts holds, but matrix positions must be maintained, that is,

$$\int_{t_0}^{t_f} u^T \frac{dv}{dt} dt = \left[u^T v\right]_{t_0}^{t_f} - \int_{t_0}^{t_f} \frac{du}{dt}^T v \, dt, \tag{5.43}$$

where $u(t)$ and $v(t)$ are column vectors.

## 5.4   Function of $n$ Independent Variables

A scalar performance index involving $n$ variables can be written as

$$J = \phi(x), \tag{5.44}$$

where $x = [x_1 \cdots x_n]^T$. The first differential is given by

$$dJ = \phi_x \, dx \tag{5.45}$$

and must vanish. Because the differentials $dx$ are independent, the matrix condition to be satisfied at an optimal point is given by

$$\phi_x = 0, \tag{5.46}$$

where $\phi_x = [\phi_{x_1} \cdots \phi_{x_n}]$ is a row vector. For a matrix to be zero, each element of the matrix must vanish.

The necessary condition for a minimum is that the second differential must be nonnegative. Hence,

$$d^2 J = d(dJ) = d(\phi_x)dx = dx^T \phi_x{}^T{}_x{}^T dx \geq 0. \tag{5.47}$$

Since the second derivative matrix is symmetric, the second transpose can be dropped. Furthermore, to simplify the notation, the following convention is adopted:

$$\phi_{xx} \triangleq \phi_x{}^T{}_x. \tag{5.48}$$

Finally, Eq. (5.47) requires that

$$\phi_{xx} \geq 0. \tag{5.49}$$

## 5.5   Function of $n$ Constrained Variables

Here, the problem is to minimize the scalar performance index

$$J = \phi(x), \tag{5.50}$$

subject to the constraints

$$\psi(x) = 0, \tag{5.51}$$

where $\psi = [\psi_1 \cdots \psi_p]^T$. If a Lagrange multiplier $\nu = [\nu_1 \cdots \nu_p]^T$ is introduced, the augmented performance index is the following:

$$J' = \phi(x) + \nu^T \psi(x) \triangleq G(x, \nu). \tag{5.52}$$

Recall that in taking differentials the multiplier $\nu$ can be left unchanged. In deriving the second differential, dependent differentials do not have to be changed.

The first differential is given by

$$dJ' = G_x dx \tag{5.53}$$

and must vanish. After the $\nu$s are chosen so that the coefficients of the dependent $dx$s vanish, then the coefficients of the independent $dx$s must vanish. Hence, the first differential and the constraint lead to the following conditions:

$$G_x = 0, \quad \psi = 0 \tag{5.54}$$

which determine $x$ and $\nu$ at the optimal point.

For the optimal point to be a minimum, the second differential when expressed in terms of independent differentials must be nonnegative. The second differential is given by

$$d^2 J' = dx^T G_{xx} dx. \tag{5.55}$$

The $dx$s are not independent but must satisfy

$$d\psi = \psi_x dx = 0. \tag{5.56}$$

Equation (5.56) is used to eliminate the dependent differentials from the quadratic form (5.55). This gives rise to a new quadratic form in terms of the independent variables, and the associated matrix must be nonnegative for a minimum.

The new quadratic form can be obtained explicitly by defining the independent variables to be $u = [u_1 \cdots u_m]^T$ and the dependent variables to be $z = [z_1 \cdots z_p]^T$, that is, $x = [u^T z^T]^T$. Then, Eqs.(5.55) and (5.56) become

$$d^2 J' = du^T G_{uu} du + dz^T G_{zu} du + du^T G_{uz} dz + dz^T G_{zz} dz \tag{5.57}$$

and

$$\psi_u du + \psi_z dz = 0. \tag{5.58}$$

If the dependent variables $z$ are chosen such that the square matrix $\psi_z$ is nonsingular (det $\psi_z \neq 0$), it can be inverted so that

$$dz = -\psi_z^{-1}\psi_u du. \tag{5.59}$$

Then, elimination of $dz$ from Eq.(5.57) leads to a quadratic form in $du$ where the enclosed matrix is symmetric. Hence, the necessary condition for a minimum becomes

$$G_{uu} - \psi_u^T \psi_z^{-T} G_{zu} - G_{uz}\psi_z^{-1}\psi_u + \psi_u^T\psi_z^{-T}G_{zz}\psi_z^{-1}\psi_u \geq 0. \tag{5.60}$$

For the case of two variables, Eq.(5.60) reduces to Eq.(3.25) as it should.

# Problems

5.1 If $c = y^T x$, where $y$ is an $n \times 1$ constant vector and $x$ is $n \times 1$, derive an expression for the derivative $c_x$. Repeat the derivation for $c = x^T y$.

5.2 Minimize the performance index

$$J = (1/2)(z^T P z + u^T Q u),$$

subject to the constraint

$$Az + Bu + C = 0.$$

Here, the dimensions of $z$, $u$, $P$, $Q$, $A$, $B$, and $C$ are $p \times 1$, $m \times 1$, $p \times p$, $m \times m$, $p \times p$, $p \times m$, and $p \times 1$, respectively. Also, $P$ and $Q$ are positive definite and symmetric, and $P$, $Q$, $A$, $B$, $C$ are constants. The variable $z$ represents the dependent variables, and the variable $u$ represents the independent variables assuming that $A$ is invertible. Show that the optimal point is defined by

$$u = -Q^{-1}B^T(AP^{-1}A^T + BQ^{-1}B^T)^{-1}C$$
$$z = -P^{-1}A^T(AP^{-1}A^T + BQ^{-1}B^T)^{-1}C$$
$$\nu = (AP^{-1}A^T + BQ^{-1}B^T)^{-1}C$$

and that the condition for a minimum is given by

$$B^T A^{-T} P A^{-1} B + Q \geq 0.$$

5.3 Apply the results of Problem 5.2 to Problem 3.4.

5.4 Apply the results of Problem 5.2 to Problem 3.5.

# Part II. Optimal Control Theory

A standard optimal control problem involves a scalar performance index, a set of differential constraints (system dynamics), and some boundary conditions. In the differential equations, the differentiated variables are called states and the undifferentiated variables are called controls. The objective is to find the control histories that drive the system from the initial conditions to the final conditions while optimizing the performance index. The unknowns are curves and points.

Optimal control theory is another name for a branch of mathematics called the calculus of variations. Both names are used interchangeably. The origin of the calculus of variations is generally credited to Newton, but it has been developed through the years by many famous mathematicians. In recent times, the need to calculate optimal performance of dynamical systems has lead to a renewed interest in this subject matter as well as a reformulation of the problem in a control notation. Neither name is entirely appropriate for the subject matter. In the calculus of variations, the primary operation is the differential, and the variation is a differential taken while holding the time variable of integration constant. The name optimal control theory seems to imply that the subject matter is control theory. While there are many optimization problems associated with control theory, the general subject matter in optimal control theory is optimization. With this understanding, the name optimal control theory is used here.

In the derivation of optimality conditions, the minimal path is assumed to be known. Then, the value of the performance index along the minimal path is compared with that along an admissible comparison path (one that satisfies all the constraints). The admissible comparison path is related to the minimal path by a system of small changes called differentials. Using a Taylor series expansion (Appendix A), the total change in the performance index can be expressed as a first-order part (first differential) plus a second-order part (second differential). Differential calculus is a calulus of small but finite changes that can be used to make a Taylor expansion one term at a time. It is developed in Chapter 6 for differential equations, boundary conditions, and integrals. A particular admissible comparison path is used to derive first and second differential conditions. It exists if the first-order differential equation and boundary condition satisfy the controllability condition (Chapter 7).

In the remainder of Part II, necessary conditions and sufficient conditions for a minimal control are derived for a sequence of optimal control problems of increasing complexity. First, in Chapter 8, the simplest optimal control problem, that is, one that does not involve any states, is investigated. Because of its simplicity, many of the fundamentals of optimal control theory such as the first differential, second differential, necessary conditions, sufficient conditions, control discontinuities, control constraints, and so on, can be discussed in one place.

In Chapters 9 through 12, the final time is assumed fixed. In Chapter 9, the first differential conditions are developed for a standard optimal control problem, and the Euler–Lagrange equations and the natural boundary conditions are derived. Also, it is shown that a first integral exists for problems that do not explicitly contain the time. Next, two tests for a minimum are derived in Chapter 10: the Weierstrass condition and the Legendre–Clebsch condition. The second differential is used in Chapter 11 to present another derivation of the Legendre–Clebsch condition and to derive the sufficient condition (the Jacobi conjugate point condition) for a minimum for the class of problems in which the strengthened Legendre–Clebsch condition is satisfied. Finally, guidance, the process of controlling a dynamical system to a prescribed final point with and without a reference path, is discussed in Chapter 12.

If the prescribed boundary conditions are not adjoined to the performance index by Lagrange multipliers, the vanishing of the first differential leads to a condition called the transversality condition. The transversality condition is combined with differentials of the prescribed boundary conditions to obtain the natural bounday conditions. On the other hand, if the prescribed boundary conditions are adjoined , the vanishing of the first differential is accomplished by choosing the Lagrange multipliers and leads directly to the natural boundary conditions. Since the latter is the procedure followed in this text, there is no transversality condition.

Optimal control problems with free final time are considered in Chapter 13. In addition to a natural boundary condition for the final time, necessary conditions and a sufficient condition similar to those of Chapter 11 are derived. Another way to deal with a free final time problem is to convert it into a "fixed final time" problem by a coordinate transformation. This transformation makes the actual final time an unknown parameter. The optimal control problem involving parameters is discussed in Chapter 14.

To investigate control discontinuities, it is necessary to be able to solve optimal control problems with free initial conditions. In Chapter 15, necessary conditions and sufficient conditions for a minimal initial point are derived. Then, first and second differential conditions for a minimal control with a discontinuity are developed in Chapter 16. The resulting Erdmann–Weierstrass corner conditions are used to join the pieces (subarcs) of a minimal path. Sufficient conditions similar to those of Chapter 13 are also obtained.

In Chapter 17, path constraints such as integral constraints, control constraints, and state constraints are investigated. In each case, the new problem can be converted into a previously considered problem and solved directly. Because the optimal path can have a wide variety of subarcs, only first differential conditions are derived. The Weierstrass and Legendre–Clebsch conditions are relied upon to indicate whether or not the optimal path is a minimum. For optimal control problems with control inequality constraints, the first differential conditions plus the Weierstrass condition are the equivalent of the Pontryagin minimum principle.

# 6

# Differentials in Optimal Control

## 6.1 Introduction

In Chapter 2, it was shown that Taylor expansions of algebraic functions can be performed one term at a time by taking differentials. Once the differential process was defined, it was used from then on to obtain the first and second differentials associated with each parameter optimization problem. The purpose of this chapter is to extend the use of differentials to do Taylor expansions associated with the optimal control problem (Ref. Bl, p. 18). The elements of the optimal control problem are ordinary differential equations, algebraic final conditions, and integrals. In general, the first-order results of the Taylor expansion are used to define the differential operation, and their correctness is verified by deriving the second-order results by taking differentials.

A fairly general optimal control problem is stated in Section 6.2. It contains ordinary differential equations, prescribed final conditions, and a performance index that has an integral. The differential associated with the differential equation is established in Section 6.3. It is shown that two neighboring paths can be related by a system of time-fixed differentials which, to connect with the existing terminology of the calculus of variations, are called variations. Next, variations are related to differentials in Section 6.4. The final conditions are algebraic and taking their differential is reviewed in Section 6.5. In Section 6.6, the differential of an integral, called Leibnitz' rule, is derived. Finally, the results are summarized in Section 6.7.

## 6.2  Standard Optimal Control Problem

To derive necessary conditions and sufficient conditions for a minimum, a somewhat general problem (see Chapter 1) is posed. The standard optimal control problem with free final time is the following: Find the control history $u(t)$ that minimizes the performance index

$$J = \phi(t_f, x_f) + \int_{t_0}^{t_f} L(t, x, u)dt, \tag{6.1}$$

subject to the differential constraints

$$\dot{x} = f(t, x, u), \tag{6.2}$$

the prescribed initial conditions

$$t_0 = t_{0_s}, \quad x_0 = x_{0_s}, \tag{6.3}$$

and the prescribed final conditions

$$\psi(t_f, x_f) = 0. \tag{6.4}$$

Here, $x$ denotes the $n \times 1$ state vector; $u$ denotes the $m \times 1$ control vector; $\phi$ and $L$ are scalars; $f$ is an $n \times 1$ vector; and $\psi$ is a $(p + 1) \times 1$ vector where $p \le n$. **Note.** All of the above functions are assumed to have as many derivatives as are needed for the theory being developed.

As stated, the unknown of this problem is the control history $u(t)$. However, only the control histories that, through the differential equation (6.2) with initial conditions (6.3) generate state histories that satisfy the final condition (6.4), are admissible, that is, in contention for being the minimum.

Some comments must be made about the geometry of the problem. First, the variable of integration $t$ (which is not always the time) is assumed to be monotonically increasing. Second, for many problems, the states and the controls are single-valued functions (Fig. 6.1) and, hence, continuous (Fig. 6.2). Third, because the states are differentiated in the state equations (6.2), they can be required to be continuous. However, there are some problems in which states can be discontinuous, for example, the mass of a multistage rocket or the velocity in an impulsive orbit transfer. Finally, because the controls are not differentiated in the state equations, they can be discontinuous. Here,

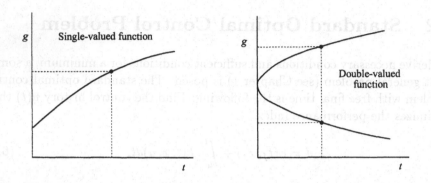

Figure 6.1: Single- and Multiple-Valued Functions

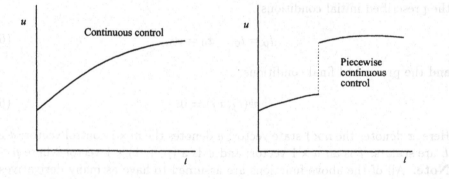

Figure 6.2: Control Continuity Requirements

they are assumed in general to be piecewise continuous (Fig. 6.2), that is, double-valued at isolated points and single-valued over the intervals.

To derive necessary conditions and sufficient conditions for a minimum, use is made of two paths: the minimal path represented by $u(t), x(t)$ and an admissible comparison path represented by $u_*(t), x_*(t)$. These paths are shown schematically in Fig. 6.3. The admissible comparison path lies in the neighborhood of the minimal path and satisfies all of the constraints, Eqs.(6.2) through (6.4). Because the admissible comparison path lies in the neighborhood of the minimal path, it can be expressed in terms of small changes from the minimal path. Then, the difference of the performance index evaluated

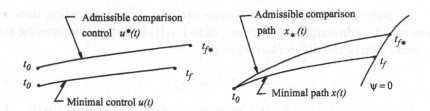

Figure 6.3: Admissible Comparison Control and Path

along the comparison path and that along the minimal path can be expanded in a Taylor series, resulting in first-order terms, second-order terms, and so on. Ultimately, the conditions for a minimum come from the vanishing of the first-order terms and the positiveness of the second-order terms.

How a comparison control/path is expressed in terms of changes from the minimal control/path requires some discussion. First, there are weak changes and strong changes. A system of weak changes is one in which the changes in the control and the state and all of its derivatives are small. On the other hand, in a system of strong changes, the change in the control can be large, but the change in the state is small. This can be accomplished by having a large change in the control over a small time interval and a small change in the control thereafter. For weak changes, the Taylor series process can be carried out by taking differentials. Strong changes are discussed in Chapter 10.

The comparison control/path can be related to the minimal control/path by a system of *time-fixed differentials* over $[t_0, t_f]$ and an ordinary differential at $t_f$. Time-fixed differentials (*variations*) are discussed in Section 6.3. Before discussing differentials associated with the final point (Section 6.5), the relationship between variations and differentials is derived in Section 6.4. The differential of the integral in Eq.(6.1) is derived in Section 6.6. Finally, a summary of the results is presented in Section 6.7.

## 6.3   Differential of the State Equation

In terms of the state differential equation (6.2), the control $u(t)$ is the independent variable, and the state $x(t)$ is the dependent variable. In other words, given $u(t)$, $x(t)$ can be determined.

Assume that the minimal control $u(t)$ and its corresponding state $x(t)$ are known. An admissible comparison control $u_*(t)$ and its corresponding state $x_*(t)$ must satisfy the differential equation

$$\dot{x}_* = f(t, x_*, u_*). \tag{6.5}$$

The comparison path lies in the neighborhood of the minimal path. This means that the total changes $\tilde{\Delta}u(t)$ and $\tilde{\Delta}x(t)$ shown in Fig. 6.4 are small, as are $\tilde{\Delta}\dot{x}(t), \tilde{\Delta}\ddot{x}(t)$, and so on. In general, the total change of something has a first-order part, a second-order part, and so on. It can be written as

$$\tilde{\Delta}(\ ) = \delta(\ ) + \frac{1}{2!}\delta^2(\ ) + \cdots. \tag{6.6}$$

The comparison control is related to the minimal control by a system of time-fixed changes defined as

$$u_* = u + \tilde{\Delta}u, \tag{6.7}$$

where $\tilde{\Delta}u$ is the small total change in the control (see Fig. 6.4). Since the control is an independent variable, $\tilde{\Delta}u$ only has a first-order part, that is,

$$\tilde{\Delta}u = \delta u. \tag{6.8}$$

Similarly, the comparison state is given by

$$x_* = x + \tilde{\Delta}x, \tag{6.9}$$

where, because the state is a dependent variable, $\tilde{\Delta}x$ is written as

$$\tilde{\Delta}x = \delta x + \frac{1}{2!}\delta^2 x + \cdots. \tag{6.10}$$

The control change $\delta u$ causes the state change $\tilde{\Delta}x$. Ultimately, $\delta x$ is related to $\delta u$, $\delta^2 x$ is related to $\delta u^2$, and so on.

The next step is to derive the differential equations for $\delta x$ and $\delta^2 x$. By combining the derivative of Eq.(6.9) with Eq.(6.2), it is seen that

$$\frac{d}{dt}\tilde{\Delta}x = \dot{x}_* - \dot{x} = f(t, x + \tilde{\Delta}x, u + \tilde{\Delta}u) - f(t, x, u). \tag{6.11}$$

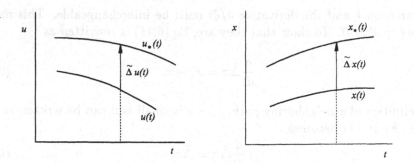

Figure 6.4: Definition of $\tilde{\Delta}u$ and $\tilde{\Delta}x$

To carry out the Taylor series expansion to second order, it is necessary to write this equation in terms of elements, that is,

$$\frac{d}{dt}\tilde{\Delta}x_k = f_k(t, x + \tilde{\Delta}x, u + \tilde{\Delta}u) - f_k(t, x, u). \qquad (6.12)$$

Then, the Taylor series expansion gives

$$\begin{aligned} \frac{d}{dt}\tilde{\Delta}x_k &= (f_k)_x\tilde{\Delta}x + (f_k)_u\tilde{\Delta}u + \frac{1}{2!}[\tilde{\Delta}x^T(f_k)_{xx}\tilde{\Delta}x + \tilde{\Delta}x^T(f_k)_{xu}\tilde{\Delta}u \\ &\quad + \tilde{\Delta}u^T(f_k)_{ux}\tilde{\Delta}x + \tilde{\Delta}u^T(f_k)_{uu}\tilde{\Delta}u] + \cdots \end{aligned} \qquad (6.13)$$

which, when combined with Eqs.(6.8) and (6.10) and like-order terms are equated, leads to

$$\frac{d}{dt}\delta x_k = (f_k)_x\delta x + (f_k)_u\delta u \qquad (6.14)$$

$$\begin{aligned} \frac{d}{dt}\delta^2 x_k &= (f_k)_x\delta^2 x + \delta x^T(f_k)_{xx}\delta x + \delta x^T(f_k)_{xu}\delta u \\ &\quad + \delta u^T(f_k)_{ux}\delta x + \delta u^T(f_k)_{uu}\delta u \end{aligned} \qquad (6.15)$$

and so on. Actually, the first-order equation is obtained by assuming that $\delta u$ is sufficiently small that second- and higher-order terms are negligible. The same procedure is applied to obtain the second-order equation, and so on.

The right-hand side of Eq.(6.14) implies that $\delta$ behaves like a differential taken while holding the time constant. To connect with the nomenclature of the calculus of variations, the time-fixed differential $\delta$ is called a variation. Now, for Eq.(6.14) to be the variation of Eq.(6.2), that is,

$$\delta\dot{x} = f_x\delta x + f_u\delta u, \qquad (6.16)$$

the variation $\delta$ and the derivative $d/dt$ must be interchangeable. This means that $\delta\dot{x} = d\delta x/dt$. To show that they are, Eq.(6.11) is rewritten as

$$\frac{d}{dt}\tilde{\Delta}x = \dot{x}_* - \dot{x}. \tag{6.17}$$

By definition of a neighboring path, $\dot{x}_* - \dot{x}$ is small and can be written as $\tilde{\Delta}\dot{x}$. Hence, Eq.(6.17) becomes

$$\frac{d}{dt}\tilde{\Delta}x = \tilde{\Delta}\dot{x}. \tag{6.18}$$

Because $\tilde{\Delta}\dot{x}$ is small, it can be expressed as

$$\tilde{\Delta}\dot{x} = \delta\dot{x} + \frac{1}{2!}\delta^2\dot{x} + \cdots. \tag{6.19}$$

Finally, combining Eqs.(6.10), (6.18), and (6.19) and equating like-order terms leads to the relations

$$\frac{d}{dt}\delta x \;=\; \delta\dot{x} \tag{6.20}$$

$$\frac{d}{dt}\delta^2 x \;=\; \delta^2\dot{x} \tag{6.21}$$

$$\vdots$$

Hence, the variation $\delta$ and the derivative $d/dt$ are interchangeable, so that Eq.(6.16) becomes

$$\frac{d}{dt}\delta x = f_x\delta x + f_u\delta u, \tag{6.22}$$

which is the matrix form of Eq. (6.14).

If the same process is applied to Eq.(6.14) using the rule for taking the variation of a row vector (Chapter 5) and using the fact that variations of independent variations are zero ($\delta(\delta u) = 0$), Eq.(6.15) is obtained.

The conclusions of this analysis are that the variation $\delta$ is a differential taken while holding the time constant, that the variation and the derivative interchange, and that variations of independent variations are zero. Also, the first-order part $\delta x$, the second-order part $\delta^2 x$, and so on that define the admissible comparison path relative to the minimal path can be obtained by taking variations of the state equation (6.2) and evaluating them on the minimal path.

## 6.4 Relationship Between $\delta$ and $d$

At the final point, there is a variation and a differential. The purpose of this section is to derive the relationship between the two. For generality, an arbitrary point $A$ on the minimal path and a neighboring point $B$ on the comparison path are considered.

Figure 6.5 shows two points $A$ and $B$ that lie on neighboring paths at different times $t$ and $t + \Delta t$, where $\Delta t$ is small. The total change $\Delta x$ between

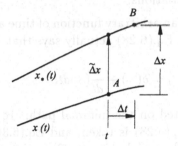

Figure 6.5: The Changes $\Delta$ and $\tilde{\Delta}$

these points is the difference

$$\Delta x(t) = x_*(t + \Delta t) - x(t). \tag{6.23}$$

Since $\Delta t$ is small, a Taylor series expansion yields

$$\Delta x(t) = \tilde{\Delta}x(t) + \dot{x}_*(t)\Delta t + \frac{1}{2!}\ddot{x}_*(t)\Delta t^2 + \cdots, \tag{6.24}$$

which, because of Eq.(6.9), can be rewritten as

$$\Delta x = \tilde{\Delta}x + \dot{x}\Delta t + \tilde{\Delta}\dot{x}\Delta t + \frac{1}{2!}\ddot{x}\Delta t^2 + \cdots. \tag{6.25}$$

The last step is to write $\Delta x$ and $\Delta t$ as

$$\Delta x = dx + \frac{1}{2!}d^2x + \cdots \tag{6.26}$$

and

$$\Delta t = dt + \frac{1}{2!}d^2t + \cdots, \tag{6.27}$$

and substitute everything into Eq.(6.25). Then, equating like-order terms gives

$$dx = \delta x + \dot{x}dt \tag{6.28}$$

and

$$d^2x = \delta^2x + \dot{x}d^2t + 2\delta\dot{x}dt + \ddot{x}dt^2, \tag{6.29}$$

which relate differentials to variations.

Since $x(t)$ represents an arbitrary function of time as far as the derivation of Eq.(6.28) is concerned, Eq.(6.28) actually says that

$$d(\ ) = \delta(\ ) + \frac{d}{dt}(\ )dt, \tag{6.30}$$

where the derivative is evaluated on the minimal path. To verify that this is the case, the differential of Eq.(6.28) is taken, and Eq.(6.30) is used to show that Eq.(6.29) results. The differential of Eq.(6.28) is given by

$$d^2x = d(\delta x) + d\dot{x}dt + \dot{x}d^2t. \tag{6.31}$$

Application of Eq.(6.30) yields

$$d(\delta x) = \delta(\delta x) + \frac{d}{dt}(\delta x)dt = \delta^2x + \delta\dot{x}dt; \tag{6.32}$$

also

$$d\dot{x} = \delta\dot{x} + \ddot{x}dt. \tag{6.33}$$

Combining these equations leads to Eq.(6.29). Hence, the relationship between $d$ and $\delta$ is given by Eq.(6.30).

For a fixed final time problem, the differential at the final point equals the variation. This is seen from Eq.(6.30) where for $dt_f = 0$

$$d(\ )_f = \delta(\ )_f. \tag{6.34}$$

For constants, Eq.(6.30) gives

$$d(Const) = \delta(Const). \tag{6.35}$$

## 6.5   Differential of the Final Condition

At the final point, the admissible comparison path must satisfy the constraint

$$\psi(t_{f_*}, x_{f_*}) = 0. \tag{6.36}$$

Then, the comparison final point is to be expressed in terms of differentials from the minimal final point, and Eq.(6.36), expanded in a Taylor series to obtain $d\psi$ and $d^2\psi$. However, because Lagrange multipliers will be used, it is only necessary to obtain $d\psi$ (Section 3.2.3). Also, because Eq.(6.4) is algebraic (a point condition), $d\psi$ can be obtained by taking the differential of Eq.(6.4). Doing so leads to

$$d\psi = \psi_{t_f}(t_f, x_f)dt_f + \psi_{x_f}(t_f, x_f)dx_f = 0. \tag{6.37}$$

At the final point where $x_f = x(t_f)$, Eq.(6.30) becomes

$$dx_f = \delta x_f + \frac{dx_f}{dt_f}\, dt_f. \tag{6.38}$$

Then, taking the derivative of $x_f$ with respect to $t_f$ leads to

$$\frac{dx_f}{dt_f} = \frac{dx(t_f)}{dt_f} = \left(\frac{dx}{dt}\right)_f = \dot{x}_f. \tag{6.39}$$

Hence, Eq.(6.38) becomes

$$dx_f = \delta x_f + \dot{x}_f dt_f. \tag{6.40}$$

Returning to Eq.(6.37), it is seen that

$$d\psi = (\psi_{t_f} + \psi_{x_f}\dot{x}_f)dt_f + \psi_{x_f}\delta x_f = 0. \tag{6.41}$$

To shorten this expression, it is rewritten as

$$d\psi = \psi' dt_f + \psi_{x_f}dx_f = 0, \tag{6.42}$$

where

$$\psi' \triangleq \frac{d\psi}{dt_f} = \psi_{t_f} + \psi_{x_f}\dot{x}_f. \tag{6.43}$$

## 6.6   Differential of the Integral

In developing the necessary conditions for a minimum, the differential of an integral having the form

$$I = \int_{t_0}^{t_f} F(t, y(t)) dt \tag{6.44}$$

must be taken. Here, $t_0$ is fixed, but $t_f$ changes with $y(t)$ as shown in Fig. 6.6. The function $y(t)$ is a vector that can represent $x$ and $u$.

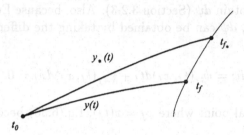

Figure 6.6: Neighboring Path with $t_0$ and $y_0$ Fixed

To obtain the formula for the differential, $I$ is evaluated on a neighboring path, and the difference of the integrals

$$\Delta I = \int_{t_0}^{t_{f*}} F(t, y_*(t)) dt - \int_{t_0}^{t_f} F(t, y(t)) dt \tag{6.45}$$

is expanded in a Taylor series to obtain the differential. The difference (6.45) can be rewritten as

$$\Delta I = \int_{t_0}^{t_f} (F_* - F) dt + \int_{t_f}^{t_f + \Delta t_f} F_* dt, \tag{6.46}$$

where

$$F_* = F(t, y_*(t)), \qquad F = F(t, y(t)). \tag{6.47}$$

The path $y_*(t)$ is related to $y(t)$ as

$$y_* = y + \tilde{\Delta} y, \tag{6.48}$$

where $\tilde{\Delta} y$ is the time-fixed total change in $y$.

Consider the second integral in Eq.(6.46). Since the region $t_f$ to $t_f + \Delta t_f$ is small, $F_*$ can be expanded in a Taylor series about $t_f$ to obtain

$$\int_{t_f}^{t_f+\Delta t_f} F_* dt = \int_{t_f}^{t_f+\Delta t_f} [(F_*)_f + (\dot{F}_*)_f(t - t_f) + \cdots] dt, \qquad (6.49)$$

where $\dot{F}_{*f} = (dF_*/dt)_f$. Integration leads to

$$\int_{t_f}^{t_f+\Delta t_f} F_* dt = (F_*)_f \Delta t_f + \frac{1}{2}(\dot{F}_*)_f \Delta t_f^2 + \cdots . \qquad (6.50)$$

Next, from Eqs.(6.47), (6.48), and $\tilde{\Delta}y = \delta y + (1/2!)\delta^2 y + \cdots$, it is seen that

$$\begin{aligned}
(F_*)_f &= F_f + (F_y)_f \tilde{\Delta} y_f + \cdots = F_f + (F_y)_f \delta y_f + \cdots \\
(\dot{F}_*)_f &= \dot{F}_f + \cdots,
\end{aligned} \qquad (6.51)$$

where

$$\left(\frac{dF_*}{dt}\right)_f = \frac{d(F_*)_f}{dt_f}. \qquad (6.52)$$

Then, with $\Delta t_f = dt_f + (1/2!)d^2 t_f + \cdots$,

$$\int_{t_f}^{t_f+\Delta t_f} F_* dt = F_f dt_f + \frac{1}{2!}F_f d^2 t_f + (F_y)_f \delta y_f dt_f + \frac{1}{2!}\dot{F}_f dt_f^2 + \cdots . \qquad (6.53)$$

The integrand of the first integral in Eq.(6.46) is expanded in terms of the time-fixed total change to obtain

$$\begin{aligned}
F_* - F &= F_y \tilde{\Delta} y + \frac{1}{2!}\tilde{\Delta}y^T F_{yy} \tilde{\Delta} y + \cdots \\
&= F_y \delta y + \frac{1}{2!}(F_y \delta^2 y + \delta y^T F_{yy} \delta y) + \cdots .
\end{aligned} \qquad (6.54)$$

Combining Eqs.(6.46), (6.53), and (6.54) leads to

$$\Delta I = dI + \frac{1}{2!}d^2 I + \cdots, \qquad (6.55)$$

where the first differential of $I$ is given by

$$dI = F_f dt_f + \int_{t_0}^{t_f} F_y \delta y \, dt \qquad (6.56)$$

and the second differential, by

$$d^2 I = F_f \delta^2 t_f + 2(F_y)_f \delta y_f \delta t_f + \dot{F}_f \delta t_f^2 + \int_{t_0}^{t_f} (F_y \delta^2 y + \delta y^T F_{yy} \delta y) dt. \qquad (6.57)$$

Eq.(6.56) can be rewritten as

$$dI = F_f dt_f + \int_{t_0}^{t_f} \delta F dt, \qquad (6.58)$$

which is the rule for the differential of an integral.

To further verify that Eq.(6.58) is the rule for taking the differential of an integral, the differential of Eq.(6.56) is taken as

$$d^2 I = dF_f dt_f + F_f d^2 t_f + (F_y)_f \delta y_f dt_f + \int_{t_0}^{t_f} (F_y \delta^2 y + \delta y^T F_{yy} \delta y) dt. \qquad (6.59)$$

Next, it is seen that

$$dF_f dt_f = (\delta F_f + \dot{F}_f dt_f) dt_f = (F_y)_f \delta y_f dt_f + \dot{F}_f dt_f^2, \qquad (6.60)$$

where $F_{f y_f} = (F_y)_f$. Then, Eq.(6.59) becomes Eq.(6.57).

Should the initial time be free, the more general formula for the differential of an integral becomes

$$dI = [F dt]_{t_0}^{t_f} + \int_{t_0}^{t_f} \delta F dt \qquad (6.61)$$

and is known as Leibnitz's rule (Ref. Gr, p.18). This equation shows that the differential of the integral in the performance index (6.1) can be obtained from differentials at $t_0$ and $t_f$ and variations over $[t_0, t_f]$.

## 6.7   Summary of Differential Properties

In the previous sections, a number of properties associated with small changes has been developed. A summary of these properties is the following:

(a) The variation $\delta$ is a time-fixed differential that satisfies the relationship

$$\delta F(t, x) = F_x(t, x) \delta x. \qquad (6.62)$$

(b) Variations and derivatives interchange, that is,

$$\delta \frac{d}{dt}( \ ) = \frac{d}{dt}\delta( \ ).$$

(6.63)

(c) Variations of independent variations are zero.

(d) The differential is related to the variation and $dt$ as

$$d( \ ) = \delta( \ ) + \frac{d}{dt}( \ ) \, dt.$$

(6.64)

For neighboring points with $dt = 0$,

$$d( \ ) = \delta( \ ).$$

(6.65)

Also, for a constant,

$$d(Const) = \delta(Const).$$

(6.66)

For consistency, variations of constants are written as differentials.

(e) The first-order part of the admissible comparison path can be obtained by taking the variation of the state equation and evaluating it on the minimal path. It satisfies the differential equation

$$\frac{d}{dt}\delta x = f_x \delta x + f_u \delta u.$$

(6.67)

The boundary conditions for this equation are obtained by taking the differentials of the prescribed boundary conditions, evaluating them on the minimal path, and relating differentials to variations. They are given by

$$t_0 = 0, \quad \delta x_0 = 0,$$

(6.68)

and

$$\psi_{x_f}\delta x_f + \psi' dt_f = 0,$$

(6.69)

where

$$\psi' \triangleq \frac{d\psi}{dt_f} = \psi_{t_f} + \psi_{x_f}\dot{x}_f.$$

(6.70)

All partial derivatives are evaluated on the minimal path.

The equations for the higher-order parts of the admissible comparison path are obtained in a similar manner, recalling that variations of independent variations are zero.

(f) The differential of an integral of the form

$$I = \int_{t_0}^{t_f} F\,dt \tag{6.71}$$

is given by Leibnitz' rule

$$dI = [F\,dt]_{t_0}^{t_f} + \int_{t_0}^{t_f} \delta F\,dt. \tag{6.72}$$

With these tools and the concept of controllability presented in the next chapter, the necessary conditions and sufficient conditions for a relative minimum can be derived. The general procedure is to adjoin the constraints to the performance index $J$ with Lagrange multipliers to form the augmented performance index $J'$. Next, the value of $J'$ along the minimum is differenced with that, $J'_*$, along an admissible comparison path that lies in the neighborhood of the minimum. Then, the difference $\Delta J' = J'_* - J'$ is expressed in terms of first-order terms (first differential), second-order terms (second differential), ..., as

$$\Delta J' = dJ' + \frac{1}{2!}d^2 J' + \cdots. \tag{6.73}$$

Since $\Delta J'$ must be positive regardless of the choice of the admissible comparison path, the first differential conditions for an optimal path are obtained from the vanishing of the first differential, that is,

$$dJ' = 0, \tag{6.74}$$

and if the second differential is positive, that is,

$$d^2 J' > 0, \tag{6.75}$$

the optimal path is a minimal path.

# 7

# Controllability

## 7.1   Introduction

In deriving the conditions for a minimum, the value of the performance index for the minimum is compared with that of an admissible comparison path. An admissible comparison path satisfies the differential constraints, the prescribed initial conditions, and the prescribed final conditions, and it lies in the neighborhood of the minimum. The admissible comparison path is composed of a first-order part, a second-order part, and so on. The notion of controllability is that a control perturbation can be found which drives the first-order or the linearized state equation from a point close to the minimal path to the first-order or linearized final conditions.

In this chapter, the controllability condition is derived for the fixed final time and the free final time problems. In doing so, the general solution of the linearized state equation is obtained in terms of the transition matrix.

## 7.2   Fixed Final Time

In the derivation of optimality conditions, use is made of the particular admissible comparison path $x_*(t)$ shown in Fig. 7.1. A small pulse in the control is applied over $[t_p, t_q]$, and it generates a small displacement from the mini-

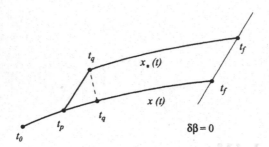

Figure 7.1: Perturbed Path

mal path. From $t_q$ to $t_f$, the admissible comparison path returns to the final
conditions.

The total change between the admissible comparison path and the
minimal path is composed of a first-order part, a second-order part, and so on.
Over $[t_p, t_f]$, the differential equation for the first-order part is given by the
linear equation

$$\frac{d}{dt}\delta x = f_x \delta x + f_u \delta u. \tag{7.1}$$

This system is said to be controllable if a solution from $t_q$ to $t_f$ exists.

The initial conditions for this solution are given by

$$t_q = t_{q_s}, \quad \delta x_q = \delta x_{q_s}, \tag{7.2}$$

where the subscript $s$ denotes a specified value. For fixed final time, the general
final conditions are given by

$$t_f = t_{f_s}, \quad \beta(x_f) = 0. \tag{7.3}$$

The differential of these conditions leads to

$$dt_f = 0, \quad d\beta = \beta_{x_f} dx_f = 0. \tag{7.4}$$

Equation (6.53) then implies that $dx_f = \delta x_f$ so that the final conditions for
Eq.(7.1) are given by

$$t_f = t_{f_s}, \quad \beta_{x_f} \delta x_f = 0. \tag{7.5}$$

In summary, Eq.(7.1) is said to be controllable if there exists a control
perturbation $\delta u(t)$ that drives the system from the initial conditions (7.2) to
the final conditions (7.5).

## 7.3  Solution of the Linear Equation

The solution of the linear differential equation (7.1) is now derived by finding its general solution and a particular solution.

By introducing the transition matrix, the solution of the homogenous equation

$$\frac{d}{dt}\delta x = f_x \delta x \tag{7.6}$$

can be obtained. It is assumed that

$$\delta x = \Phi(t, t_q)\delta x_q, \tag{7.7}$$

where $\Phi$ is the $n \times n$ transition matrix, and that

$$\Phi_q = \Phi(t_q, t_q) = I \tag{7.8}$$

to guarantee satisfaction of the initial conditions. Substitution of (7.7) into (7.6) allows $\Phi$ to be chosen as

$$\dot{\Phi} = f_x \Phi. \tag{7.9}$$

Hence, if $\Phi$ satisfies the differential equation (7.9) and the boundary condition (7.8), (7.7) is the general solution of (7.6).

To get the solution of the forced equation (7.1), the differential equation for $\Phi^{-1}$, which always exists, is needed. By definition,

$$\Phi\Phi^{-1} = I, \tag{7.10}$$

and differentiation with respect to time leads to

$$\frac{d}{dt}\Phi^{-1} = -\Phi^{-1}f_x \tag{7.11}$$

in view of Eq.(7.9). Also, from Eqs.(7.8) and (7.10),

$$\Phi_q^{-1} = I. \tag{7.12}$$

Now, multiply Eq.(7.1) on the left by $\Phi^{-1}$ as follows:

$$\Phi^{-1}d\delta x/dt = \Phi^{-1}f_x\delta x + \Phi^{-1}f_u\delta u. \tag{7.13}$$

With Eq.(7.11), this equation can be rewritten as

$$\frac{d}{dt}(\Phi^{-1}\delta x) = \Phi^{-1}f_u\delta u \tag{7.14}$$

and integrated to obtain the solution of Eq.(7.1) as

$$\delta x = \Phi\delta x_q + \Phi\int_{t_q}^{t}\Phi^{-1}f_u\delta u\ dt. \tag{7.15}$$

The transition matrix $\Phi$ denotes $\Phi(t, t_q)$, and the integral term is the particular solution.

## 7.4   Controllability Condition

To show that the system (7.1) is controllable, it is necessary to show that a $\delta u$ can be found which takes the system from the initial condition (7.2) to the final condition (7.5). The final state $\delta x_f$ is related to the control $\delta u$ and the perturbed state $\delta x_q$ by Eq.(7.15). Use of this equation in Eq.(7.5) leads to

$$\beta_{x_f}\ \Phi_f\delta x_q + \beta_{x_f}\ \Phi_f\int_{t_q}^{t_f}\Phi^{-1}f_u\delta u\ dt = 0. \tag{7.16}$$

Since there are many controls that satisfy this equation, a unique control is sought that minimizes the performance index

$$J = \int_{t_q}^{t_f}\delta u^T\ \delta u\ dt. \tag{7.17}$$

While this is an optimization problem, it can be solved without using optimization theory. Another approach is that of linear system theory (Ref. Ch) and the controllability grammian. It involves the use of linearly independent functions.

The constraint (7.16) is multiplied by the constant $-2d\nu^T$ and added to the performance index (7.17) to form

$$\begin{aligned}J' &= -2d\nu^T\beta_{x_f}\ \Phi_f\delta x_q\\ &+ \int_{t_g}^{t_f}(\delta u^T\delta u - 2d\nu^T\beta_{x_f}\Phi_f\Phi^{-1}f_u\delta u)\ dt.\end{aligned} \tag{7.18}$$

Next, the term

$$dv^T \beta_{x_f} \Phi_f \Phi^{-1} f_u f_u{}^T \Phi^{-T} \Phi_f{}^T \beta_{x_f}^T dv \qquad (7.19)$$

is added and subtracted under the integral so that the terms involving the control variation form a perfect square, that is,

$$
\begin{aligned}
J' \;=\; & -2dv^T \beta_{x_f} \, \Phi_f \delta x_q \\[4pt]
& - \int_{t_q}^{t_f} dv^T \beta_{x_f} \Phi_f \Phi^{-1} f_u f_u{}^T \Phi^{-T} \Phi_f{}^T \beta_{x_f}^T dv \; dt \\[4pt]
& + \int_{t_q}^{t_f} [\delta u - f_u{}^T \Phi^{-T} \Phi_f{}^T \beta_{x_f}^T dv]^T [\delta u - f_u{}^T \Phi^{-T} \Phi_f{}^T \beta_{x_f}^T dv] \; dt.
\end{aligned}
\qquad (7.20)
$$

Since the control term is squared, $J'$ is a minimum with respect to $\delta u$ if

$$\delta u = f_u{}^T \Phi^{-T} \Phi_f{}^T \beta_{x_f}^T dv. \qquad (7.21)$$

To determine the quantity $dv$, Eq. (7.21) is substituted into the constraint (7.16) leading to

$$Q dv + r = 0, \qquad (7.22)$$

where

$$
\begin{aligned}
Q &= \beta_{x_f} \, \Phi_f \int_{t_q}^{t_f} \Phi^{-1} f_u f_u{}^T \Phi^{-T} \; dt \; \Phi_f{}^T \beta_{x_f}^T \\
r &= \beta_{x_f} \, \Phi_f \delta x_q.
\end{aligned}
\qquad (7.23)
$$

To be able to solve for $dv$, the inverse of $Q$ must exist, which is the controllability condition. If it is assumed that $Q^{-1}$ exists, $dv$ is given by

$$dv = -Q^{-1} r = -Q^{-1} \beta_{x_f} \, \Phi_f \delta x_q. \qquad (7.24)$$

Finally, a control perturbation that satisfies $\beta_{x_f} \delta x_f = 0$ is given by

$$\delta u = U(t) \delta x_q, \qquad (7.25)$$

where

$$U(t) = -f_u{}^T \Phi^{-T} \Phi_f{}^T \beta_{x_f}^T Q^{-1} \beta_{x_f} \Phi_f. \qquad (7.26)$$

The corresponding state perturbation follows from Eq. (7.15) as

$$\delta x = X(t) \delta x_q, \qquad (7.27)$$

where
$$X(t) = \Phi + \Phi \int_{t_q}^{t} \Phi^{-1} f_u U \ dt. \tag{7.28}$$

In conclusion, the linear state equation (7.1) is controllable if the controllability matrix

$$Q = \beta_{x_f} \ \Phi_f \int_{t_q}^{t_f} \Phi^{-1} f_u f_u{}^T \Phi^{-T} \ dt \ \Phi_f{}^T \beta_{x_f}^T \tag{7.29}$$

is invertible.

## 7.5  Examples

Consider a linear system whose scalar differential equation and initial conditions are given by

$$\frac{d}{dt} \delta x = \delta u, \quad t_q = t_{q_s}, \quad \delta x_q = \delta x_{q_s}, \tag{7.30}$$

and whose final conditions are to be

$$t_f = t_{f_s}, \quad \delta x_f = 0. \tag{7.31}$$

Is this system controllable? Comparing Eqs.(7.30) and (7.1) shows that $f_x = 0$ and $f_u = 1$. Hence, $\Phi = 1$. Also, comparison of Eqs.(7.31) and (7.5) leads to $\beta_{x_f} = 1$. Then, the controllability matrix Q becomes

$$Q = t_f - t_q. \tag{7.32}$$

Since $Q^{-1}$ exists, the system is controllable.

Next, is the linear system

$$\frac{d}{dt}\delta x_1 = \delta x_2$$
$$\frac{d}{dt}\delta x_2 = \delta u \tag{7.33}$$

controllable to the final conditions

$$t_f = t_{f_s}, \quad \delta x_{1f} = 0, \quad \delta x_{2f} = 0 \ ? \tag{7.34}$$

For this problem,

$$f_x = \begin{bmatrix} 0 & 1 \\ 0 & 0 \end{bmatrix}, \quad f_u = \begin{bmatrix} 0 \\ 1 \end{bmatrix}, \quad \beta_{x_f} = \begin{bmatrix} 1 & 0 \\ 0 & 1 \end{bmatrix}. \tag{7.35}$$

Hence, by solving the differential equation (7.9) subject to the initial conditions (7.8), the transition matrix becomes

$$\Phi = \begin{bmatrix} 1 & t - t_q \\ 0 & 1 \end{bmatrix}. \tag{7.36}$$

Then, the controllability matrix becomes

$$Q = \begin{bmatrix} \frac{\Delta t^3}{3} & \frac{\Delta t^2}{2} \\ \frac{\Delta t^2}{2} & \Delta t \end{bmatrix}, \tag{7.37}$$

where $\Delta t = t_f - t_q$. Since $Q$ is positive definite (det $Q \neq 0$), its inverse exists, and the system is controllable.

## 7.6 Controllability: Free Final Time

If the final time is free, the final condition is written as $\psi(t_f, x_f) = 0$ and by taking the differential leads to

$$d\psi = \psi_{t_f} dt_f + \psi_{x_f} dx_f = 0. \tag{7.38}$$

Then, by using Eq.(6.53), this result can be rewritten as

$$\psi' dt_f + \psi_{x_f} \delta x_f = 0, \tag{7.39}$$

where by definition

$$\psi' = \psi_{t_f} + \psi_{x_f} \dot{x}_f = \psi_{t_f} + \psi_{x_f} f_f. \tag{7.40}$$

Since the final time is free, at least one element of $\psi$ must exist such that Eq.(7.39) can be solved for $dt_f$. If this element is called $\sigma$ and the remaining elements are called $\beta$, $\psi$ can be rewritten as

$$\psi = \begin{bmatrix} \sigma \\ \beta \end{bmatrix}, \tag{7.41}$$

and Eq.(7.39) becomes

$$\sigma' dt_f + \sigma_{x_f} \delta x_f = 0$$
$$\beta' dt_f + \beta_{x_f} \delta x_f = 0.$$

(7.42)

Solving the $\sigma$ equation for $dt_f$ leads to

$$dt_f = -\frac{\sigma_{x_f}}{\sigma'} \delta x_f.$$

(7.43)

so that the $\beta$ equation becomes

$$\left( \beta_{x_f} - \frac{\beta'}{\sigma'} \sigma_{x_f} \right) \delta x_f = 0.$$

(7.44)

Now, the controllability condition is the same as that for fixed final time with $\beta_{x_f}$ replaced by

$$\beta_{x_f} - \frac{\beta'}{\sigma'} \sigma_{x_f}.$$

(7.45)

For the fixed final time problem, $\sigma = t_f - t_{f_s}$, so that $\sigma' = 1$ and $\sigma_{x_f} = 0$. Then, Eq.(7.44) reduces to Eq.(7.5).

## 7.7　Navigation Problem

The differential equations and boundary conditions that define the navigation problem of Section 1.4.3 are given by

$$\dot{x} = V \cos \theta$$

(7.46)

$$\dot{y} = V \sin \theta + w$$

(7.47)

$$t_0 = 0, \quad x_0 = 0, \quad y_0 = 0$$

(7.48)

$$x_f = 1, \quad y_f = 0.$$

(7.49)

Note that the final time is unspecified. If it is assumed that the minimal control is

$$\sin \theta = -\frac{w}{V},$$

(7.50)

these equations can be solved as

$$x = \sqrt{V^2 - w^2}\, t, \quad y = 0, \quad t_f = \frac{1}{\sqrt{V^2 - w^2}}. \tag{7.51}$$

The first-order part of a path that lies in the neighborhood of the minimal path must satisfy the linear differential equations

$$\frac{d}{dt}\delta x = -V \sin\theta\, \delta\theta$$
$$\frac{d}{dt}\delta y = \phantom{-}V \cos\theta\, \delta\theta \tag{7.52}$$

which are obtained by taking the variation of Eqs.(7.46) and (7.47) and interchanging $\delta$ and $d/dt$. The first-order parts of the final conditions are obtained by taking the differential ($t_f$ is free) of Eqs.(7.49), that is,

$$dx_f = 0$$
$$dy_f = 0. \tag{7.53}$$

In terms of variations, these results become

$$\delta x_f + \dot{x}_f dt_f = 0$$
$$\delta y_f + \dot{y}_f dt_f = 0 \tag{7.54}$$

or

$$\delta x_f + (V \cos\theta_f)dt_f = 0$$
$$\delta y_f + (V \sin\theta_f + w)dt_f = 0. \tag{7.55}$$

Relative to the minimal path (7.50), these final conditions take the form

$$\delta x_f + \sqrt{V^2 - w^2}\, dt_f = 0$$
$$\delta y_f = 0. \tag{7.56}$$

Starting from a displacement from the minimal path, that is,

$$t_q = t_{q_s}, \quad \delta x_q = \delta x_{q_s}, \quad \delta y_q = \delta y_{q_s}, \tag{7.57}$$

is the linear system (7.52) controllable to the final conditions (7.56)? To assess controllability, one final condition is solved for $dt_f$ which is then eliminated from the other. In this case, the first final condition can be solved for $dt_f$, but it is not contained in the second. The equations corresponding to Eqs.(7.43) and (7.44) are

$$dt_f = -\frac{\delta x_f}{\sqrt{V^2 - w^2}}, \quad [\,0\ 1\,]\begin{bmatrix} \delta x_f \\ \delta y_f \end{bmatrix} = 0. \tag{7.58}$$

Hence, the controllability condition (7.29) is applied with $\beta_{x_f}$ given by $[\,0\ 1\,]$ and

$$f_x = \begin{bmatrix} 0 \\ 0 \end{bmatrix}, \quad f_u = \begin{bmatrix} w \\ \sqrt{V^2 - w^2} \end{bmatrix}. \tag{7.59}$$

Because of the form of $f_x$, the transition matrix is $\Phi = I$ so that the controllability matrix becomes

$$Q = [\,0\ 1\,]\int_{t_q}^{t_f} \begin{bmatrix} w \\ \sqrt{V^2 - w^2} \end{bmatrix}[w\ \sqrt{V^2 - w^2}]dt\begin{bmatrix} 0 \\ 1 \end{bmatrix} \tag{7.60}$$

or

$$Q = (V^2 - w^2)(t_f - t_q). \tag{7.61}$$

Hence, since $Q$ is invertible, the system is controllable.

# Problems

7.1 Consider the first-order system

$$\delta \dot{x}_1 = \delta u$$

$$\delta \dot{x}_2 = \delta u$$

and the boundary conditions

$$t_q = t_{q_s}, \quad \delta x_{1_q} = \delta x_{1_{q_s}}, \quad \delta x_{2_q} = \delta x_{2_{q_s}}$$

and

$$t_f = t_{f_s}, \quad \delta x_{1_f} = 0, \quad \delta x_{2_f} = 0.$$

Show that the system is not full-state controllable, that is, show that $\Phi = I$ and that

$$Q = \begin{bmatrix} \Delta t & \Delta t \\ \Delta t & \Delta t \end{bmatrix}, \quad \Delta t = t_f - t_q.$$

Since $Q$ is singular, its inverse does not exist.

7.2 Show that the system in Problem 7.1 is controllable if the final condition on the state perturbation is only $\delta x_{1_f} = 0$. Here, $\psi_{x_f} = [1\ 0]$. Consider also the case where only $\delta x_{2_f} = 0$ is required.

7.3 Finish Section 7.7 by solving for $\delta\theta, \delta x$, and $\delta y$ over the portion of the path from $t_q$ to $t_f$.

7.4 If the minimal control in Section 7.7 is assumed to be $\theta = 0$, show that the controllability matrix is given by

$$Q = V^2(t_f - t_q).$$

# 8

# Simplest Optimal Control Problem

## 8.1 Introduction

The simplest optimal control problem is one that does not involve any states. Such a problem arises when it is possible to eliminate the states between the state differential equations, the performance index, and the prescribed boundary conditions. An example is the aircraft navigation problem discussed in Section 1.4.3.

In this chapter, necessary conditions and sufficient conditions to be satisfied by the optimal control are developed. First, the problem with fixed final time is considered for the case where the optimal control is assumed to be continuous. Next, the final time is allowed to vary. Then, the continuity requirements are relaxed to allow a discontinuous optimal control. Finally, integral constraints, control equality constraints, and control inequality constraints are discussed.

One reason for studying this problem is to provide a simple introduction to optimal control theory. Another is to get some idea of what conditions exist for more complicated problems.

## 8.2   General Problem with No States

The general optimal control problem (Section 1.4.4) with no states is the following: Find the control history $u(t)$ that minimizes the performance index

$$J = \phi(t_f) + \int_{t_0}^{t_f} L(t, u)dt, \tag{8.1}$$

subject to the prescribed initial condition

$$t_0 = t_{0_s}, \tag{8.2}$$

the prescribed final condition

$$t_f = t_{f_s}, \text{ or } t_f \text{ free }, \tag{8.3}$$

the integral constraint

$$\int_{t_0}^{t_f} M(t, u)dt = K, \tag{8.4}$$

the control variable equality constraint

$$C(t, u) = 0, \tag{8.5}$$

and the control variable inequality constraint

$$\bar{C}(t, u) \leq 0. \tag{8.6}$$

Here, $u$ is an $m \times 1$ vector, $\phi$ and $L$ are scalars, and $M$, $K$, $C$, and $\bar{C}$ are vectors.

In general, the procedure for finding the conditions to be satisfied by the optimal control is straightforward. First, assume the existence of the minimal control. Then, form the difference $\Delta J$ of the performance index value for an admissible comparison control and that for the minimal control; an *admissible comparison control* is one that lies in the neighborhood of the minimal control and satisfies all the constraints. Using differentials, develop the first and second differentials of $\Delta J$. The vanishing of the first differential leads to equations for determining an optimal control, and the nonnegativeness or positiveness of the second differential leads to necessary conditions or sufficient conditions for a minimal control. While this process is for weak variations (small changes in the control), strong variations (large changes in the control) can be used to develop an additional condition for a minimum.

## 8.3   Fixed Final Time and Continuous Optimal Control

The problem is to find conditions for determining the control $u(t)$ that minimizes

$$J = \int_{t_0}^{t_f} L(t,u)dt, \tag{8.7}$$

where $t_0$ and $t_f$ are fixed and $u$ is an $m \times 1$ vector of independent controls. The integrand is assumed to have as many derivatives as are needed to carry out the analysis. Controls cannot be required to be continuous, but the class of problems where the optimal control is continuous can be investigated. Even though the optimal control is continuous, admissible comparison controls can be discontinuous.

For weak variations, the difference of the values of $J$ for an admissible comparison control $u_*(t)$ and the minimal control $u(t)$ (see Fig. 8.1) can be expressed as

$$\Delta J = dJ + \frac{1}{2!}d^2J + \cdots. \tag{8.8}$$

where the first differential is

$$dJ = \int_{t_0}^{t_f} L_u \delta u \, dt \tag{8.9}$$

and the second differential is

$$d^2J = \int_{t_0}^{t_f} (\delta u^T L_{uu} \delta u + L_u \delta^2 u) dt. \tag{8.10}$$

### 8.3.1   First Differential Condition

Because the first differential is linear in $\delta u$ and because $\delta u$ is arbitrary ($u_*$ does not have to satisfy any constraints), it seems reasonable that the first differential should vanish and that this should be accomplished by $L_u = 0$. To prove this, assume that $u(t)$ is the minimum but that $L_u \neq 0$. Since the first differential is not zero, the total change in the performance index can be approximated by the first differential, that is, $\Delta J \cong dJ$. Even though the minimal control is assumed to be continuous, controls in general can be discontinuous so that

# Springer Texts in Statistics

*(continued after index)*

# Springer Texts in Statistics

*Advisors:*
George Casella   Stephen Fienberg   Ingram Olkin

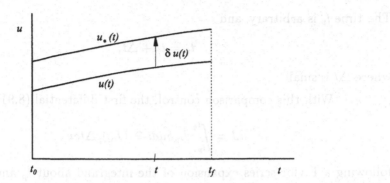

Figure 8.1: Admissible Comparison Control, Scalar Control

the admissible comparison control can be continuous or discontinuous. Hence,
consider the admissible comparison control shown in Fig. 8.2. This admissible
comparison control is a pulse defined as

$$
\begin{aligned}
t_0 \le t \le t_p, \quad & \delta u - 0 \\
t_p \le t \le t_q, \quad & \delta u = \text{Const} = \delta v \\
t_q \le t \le t_f, \quad & \delta u = 0.
\end{aligned}
$$
(8.11)

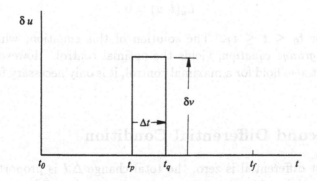

Figure 8.2: Weak Pulse, Scalar Control

The time $t_p$ is arbitrary, and

$$t_q = t_p + \Delta t, \tag{8.12}$$

where $\Delta t$ is small.

With this comparison control, the first differential (8.9) becomes

$$dJ = \int_{t_p}^{t_q} L_u \delta u \, dt \cong (L_u)_p \Delta t \delta v \tag{8.13}$$

following a Taylor series expansion of the integrand about $t_p$ and making $\Delta t$ sufficiently small that higher-order terms are negligible. Since $(L_u)_p \neq 0$, the elements of $\delta v$ can be chosen to have the same signs as the corresponding elements of $(L_u)_p$ so that $dJ > 0$ and, hence, $\Delta J \cong dJ > 0$. However, by changing the sign of $\delta v$,

$$dJ = (L_u)_p(-\delta v)\Delta t = -(L_u)_p \delta v \Delta t < 0. \tag{8.14}$$

Hence, the sign of $\Delta J$ becomes negative. Since $\Delta J$ is not positive for all admissible comparison controls, $u(t)$ cannot be a minimum. This is a contradiction so that

$$(L_u)_p = 0 \tag{8.15}$$

must hold. Since $t_p$ is arbitrary,

$$L_u(t, u) = 0 \tag{8.16}$$

must hold for $t_0 \leq t \leq t_f$. The solution of this equation, which is called the *Euler–Lagrange equation*, yields the optimal control. However, since this equation must also hold for a maximal control, it is only necessary for a minimal control.

## 8.3.2   Second Differential Condition

Since the first differential is zero, the total change $\Delta J$ is proportional to the second differential and must be nonnegative for all admissible comparison controls for $u(t)$ to be a minimum. For $L_u = 0$ and the system of weak variations (8.11), the second differential (8.10) leads to

$$d^2 J = \int_{t_p}^{t_q} \delta u^T L_{uu} \delta u \, dt \cong \delta v^T (L_{uu})_p \delta v \Delta t. \tag{8.17}$$

The quantity $\delta v^T (L_{uu})_p \delta v$ is a quadratic form. If $(L_{uu})_p$ is positive definite, the quadratic form is positive, $d^2 J > 0$, and $\Delta J > 0$. Hence, since $t_p$ is arbitrary, a sufficient condition for a minimum is that

$$L_{uu}(t, u) > 0 \qquad (8.18)$$

for all $t_0 \le t \le t_f$.

On the other hand, it is possible for $L_{uu}$ to vanish, and a minimum still exist. Consider $L = u^4$. Hence, a necessary condition for a minimum is that

$$L_{uu}(t, u) \ge 0 \qquad (8.19)$$

along the optimal path. This is the *Legendre–Clebsch condition* for a weak relative minimum.

### 8.3.3  Summary

In summary, the necessary conditions for a weak minimal control of the performance index (8.7) are the Euler–Lagrange equation

$$L_u(t, u) = 0 \qquad (8.20)$$

and the Legendre–Clebsch condition

$$L_{uu}(t, u) \ge 0 \qquad (8.21)$$

at each time $t_0 \le t \le t_f$. Also, the strengthened Legendre–Clebsch condition

$$L_{uu}(t, u) > 0, \qquad (8.22)$$

at each time instant, is sufficient for the solution of Eq.(8.20) to be a minimum. Hence, if $u(t)$ satisfies Eqs.(8.20) and (8.22), it is a weak minimal control, and $L(t, u)$ is a local minimum with respect to $u$.

### 8.3.4  Strong Variations

These conditions have been developed using weak variations, that is, $\delta u$ is small. Consider the system of strong variations shown in Fig. 8.3. Here,

$$
\begin{aligned}
t_0 \le t \le t_p, & \quad u_* - u = 0 \\
t_p \le t \le t_q, & \quad u_* - u = \text{Const} \\
t_q \le t \le t_f, & \quad u_* - u = 0,
\end{aligned}
\qquad (8.23)
$$

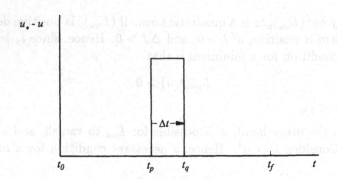

Figure 8.3: Strong Pulse, Scalar Control

where $t_p$ is arbitrary, $t_q = t_p + \Delta t$, $\Delta t$ is small, and $u_* - u$ is not small.

For this system of variations, $\Delta J$ is given by

$$\Delta J = \int_{t_p}^{t_p + \Delta t} [L(t, u_*) - L(t, u)]dt. \tag{8.24}$$

A Taylor series expansion in time of the integrand followed by integration leads to

$$\Delta J = [L(t, u_*) - L(t, u)]_p \Delta t \tag{8.25}$$

which must be positive for all $u_* \neq u$ for $u$ to be a minimum. Since $t_p$ is arbitrary, a necessary condition for a strong relative minimum is

$$L(t, u_*) - L(t, u) > 0 \tag{8.26}$$

for all $u_* \neq u$ at every time instant. This necessary condition is known as the *Weierstrass condition* and says that along a minimal path $L$ must be an absolute minimum with respect to the control.

Note that the Weierstrass condition reduces to the Legendre–Clebsch condition when $u_* = u + \delta u$, where $\delta u$ is small. This is accomplished by expanding $L(t, u + \delta u)$ in a Taylor series expansion and noting that $L_u = 0$.

## 8.4   Examples

As a simple example, find $u(t)$ that minimizes the performance index

$$J = \int_{t_0}^{t_f} u^2 dt, \tag{8.27}$$

where $L = u^2$. The Euler–Lagrange equation (8.20) leads to

$$L_u = 2u = 0 \ \Rightarrow \ u = 0 \tag{8.28}$$

and since

$$L_{uu} = 2 > 0, \tag{8.29}$$

$u = 0$ is a weak relative minimum. Finally, the Weierstrass condition (8.26), that is,

$$L(u_*) - L(u) = u_*^2 - u^2 = u_*^2 > 0, \tag{8.30}$$

for all $u_* \neq u$, is satisfied.

Consider the navigation problem for the case where the aircraft is only required to reach the line $x_f = 1$ (see Section 1.4.3). Here, the performance index is given by

$$J = \int_{x_0}^{x_f} \frac{dx}{\cos \theta}, \tag{8.31}$$

where $\theta(x)$ is the control and $L = 1/\cos \theta$. The Euler–Lagrange equation (8.20) leads to

$$L_\theta = \frac{\sin \theta}{\cos^2 \theta} = 0 \ \Rightarrow \ \theta = 0 \tag{8.32}$$

and since the sufficient condition (8.22) is satisfied, that is,

$$L_{\theta\theta} = \frac{1 + \sin^2 \theta}{\cos^3 \theta} = 1 > 0, \tag{8.33}$$

the solution $\theta = 0$ is a weak minimum. The Weierstrass condition for $\theta = 0$ becomes

$$\frac{1 - \cos \theta_*}{\cos \theta_*} > 0 \tag{8.34}$$

for all $\theta_* \neq 0$. This condition is not satisfied for $\pi/2 < \theta_* < 3\pi/2$. However, $-\pi/2 \leq \theta_* \leq \pi/2$ is an implied constraint which if included in the statement of the problem allows the Weierstrass condition to be satisfied (see Section 8.9).

As a final example, minimize the performance index

$$J = \int_{t_0}^{t_f} u^2(1-u)dt. \tag{8.35}$$

The Euler–Lagrange equation and the Legendre–Clebsch condition indicate a minimum at $u = 0$. However, the Weierstrass condition leads to

$$u_*^2(1-u_*) > 0 \tag{8.36}$$

for all $u_* \neq 0$. This condition is not satisfied because $u_*^3$ dominates the left-hand side for large $u_*$, and it is positive for positive $u_*$ and negative for negative $u_*$. It is obvious from Eq.(8.35) that a negative performance index can be achieved by choosing $u = \text{Const} > 1$, and $J = -\infty$ occurs for $u = \infty$. This is a nonanalytical minimum, but its existence is indicated by the Weierstrass condition.

## 8.5   Free Final Time and Continuous Optimal Control

If the final time is free, the performance index can have the more general form

$$J = \phi(t_f) + \int_{t_0}^{t_f} L(t, u)dt. \tag{8.37}$$

Then, finding $u(t)$ and $t_f$ that minimize $J$ is the optimal control problem.

Use of Leibniz' rule leads to the following expression for the first differential:

$$dJ = (\phi_{t_f} + L_f)dt_f + \int_{t_0}^{t_f} L_u \delta u \, dt \tag{8.38}$$

since $t_0$ is fixed. First, it is assumed that the first differential does not vanish so that $\Delta J \cong dJ$ and must be positive regardless of the choice of the admissible comparison control. Next, it is observed that $dt_f$ and $\delta u$ are independent quantities, that is, their values can be chosen arbitrarily. For the class of admissible controls where $dt_f = 0$ (Fig. 8.1), the first differential reduces to

$$dJ = \int_{t_0}^{t_f} L_u \delta u \, dt. \tag{8.39}$$

Then, by applying the same arguments as in Section 8.3, it is seen that

$$L_u(t, u) = 0 \tag{8.40}$$

must hold for all $t_0 \leq t \leq t_f$. Since $L_u = 0$ , the first differential for arbitrary comparison controls (Fig. 8.2) becomes

$$dJ = (\phi_{t_f} + L_f)dt_f. \tag{8.41}$$

If the coefficient of $dt_f$ is not zero, it has some value. Since $dt_f$ can be positive

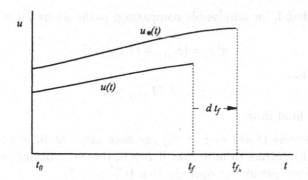

Figure 8.4: Admissible Comparison Control for $dt_f \neq 0$, Scalar Control

or negative, the sign of $dJ$ and, hence, $\Delta J$, changes. This cannot be if $u(t)$ is a minimum, so the following condition must be satisfied:

$$\phi_{t_f} + L_f = 0. \tag{8.42}$$

The Euler–Lagrange equation (8.40) determines the optimal control $u(t)$, and Eq.(8.42), called a *natural boundary condition*, determines the optimal value of $t_f$. At this point, the control is only optimal, and the second differential must be investigated to determine whether or not it is a minimum.

By taking the differential of Eq.(8.37) according to Leibniz' rule and invoking the first differential conditions, the second differential is given by

$$
\begin{aligned}
d^2 J \;=\; & [\phi_{t_f t_f} + (L_f)_{t_f}]dt_f^2 + (L_f)_{u_f}du_f dt_f \\
& + (L_u)_f \delta u_f dt_f + \int_{t_0}^{t_f} \delta u^T L_{uu} \delta u \, dt.
\end{aligned}
\tag{8.43}
$$

Since $L_u = 0$ for the optimal control and $(L_f)_{u_f} = (L_u)_f$, the second differential becomes

$$d^2 J = [\phi_{t_f t_f} + (L_f)_{t_f}]dt_f^2 + \int_{t_0}^{t_f} \delta u^T L_{uu} \delta u \, dt \qquad (8.44)$$

and must be nonnegative for a minimal control. For admissible comparison controls where $dt_f = 0$, the second differential reduces to that for a fixed final time and, hence, leads to the Legendre–Clebsch condition

$$L_{uu}(t, u) \geq 0. \qquad (8.45)$$

On the other hand, for admissible comparison paths where $\delta u = 0$, Eq. (8.44) reduces to

$$d^2 J = [\phi_{t_f t_f} + (L_f)_{t_f}]dt_f^2 \qquad (8.46)$$

and requires that

$$\phi_{t_f t_f} + (L_f)_{t_f} \geq 0 \qquad (8.47)$$

for a minimal final time.

Equations (8.44) and (8.46) are necessary conditions for a minimal control. It is interesting to note that if $L = 0$, these conditions reduce to those for unconstrained parameter optimization (Chapter 2).

It is easily seen that

$$L_{uu}(t, u) > 0 \qquad (8.48)$$

and

$$\phi_{t_f t_f} + (L_f)_{t_f} > 0 \qquad (8.49)$$

are sufficient conditions for a minimal control.

Also, since the Weierstrass condition (8.20) is a local condition, it is valid here. In other words, its derivation is not affected by free final time.

## 8.6 Discontinuous Optimal Control

The previous developments have been for the case where the optimal control is continuous. Here, the optimal control is assumed to be discontinuous. Only one jump in the control is necessary for deriving the conditions; if more jumps occur, the same conditions are applied at each discontinuity. Then, the final time is assumed fixed. If it is free, the conditions derived in Section 8.5 apply.

The effect of a jump in the control is that $L$ has two values at the time of the discontinuity. To avoid this, the integral in the performance index is divided into two parts as follows:

$$J = \int_{t_0}^{t_c} L dt + \int_{t_c}^{t_f} L dt, \tag{8.50}$$

where $t_c$ is the time of the jump. In what follows, the subscript $c-$ denotes a quantity evaluated with the control just before the corner, and the subscript $c+$ denotes a quantity evaluated with the control just after the corner.

Taking the differential of Eq.(8.50) leads to

$$\begin{aligned} dJ &= [L dt]_{t_0}^{t_{c-}} + \int_{t_0}^{t_c} L_u \delta u\, dt \\ &+ [L dt]_{t_{c+}}^{t_f} + \int_{t_c}^{t_f} L_u \delta u\, dt. \end{aligned} \tag{8.51}$$

The quantities $dt_0$ and $dt_f$ are zero, and since no timewise discontinuity can exist in $t_c$, $dt_{c-} = dt_{c+} = dt_c$. Hence, the first differential becomes

$$dJ = (L_{c-} - L_{c+})dt_c + \int_{t_0}^{t_c} L_u \delta u\, dt + \int_{t_c}^{t_f} L_u \delta u\, dt, \tag{8.52}$$

where, for example, $L_{c-} = L(t_c, u_{c-})$. Admissible comparison controls for $dt_c = 0$ and $dt_c \neq 0$ are shown in Figs. 8.3 and 8.4. Consider an admissible comparison control where $dt_c = 0$ and $\delta u = 0$ over $[t_0, t_c]$. The first differential reduces to

$$dJ = \int_{t_c}^{t_f} L_u \delta u\, dt \tag{8.53}$$

and following the approach of Section 8.3 leads to

$$L_u(t, u) = 0, \quad t_c \leq t \leq t_f. \tag{8.54}$$

Then, the first differential reduces to

$$dJ = (L_{c-} - L_{c+})dt_c + \int_{t_0}^{t_c} L_u \delta u\, dt \tag{8.55}$$

even if $\delta u \neq 0$ over $[t_c, t_f]$. Now, consider an admissible comparison control where $dt_c = 0$. The first differential becomes

$$dJ = \int_{t_0}^{t_c} L_u \delta u\, dt \tag{8.56}$$

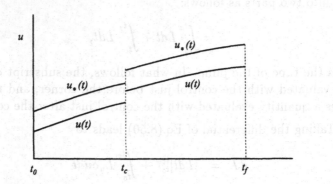

Figure 8.5: Admissible Comparison Control for $dt_c = 0$, Scalar Control

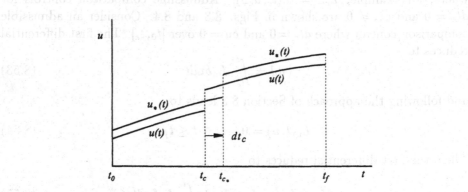

Figure 8.6: Admissible Comparison Control for $dt_c \neq 0$, Scalar Control

and implies that

$$L_u(t, u) = 0, \quad t_0 \le t \le t_c. \tag{8.57}$$

Finally, the first differential for arbitrary comparison controls $(dt_c \ne 0, \ \delta u \ne 0)$ is

$$dJ = (L_{c-} - L_{c+})dt_c. \tag{8.58}$$

Since $dt_c$ can be positive or negative, the condition

$$L(t_c, u_{c-}) = L(t_c, u_{c+}) \tag{8.59}$$

must hold and determines the optimal time for the discontinuity. This condition requires that $L$ be continuous at the jump.

In summary, the first differential conditions for a minimum are

$$L_u(t, u) = 0, \quad t_0 \le t \le t_f, \tag{8.60}$$

which yields the optimal control, and

$$L(t_c, u_{c-}) = L(t_c, u_{c+}) \tag{8.61}$$

at any jump in the optimal control. In applying these conditions, something must indicate that a jump can exist. This could happen when Eq.(8.60) yields multiple solutions. Then, the optimal location of the jump is determined by Eq.(8.61), which is called the *Erdmann–Weierstrass corner condition*. These conditions only identify an optimal control so the second differential must be used to verify that the optimal control is a minimum.

The second differential is obtained by taking the differential of the first differential (8.52) according to Leibniz' rule. If the first differential conditions are used to eliminate the $\delta^2 u$ and $d^2 t_f$ terms, it is given by

$$\begin{aligned} d^2 J &= [(L_{c-})_{t_c} - (L_{c+})_{t_c}]\, dt_c^2 + (L_{c-})_{u_{c-}}\, du_{c-}\, dt_c - (L_{c+})_{u_{c+}}\, du_{c+}\, dt_c \\[1mm] &\quad + [L_u \delta u dt]_{t_0}^{t_{c-}} + \int_{t_0}^{t_c} \delta u^T L_{uu} \delta u dt \\[1mm] &\quad + [L_u \delta u dt]_{t_{c+}}^{t_f} + \int_{t_c}^{t_f} \delta u^T L_{uu} \delta u dt. \end{aligned} \tag{8.62}$$

In this equation, $dt_0 = dt_f = 0$, $dt_{c-} = dt_{c+} = dt_c$, and since $L_u = 0$, $(L_{c-})_{u_{c-}} = (L_{c+})_{u_{c+}} = 0$. Hence, the second differential reduces to

$$d^2 J = [(L_{c-})_{t_c} - (L_{c+})_{t_c}]\, dt_c^2 + \int_{t_0}^{t_c} \delta u^T L_{uu} \delta u dt + \int_{t_c}^{t_f} \delta u^T L_{uu} \delta u dt \tag{8.63}$$

and must be nonnegative for a minimal control. By considering a sequence of admissible comparison controls identical with that used in the analysis of the first differential, Eq.(8.63) yields the following necessary conditions:

$$
\begin{aligned}
L_{uu} &\geq 0, \quad t_0 \leq t \leq t_c \\
L_{uu} &\geq 0, \quad t_c \leq t \leq t_f \\
(L_{c-})_{t_c} - (L_{c+})_{t_c} &\geq 0
\end{aligned}
\tag{8.64}
$$

or, equivalently,

$$
L_{uu}(t, u) \geq 0, \quad (L_{c-})_{t_c} - (L_{c+})_{t_c} \geq 0.
\tag{8.65}
$$

It is obvious that

$$
L_{uu}(t, u) > 0 \, , \quad (L_{c-})_{t_c} - (L_{c+})_{t_c} > 0
\tag{8.66}
$$

are sufficient conditions for a minimal control. Also, the Weierstrass condition (8.26) must hold because it is a local condition. At the corner, where $L_{c-} - L_{c+} = 0$ by Eq.(8.61), the Weierstrass condition is satisfied by equality if $u = u_{c-}$ and $u_* = u_{c+}$. Hence, the Weierstrass condition now says that $L$ must be an absolute minimum with respect to the control for all $u_* \neq u$ and $u_* \neq u_{c+}$ if a corner exists.

# 8.7   Integral Constraint

Consider the problem of finding the control $u(t)$ that minimizes the performance index

$$
J = \int_{t_0}^{t_f} L(t, u) \, dt,
\tag{8.67}
$$

subject to the integral constraint

$$
\int_{t_0}^{t_f} M(t, u) \, dt = K,
\tag{8.68}
$$

where $t_0$ and $t_f$ are fixed and $K$ is a constant. The integral constraint does not severely restrict the control because there are an infinite number of $u(t)$ that satisfy (8.68). For the same reason, many integral constraints can be imposed. Only the case of a single integral constraint is considered here. For additional constraints, the procedure is the same. Finally, $t_f$ is assumed fixed and the optimal control is assumed to be continuous in order to concentrate on the integral constraint.

The integral constraint is like a point condition and, hence, is adjoined to the performance index with a constant Lagrange multiplier $\lambda$ to form the augmented performance index

$$J' = \int_{t_0}^{t_f} (L + \lambda M)\, dt - \lambda K. \tag{8.69}$$

Taking the differential of $J'$ leads to

$$dJ' = \int_{t_0}^{t_f} H_u \delta u \, dt, \tag{8.70}$$

where the quantity $H$ is defined as

$$H = L + \lambda M. \tag{8.71}$$

## 8.7.1  First Differential Condition

It seems reasonable that $H_u$ should be zero along the minimal path. However, in the proof, admissible comparison controls are not arbitrary but must satisfy

$$\int_{t_0}^{t_f} M_u \delta u \, dt = 0, \tag{8.72}$$

which is obtained by taking the differential of Eq.(8.68).

Consider the admissible comparison control defined as follows:

$$
\begin{array}{ll}
t_0 \le t \le t_p, & \delta u = 0 \\
t_p \le t \le t_q, & \delta u = \delta v = \text{Const} \\
t_q \le t \le t_f, & \delta u \text{ such that Eq.(8.72) is satisfied,}
\end{array}
\tag{8.73}
$$

where $t_q = t_p + \Delta t$. For this control, Eq.(8.71) becomes

$$(M_u)_p \Delta t \delta v + \int_{t_q}^{t_f} M_u \delta u \, dt = 0, \tag{8.74}$$

and there are many $\delta u$ over $[t_q,\ t_f]$ that satisfy this equation. To obtain a unique $\delta u$, find the $\delta u$ that minimizes

$$\bar{J} = \int_{t_q}^{t_f} \delta u^T \delta u \, dt \tag{8.75}$$

subject to the constraint (8.74). The derivation is similar to that for control-lability presented in Chapter 7. First, multiply (8.74) by the constant $-2d\nu^T$ and add it to Eq.(8.75) to obtain

$$\bar{J} = -2d\nu^T(M_u)_p\Delta t\delta v + \int_{t_q}^{t_f}(\delta u^T\delta u - 2d\nu^T M_u\delta u)\ dt. \qquad (8.76)$$

To make the integrand a perfect square, add and subtract the term $d\nu^T M_u M_u^T d\nu$ so that

$$\begin{aligned}
\bar{J} &= -2d\nu^T(M_u)_p\Delta t\delta v - \int_{t_q}^{t_f} d\nu^T M_u M_u^T d\nu\ dt \\
&\quad + \int_{t_q}^{t_f}(\delta u - M_u^T d\nu)^T(\delta u - M_u^T d\nu)\ dt.
\end{aligned} \qquad (8.77)$$

Because the control term is squared, $\bar{J}$ is minimized with respect to $\delta u$ when

$$\delta u = -M_u^T d\nu. \qquad (8.78)$$

Substitution of $\delta u$ into the constraint (8.74) gives

$$d\nu = -Q^{-1}(M_u)_p\Delta t\delta v, \qquad (8.79)$$

where

$$Q = \int_{t_q}^{t_f} M_u M_u^T\ dt \qquad (8.80)$$

and $Q^{-1}$ exists because $Q > 0$. Finally, a $\delta u$ over $[t_q,\ t_f]$ that satisfies Eq.(8.74) is given by

$$\delta u = U\Delta t\delta v, \qquad (8.81)$$

where

$$U(t) = -M_u^T Q^{-1}(M_u)_p. \qquad (8.82)$$

For the admissible comparison control (8.73), the first differential (8.70) becomes

$$dJ' = \int_{t_p}^{t_q} H_u\delta v\ dt + \int_{t_p}^{t_q} H_u U\Delta t\delta v\ dt \qquad (8.83)$$

and reduces to

$$dJ' = P\Delta t\delta v, \qquad (8.84)$$

where

$$P = (H_u)_p + \int_{t_q}^{t_f} H_u U\ dt. \qquad (8.85)$$

The proof that $H_u = 0$ is by contradiction. Assume that $u(t)$ is the minimal control but that $H_u \neq 0$. Since the first differential does not vanish, $\Delta J'$ can be approximated by $dJ'$, that is, $\Delta J' \cong P\Delta t \delta v$. At least one element of $P$ must be nonzero, otherwise the control cannot change the performance index. Now, choose $\delta v$ such that $\Delta J' > 0$. Then, by changing the sign of $\delta v$, $\Delta J' = P\Delta t(-\delta v) = -P\Delta t \delta v < 0$. Since $\Delta J'$ must be $> 0$ regardless of the choice of $\delta v$, $u(t)$ is not a minimum. Since this is a contradiction, it is concluded that $H_u$ must be zero. Hence, a necessary condition for a minimum is the Euler-Lagrange equation

$$H_u(t, u, \lambda) = 0. \tag{8.86}$$

It must be satisfied at every point of the minimal path. This condition implies that $u = u(t, \lambda)$, and substitution into the constraint (8.68) leads to a nonlinear algebraic equation that can be solved for $\lambda = \lambda(K)$.

## 8.7.2    Second Differential Conditions

To prove that the control history satisfying Eq.(8.86) is a minimum, the second differential

$$d^2 J' = \int_{t_0}^{t_f} \delta u^T H_{uu} \delta u \, dt \tag{8.87}$$

is investigated. For the admissible comparison control (8.73), the second differential becomes

$$d^2 J' = \int_{t_p}^{t_q} \delta v^T H_{uu} \delta v \, dt + \int_{t_q}^{t_f} \delta v^T \Delta t U^T H_{uu} U \Delta t \delta v \, dt \tag{8.88}$$

or

$$d^2 J' = \delta v^T (H_{uu})_p \delta v + \delta v^T \Delta t \int_{t_q}^{t_f} U^T H_{uu} U \, dt \, \Delta t \delta v. \tag{8.89}$$

The second term can be made negligible with respect to the first by making $\Delta t$ as small as necessary. Hence,

$$d^2 J' \cong \delta v^T (H_{uu})_p \delta v \tag{8.90}$$

and implies that $(H_{uu})_p$ must be $\geq 0$ for a minimum. Since $t_p$ is arbitrary, the necessary condition

$$H_{uu}(t, u, \lambda) \geq 0 \tag{8.91}$$

must be satisfied over $[t_0, t_f]$ for $u(t)$ to be a minimum.

From Eq.(8.87) it is apparent that

$$H_{uu}(t, u, \lambda) > 0 \qquad (8.92)$$

is the sufficient condition for a minimum since $d^2J' > 0$ for all $\delta u$.

### 8.7.3   Strong Variations

Consider the case of strong variations similar to that of Eq.(8.73) with $u_* - u$ not small over $[t_p, t_q]$. For this system of variations, the total change in the augmented performance index becomes

$$\Delta J' = \int_{t_p}^{t_q} [H(t, u_*, \lambda_*) - H(t, u, \lambda)]\, dt$$
$$+ \int_{t_q}^{t_f} [H(t, u_*, \lambda_*) - H(t, u, \lambda)]\, dt - \lambda_* K + \lambda K. \qquad (8.93)$$

Since the integral constraint is satisfied by the choice of the admissible comparison control, the multiplier $\lambda_*$ can have any value. It is chosen to be $\lambda_* = \lambda$. Next, over $[t_q, t_f]$, $u_* = u + \delta u$ so that a Taylor expansion is possible. Then, since $t_q = t_p + \Delta t$, $\Delta J'$ becomes

$$\Delta J' = (H_* - H)_p\, \Delta t + \int_{t_q}^{t_f} H_u \delta u\, dt. \qquad (8.94)$$

Since $H_u = 0$ over $[t_q, t_f]$, $\Delta J'$ can be rewritten as

$$\Delta J' = [H_* - H]_p \Delta t \qquad (8.95)$$

and leads to $(H_* - H)_p \geq 0$. Finally, since $\Delta J' > 0$ for a minimum and $t_p$ is arbitrary, it follows that

$$H(t, u_*, \lambda) - H(t, u, \lambda) > 0 \qquad (8.96)$$

must hold over $[t_0, t_f]$ for all $u_* \neq u$ and is a necessary condition for a minimum. It says that $H$ must be an absolute minimum with respect to the control.

Note that the Weierstrass condition reduces to the Legendre–Clebsch condition for weak variations, that is, $u_* = u + \delta u$.

## 8.7.4   Navigation Problem

As an example, consider the navigation problem defined in Chapter 1. The optimal control problem is to find the steering angle $\theta(t)$ that minimizes the integral

$$J = \int_{x_0}^{x_f} \frac{dx}{\cos \theta} \qquad (8.97)$$

in going from $x_0 = 0, y_0 = 0$ to $x_f = 1, y_f = 0$, where the conditions on $y$ are represented by the integral constraint

$$\int_{x_0}^{x_f} \frac{V \sin \theta + w}{\cos \theta} dx = 0. \qquad (8.98)$$

Here,

$$L = \frac{1}{\cos \theta}, \quad M = \frac{V \sin \theta + w}{\cos \theta}, \quad K = 0, \qquad (8.99)$$

and

$$H = \frac{1}{\cos \theta} + \lambda \frac{V \sin \theta + w}{\cos \theta}. \qquad (8.100)$$

The first differential condition (8.86) gives

$$H_\theta = \frac{\sin \theta + \lambda[V + w \sin \theta]}{\cos^2 \theta} = 0 \qquad (8.101)$$

and implies that $\theta = $ Const. Next, integration of the constraint (8.98) leads to

$$\frac{V \sin \theta + w}{\cos \theta}(x_f - x_0) = 0 \qquad (8.102)$$

or

$$\sin \theta = -\frac{w}{V}. \qquad (8.103)$$

Finally, Eq.(8.101) yields the multiplier $\lambda$, that is,

$$\lambda = \frac{w}{V^2 - w^2}. \qquad (8.104)$$

The second differential condition (8.91) or $H_{\theta\theta} \geq 0$ leads to

$$\cos^3 \theta \; H_{\theta\theta} = \cos^2 \theta + \lambda w \cos^2 \theta + 2 \sin^2 \theta + 2\lambda \sin \theta[V + w \sin \theta] \geq 0. \qquad (8.105)$$

When combined with Eqs.(8.103), (8.104), and

$$\cos \theta = \sqrt{1 - \left(\frac{w}{V}\right)^2}, \qquad (8.106)$$

the final result is

$$H_{\theta\theta} = 1 > 0. \tag{8.107}$$

Finally, the Weierstrass condition (8.96) requires that

$$\frac{1}{\cos\theta_*} + \lambda\left[\frac{V\sin\theta_* + w}{\cos\theta_*}\right] - \frac{1}{\cos\theta} - \lambda\left[\frac{V\sin\theta + w}{\cos\theta}\right] > 0. \tag{8.108}$$

Combined with Eq.(8.103) and Eq.(8.104) in the form

$$\lambda = -\frac{\sin\theta}{\cos^2\theta}, \tag{8.109}$$

Eq.(8.108) requires that

$$\frac{1 - \cos(\theta_* - \theta)}{\cos^2\theta\cos\theta_*} > 0 \tag{8.110}$$

for all $\theta_* \neq \theta$. The Weierstrass condition is not satisfied because $\cos\theta_*$ can be negative. However, $-\pi/2 \le \theta \le \pi/2$ is an implied constraint. If it were included in the problem, the Weierstrass condition would be satisfied. See Section 8.9.

## 8.8   Control Equality Constraint

Consider the fixed final time problem for the case where the control variables are not independent but must satisfy an $r \times 1$ vector of constraints having the form

$$C(t, u) = 0. \tag{8.111}$$

This means that of the $m$ control variables only $m - r$ are independent. Once again the optimal control is assumed to be continuous.

Each constraint is multiplied by a Lagrange multiplier function $\mu_k(t)$, integrated over the time interval $[t_0, t_f]$, and added to the performance index (8.7). The augmented performance index can be written in matrix form as

$$J' = \int_{t_0}^{t_f}(L + \mu^T C)dt, \tag{8.112}$$

where $\mu$ is a $r \times 1$ vector of Lagrange multipliers.

Taking the differential of $J'$ leads to

$$dJ' = \int_{t_0}^{t_f} [(L_u + \mu^T C_u)\delta u + \delta\mu^T C]dt, \qquad (8.113)$$

where the $\delta\mu$ term vanishes because its coefficient is zero. Since the components of $\delta u$ are not independent, it is not possible to set their coefficients equal to zero. Instead, the $r$ Lagrange multipliers $\mu(t)$ are chosen such that the coefficients of the $r$ dependent components of $\delta u$ vanish. Then, the remaining $m-r$ components can be considered as independent, and their coefficients must vanish. In all, $m$ conditions result, that is,

$$L_u + \mu^T C_u = 0. \qquad (8.114)$$

Equations (8.111) and (8.114) constitute $r+m$ equations in the $m+r$ unknowns $u(t)$ and $\mu(t)$.

The second differential is obtained by taking the differential of Eq.(8.113). First, the term $\mu^T C_u$ is written as $\mu_k(C_k)_u$, where the repeated index implies a summation. Then, in view of the first differential conditions the second differential is given by

$$d^2 J' = \int_{t_0}^{t_f} \{\delta u^T [L_{uu} + \mu_k(C_k)_{uu}]\delta u + 2\delta\mu^T C_u \delta u\}dt, \qquad (8.115)$$

where the components of $\delta u$ are not independent but must be consistent with Eq.(8.111). To first order, this relationship is

$$C_u \delta u = 0 \qquad (8.116)$$

and causes the $\delta\mu$ term in (8.115) to vanish. Once the dependent variations are obtained in terms of the independent variations with Eq.(8.116) and are eliminated from Eq.(8.114), a lower-order quadratic form results and must be nonnegative for a minimal control. An explicit condition can be obtained by identifying the dependent and independent variables along the lines of Section 5.5. A sufficient condition for a minimum is that the reduced quadratic form be positive definite.

The Weierstrass condition for this problem is that

$$[L(t, u_*) + \mu^T C(t, u_*)] - [L(t, u) + \mu^T C(t, u)] > 0 \qquad (8.117)$$

for all $u_* \neq u$ that satisfy the constraint (8.111). Hence, the strong variation condition reduces to

$$L(t, u_*) - L(t, u) > 0, \qquad (8.118)$$

where the dependent $u_*$s are required to satisfy the constraint

$$C(t, u_*) = 0 \tag{8.119}$$

and the independent $u_*$s are allowed to vary between plus and minus infinity.

## 8.9  Control Inequality Constraint

Consider the fixed final time problem for the case where the optimal control must satisfy an inequality constraint of the form

$$\bar{C}(t, u) \leq 0. \tag{8.120}$$

To simplify the development, assume that there is only one control and only one control inequality constraint.

The constraint (8.120) is converted into an equality constraint through the introduction of a slack variable $\alpha(t)$ defined by (Ref. Va)

$$\bar{C}(t, u) = -\alpha^2, \tag{8.121}$$

and $\alpha$ is required to be real. To apply the results of the previous section, it is observed that

$$C = \bar{C} + \alpha^2 = 0 \tag{8.122}$$

and that there are now two controls:

$$u_1 = u, \quad u_2 = \alpha. \tag{8.123}$$

The first differential conditions (8.114) become

$$L_u + \mu \bar{C}_u = 0. \tag{8.124}$$

and

$$2\mu\alpha = 0. \tag{8.125}$$

The latter implies that the solution is possibly composed of multiple parts (Fig. 8.5): parts where $\mu = 0$ that are off the boundary ($\bar{C} < 0$) and parts where $\alpha = 0$ that are on the boundary ($\bar{C} = 0$). For $\mu = 0$, the Eq.(8.124) reduces to $L_u = 0$ as it should, and determines the control $u$. Then, Eq.(8.121) gives $\alpha$. For $\alpha = 0$, the control $u$ is obtained from $\bar{C} = 0$, and $\mu$ is obtained from Eq.(8.124).

To prove that the optimal control is a minimum, the second differential must be examined. It is given by

$$d^2 J' = \int_{t_0}^{t_f} [(L_{uu} + \mu \bar{C}_{uu})\delta u^2 + 2\mu\delta\alpha^2]dt \qquad (8.126)$$

where the variations are not independent but from (8.116) must satisfy

$$\bar{C}_u \delta u + 2\alpha\delta\alpha = 0. \qquad (8.127)$$

If $\bar{C}_u \neq 0$, the dependent variation $\delta u$ can be eliminated from the second differential leading to the following condition for a minimum:

$$(2\alpha/\bar{C}_u)^2(L_{uu} + \mu \bar{C}_{uu}) + 2\mu \geq 0. \qquad (8.128)$$

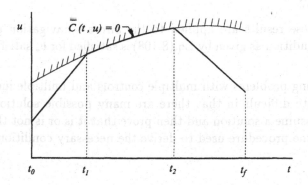

Figure 8.7: Control Inequality Constraint, Scalar Control

For the off-boundary solution $\mu = 0$, the requirement that the second differential be nonnegative leads to $L_{uu} \geq 0$. For the on-boundary solution $\alpha = 0$, so that the second differential yields $\mu \geq 0$. Hence, along that part of the control where the control inequality constraint is on the boundary, the corresponding Lagrange multiplier must be nonnegative.

The above conditions are necessary conditions. Sufficient conditions for a minimum are $L_{uu} > 0$ on off-boundary segments ($\mu = 0$) and $\mu > 0$ for on-boundary segments ($\alpha = 0$). Note that $\alpha = 0$ leads to $\delta u = 0$.

The solution process is to compute the control from $L_u = 0$. If $\bar{C} \leq 0$, this control is the optimum. If $\bar{C} > 0$, then the control that makes $\bar{C} = 0$ is the optimum. Next, if $L_{uu} > 0$ over the off-boundary segment(s) and $\mu > 0$ on the on-boundary segment(s), the control is a minimum.

The necessary condition for a strong minimum is that

$$L(t, u_*) - L(t, u) > 0 \tag{8.129}$$

for all $u_* \neq u$ that satisfy

$$\bar{C}(t, u_*) = -\alpha_*^2, \tag{8.130}$$

or for all $u_* \neq u$ that satisfy

$$\bar{C}(t, u_*) < 0. \tag{8.131}$$

This extended Weierstrass condition is also known as Pontryagin's minimum principle. Note that $l$ must be an absolute minimum with respect to the admissible controls.

If these results are applied to the aircraft navigation problem, the Weierstrass condition as given by Eq.(8.108) is satisfied for $\theta_*$ satisfying $-\pi/2 \leq \theta_* \leq \pi/2$.

Solving problems with multiple controls and multiple inequality constraints is quite difficult in that there are many possible solutions. One approach is to assume a solution and then prove that it is or is not the minimum. This is the same procedure used to derive the necessary conditions in the first place.

# Problems

8.1 Another way to solve the free final time problem is to introduce the transformation $\tau = t/t_f$. Since $\tau_0 = 0$ and $\tau_f = 1$, the problem becomes a fixed final time problem; however, the final time now appears as a parameter. The performance index in the new time is given by

$$J = \phi(t_f) + \int_{\tau_0}^{\tau_f} t_f L(t_f \tau, u) d\tau.$$

Derive the first and second differential conditions for a minimum and show that they can be manipulated into the corresponding conditions presented in Section 8.3.

8.2 Consider the problem where the controls are not independent but must satisfy the constraints $C(t, u) = 0$, where $C$ is an $r \times 1$ matrix. Develop the first and second differential conditions for a minimum for the case where a jump occurs in the control.

8.3 Consider the optimal control problem where the control equality constraint $C(t, u) = 0$ must be satisfied over the subinterval $[t_1, t_2]$ of the fixed time interval $[t_0, t_f]$. Assuming that $t_1$ and $t_2$ are fixed, derive the first and second differential conditions for a minimum.

# 9

# Fixed Final Time: First Differential

## 9.1 Introduction

First differential conditions are developed for a standard optimal control problem with fixed final time. The conditions are the Euler–Lagrange equations and the natural boundary conditions. These equations are to be solved for the optimal controls and states. For problems not explicitly containing the time, a first integral is shown to exist. Several examples are presented to illustrate the application of the theory.

## 9.2 Preliminary Remarks

In this chapter and Chapter 11, necessary conditions and sufficient conditions for a minimum are derived for the following optimal control problem with fixed final time: Find the control history $u(t)$ that minimizes the scalar performance index

$$J = \phi(x_f) + \int_{t_0}^{t_f} L(t, x, u) \, dt, \qquad (9.1)$$

subject to the differential constraints

$$\dot{x} = f(t, x, u), \qquad (9.2)$$

the prescribed initial conditions

$$t_0 = t_{0_s}, \quad x_0 = x_{0_s}, \tag{9.3}$$

and the prescribed final conditions

$$t_f = t_{f_s}, \quad \beta(x_f) = 0. \tag{9.4}$$

The quantity $u(t)$ is an $m \times 1$ vector of control variables; $x(t)$ is an $n \times 1$ vector of state variables; $f$ is an $n \times 1$ vector, and $\beta$ is a $p \times 1$ vector, where $p \leq n$. Also, the subscript $s$ denotes a specific value. The conditions derived in this chapter are valid for the case where there are no prescribed final conditions, that is, no $\beta$.

The general procedure for developing conditions for a minimum is to (a) adjoin the constraints to the performance index to form the augmented performance index $J'$, (b) form the difference $\Delta J'$ of the value of the augmented performance index along an admissible comparison path (one that satisfies all constraints) with its value along the minimal path, (c) relate the admissible comparison path to the minimal path by a system of small control changes and small state changes, and (d) expand in a Taylor series to obtain the first differential and the second differential. Here, the first and second differentials are obtained by taking differentials (Chapter 6), and the Taylor series approach is presented in Appendix A. The total change in the augmented performance index is given by

$$\Delta J' = dJ' + \frac{1}{2!}d^2 J' + \cdots. \tag{9.5}$$

The vanishing of the first differential $(dJ' = 0)$ leads to the Euler–Lagrange equations and the natural boundary conditions, and subsequently the first integral, if it exists. Then, the requirement that the second differential be non-negative $(d^2 J' \geq 0)$ for a minimum yields the Legendre–Clebsch condition and the conjugate point condition (Chapter 11).

Before investigating the second differential, two tests are developed for a minimum (Chapter 10): the Weierstrass condition and the Legendre–Clebsch condition. The Weierstrass condition is based on large control changes and small state changes, and the Legendre–Clebsch condition is a consequence of passing to small control changes.

**Note.** During the next few chapters, the minimal control is assumed to be continuous. However, since the control cannot be required to be continuous, admissible comparison controls can be discontinuous. Problems with discontinuous optimal controls are covered in Chapter 16.

## 9.3   First Differential Conditions

To develop the necessary conditions, the constraints are adjoined to the performance index by means of Lagrange multipliers. The differential constraints (9.2) must be satisfied at every point of the minimal path. Hence, each differential constraint $f_k - \dot{x}_k = 0$ is multiplied by a Lagrange multiplier $\lambda_k(t)$, integrated over $[t_0, t_f]$, and added to the performance index. Also, each prescribed final condition $\beta_k = 0$ is multiplied by a constant Lagrange multiplier $\nu_k$ and added to the performance index. The augmented performance index is then given by

$$J' = G(x_f, \nu) + \int_{t_0}^{t_f} [H(t, x, u, \lambda) - \lambda^T \dot{x}] \, dt, \qquad (9.6)$$

where $G$ is the endpoint function and $H$ is the Hamiltonian defined as

$$G = \phi(x_f) + \nu^T \beta(x_f) \qquad (9.7)$$

$$H = L(t, x, u) + \lambda^T f(t, x, u). \qquad (9.8)$$

None of the constraints on $t_0, x_0$, and $t_f$ has been adjoined with Lagrange multipliers because it is easier just to hold $t_0, x_0$, and $t_f$ fixed during the differential process.

The differential of the augmented performance index (9.6) is given by

$$dJ' = G_{x_f} dx_f + G_\nu d\nu + \int_{t_0}^{t_f} [H_x \delta x + H_u \delta u + (H_\lambda - \dot{x}^T)\delta\lambda - \lambda^T \delta\dot{x}] dt, \quad (9.9)$$

where Leibniz' rule (Section 6.7) has been used with $dt_0 = 0$ and $dt_f = 0$. First, $dx_f = \delta x_f$ since $dt_f = 0$. Then, note that the coefficients of $d\nu$ and $\delta\lambda$ are zero because $G_\nu = \beta^T = 0$ and $H_\lambda = f^T$ so that $H_\lambda - \dot{x}^T = 0$. Since $\delta\dot{x} = d\,\delta x/dt$ (Section 6.7), the last term in the integrand can be integrated by parts (Section 5.3.2) to obtain

$$dJ' = G_{x_f} \delta x_f - [\lambda^T \delta x]_{t_0}^{t_f}$$
$$+ \int_{t_0}^{t_f} [(H_x + \dot{\lambda}^T)\delta x + H_u \delta u] dt \,. \qquad (9.10)$$

Then, since the state is prescribed, $\delta x_0 = 0$, and Eq. (9.10) becomes

$$dJ' = (G_{x_f} - \lambda_f^T)\delta x_f$$
$$+ \int_{t_0}^{t_f} [(H_x + \dot{\lambda}^T)\delta x + H_u \delta u] \, dt. \qquad (9.11)$$

Ultimately, the first differential can be expressed in terms of independent variations whose coefficients must vanish for $u(t)$ to be the minimal control. This means that the first differential must vanish regardless of the choice of the admissible comparison control. An *admissible comparison control* is one that causes the differential equation

$$\frac{d}{dt}\delta x = f_x \delta x + f_u \delta u \tag{9.12}$$

to generate a $\delta x$ which satisfies the initial conditions

$$t_0 = t_{0_s}, \quad \delta x_0 = 0, \tag{9.13}$$

and the final conditions

$$t_f = t_{f_s}, \quad \beta_{x_f} \delta x_f = 0. \tag{9.14}$$

These equations are obtained by taking the variation of Eq.(9.2) and by taking the differential of Eqs.(9.3) and (9.4) and converting to variations by Eq.(6.64).

The next step in deriving the first differential conditions for a minimum is to choose the Lagrange multipliers to eliminate $\delta x$ which is integrally related to $\delta u$ (Eq. 7.15). Hence, $\lambda(t)$ is chosen to satisfy the differential equation

$$\dot{\lambda} = -H_x{}^T \tag{9.15}$$

and the final condition

$$\lambda_f = G_{x_f}^T. \tag{9.16}$$

These choices reduce the first differential to

$$dJ' = \int_{t_0}^{t_f} H_u \delta u dt. \tag{9.17}$$

However, the elements of $\delta u$ are not independent because $\delta x$, which is driven by $\delta u$, must satisfy $\beta_{x_f} \delta x_f = 0$.

For the first differential to vanish regardless of the choice of the admissible comparison control $\delta u$, it seems reasonable to expect that $H_u = 0$ along the minimal path. It is certainly true that if $H_u = 0$ the first differential vanishes regardless of the choice of $\delta u$. The question is whether $H_u$ must vanish or not.

## 9.3.1   No Final State Constraints

To prove that $H_u$ must vanish, consider first the case where there are no final state constraints (no $\beta$). Now, assume that $u(t)$ is the minimal control but that $H_u(t) \neq 0$. Next, consider an admissible comparison control where $\delta u$ is a small pulse over the small time interval $\Delta t = t_q - t_p$. The pulse is defined by the equations

$$
\begin{array}{lll}
t_0 \leq t \leq t_p, & \delta u(t) = 0 \\
t_p \leq t \leq t_q, & \delta u(t) = \text{Const} \triangleq \delta v & \qquad (9.18) \\
t_q \leq t \leq t_f, & \delta u(t) = 0
\end{array}
$$

and is shown in Fig. 9.1. For this admissible comparison control, the first differential (9.17) becomes

$$
dJ' = \int_{t_p}^{t_q} H_u dt \delta v. \qquad (9.19)
$$

Next, $H_u(t)$ is expanded in a Taylor series about $t_p$ as $H_u = (H_u)_p + (\dot{H}_u)_p(t -$

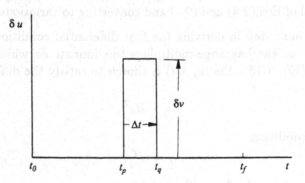

Figure 9.1: Small Pulse, Scalar Control

$t_p) + \cdots$, the integration is performed, and $\Delta t$ is assumed to be sufficiently small that higher-order terms are negligible. The result is the following:

$$
dJ' = (H_u)_p \Delta t \delta v. \qquad (9.20)
$$

Now, for the first differential to vanish regardless of the choice of $\delta v$ (plus or minus), it is necessary that $(H_u)_p = 0$. Finally, since the time $t_p$ is arbitrary, it is necessary that

$$
H_u = 0 \qquad (9.21)
$$

at every point of the minimal path.

## 9.3.2   With Final State Constraints

For the case where there are final state constraints, it seems reasonable to expect that $H_u = 0$ still must hold along the minimal path. To prove that this is true, it is assumed that $u(t)$ is the minimal control but that $H_u \neq 0$, and the proof is made by contradiction.

Consider the following admissible comparison control:

$$
\begin{aligned}
t_0 &\leq t \leq t_p, & \delta u(t) &= 0 \\
t_p &\leq t \leq t_q, & \delta u(t) &= \text{Const} \triangleq \delta v \\
t_q &\leq t \leq t_f, & \delta u(t) &\text{ such that } \beta_{x_f} \delta x_f = 0,
\end{aligned}
\tag{9.22}
$$

which is shown in Fig. 9.2. This is a pulse at $t_p$ followed by a control perturba-

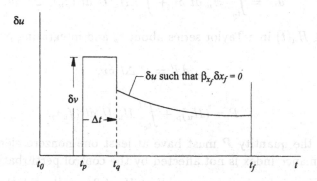

Figure 9.2: Small Pulse, Scalar Control

tion history that satisfies the perturbed final condition. Finally, the elements of $\delta v$ can be declared independent because the final condition $\beta_{x_f} \delta x_f = 0$ can be satisfied over $[t_q, t_f]$ by requiring the system (9.12) to be controllable to the final constraint surface.

Integration of Eq.(9.12) over $[t_p, t_q]$ leads to

$$
\delta x_q - \delta x_p = \int_{t_p}^{t_q} (f_x \delta x + f_u \delta v)\, dt.
\tag{9.23}
$$

If the integrand is viewed as a function of time, it can be expanded in a Taylor series about $t_p$ and integrated. Neglecting higher-order terms in $\Delta t$ leads to

$$\delta x_q = (f_u)_p \, \Delta t \, \delta v \qquad (9.24)$$

since $\delta x_p = 0$. Over $[t_q, t_f]$, a control perturbation that drives the system (9.12) from $t_q$, $\delta x_q$ to $t_f$, $\beta_{x_f} \delta x_f = 0$ is given in Section 7.4 to be $\delta u = U(t) \, \delta x_q$. In view of Eq.(9.24), this control perturbation becomes

$$\delta u = U \, (f_u)_p \, \Delta t \, \delta v. \qquad (9.25)$$

At this point, the first differential (9.18) is written as

$$dJ' = \int_{t_0}^{t_p} H_u \, \delta u \, dt + \int_{t_p}^{t_q} H_u \, \delta u \, dt + \int_{t_q}^{t_f} H_u \, \delta u \, dt. \qquad (9.26)$$

For the admissible comparison control (9.22), the first differential becomes

$$dJ' = \int_{t_p}^{t_q} H_u \, dt \, \delta v + \int_{t_q}^{t_f} H_u \, U \, dt \, (f_u)_p \, \Delta t \, \delta v. \qquad (9.27)$$

Expanding $H_u(t)$ in a Taylor series about $t_p$ and integrating over $[t_p, \, t_q]$ leads to

$$dJ' = P \, \Delta t \, \delta v, \qquad (9.28)$$

where

$$P = (H_u)_p + \int_{t_q}^{t_f} H_u \, U \, dt \, (f_u)_p. \qquad (9.29)$$

Note that the quantity $P$ must have at least one nonzero element; otherwise the performance index is not affected by the control perturbation.

Since it has been assumed that $H_u \neq 0$ over $[t_p, \, t_f]$, the $m \times 1$ vector $P$ has some value so that $dJ' \neq 0$. Because the first differential does not vanish, the total change in the augmented performance index can be approximated by the first-order change, that is, $\Delta J' \cong dJ'$. Since the elements of $\delta v$ are independent, choose each element such that $P \delta v > 0$. Then, changing the sign of $\delta v$ leads to $P \delta v < 0$. In other words, $dJ'$ can be made positive or negative. Therefore, because $H_u \neq 0$, $\Delta J'$ can be positive or negative. Since $\Delta J'$ must be greater than zero for a minimum regardless of the choice of the admissible comparison path (regardless of the choice of $\delta u$), this is a contradiction. Hence,

$$H_u = 0 \qquad (9.30)$$

must hold at every point of a minimal path.

## 9.4  Summary

The fixed final time optimal control problem is to find the control history $u(t)$ that minimizes the scalar performance index (9.1), subject to the differential constraints (9.2), the prescribed initial conditions (9.3), and the prescribed final conditions (9.4). First, the endpoint function (9.7) and the Hamiltonian (9.8) are formed. Then, the optimal state $x(t)$, Lagrange multiplier $\lambda(t)$, and control $u(t)$ are obtained from the solution of the equations

$$\dot{x} = f(t, x, u) \tag{9.31}$$

$$\dot{\lambda} = -H_x^T(t, x, u, \lambda) \tag{9.32}$$

$$0 = H_u(t, x, u, \lambda) \tag{9.33}$$

subject to the boundary conditions

$$t_0 = t_{0_s}, \quad x_0 = x_{0_s} \tag{9.34}$$

$$t_f = t_{f_s}, \quad \beta(x_f) = 0 \tag{9.35}$$

$$\lambda_f = G_{x_f}^T(x_f, \nu). \tag{9.36}$$

Eqations (9.31) through (9.33) are the *Euler–Lagrange equations*. The final conditions for the Lagrange multipliers (9.36) are called the *natural boundary conditions*. All of the conditions are called the *optimality conditions*. Note that there are $n + n + m$ equations for obtaining the $n$ $x$s, the $n$ $\lambda$s, and the $m$ $u$s. Also, there are $p + n$ final conditions for determining the $p$ $\nu$s and the $n$ $x_f$s. While Eqs.(9.31) through (9.36) are valid for a minimal control, they are also valid for a maximal control. Hence, these equations represent necessary conditions for a minimal control. For the case where there are no final state constraints (no $\beta$), Eq.(9.36) becomes $\lambda_f = \phi_{x_f}^T$.

Equations (9.31) through (9.33) have other names. Equation (9.31) is also called the *state equation*; Eq.(9.32), the *costate equation* or the *adjoint equation*; and Eq.(9.33), the *control equation*.

It might seem that the multiplier $\nu$ is not playing any role in this process. However, choosing some of the $\lambda_f$s is the same as choosing the $\nu$s.

Finally, if the problem to be solved is a maximization problem, then $J$ is replaced by $-J$, and the problem is solved as a minimization problem.

## 9.5   First Integral

Consider the total time derivative of the Hamiltonian $H(t, x, u, \lambda)$ along the minimal path, that is,

$$\frac{dH}{dt} = H_t + H_x \dot{x} + H_u \dot{u} + H_\lambda \dot{\lambda}. \tag{9.37}$$

Since

$$\dot{x} = f, \quad H_u = 0, \quad H_\lambda = f^T, \quad \dot{\lambda} = -H_x^T, \tag{9.38}$$

the derivative becomes

$$\frac{dH}{dt} = H_t. \tag{9.39}$$

As a consequence, if $H$ does not contain $t$ explicitly, that is, if $H_t = 0$, the first integral

$$H = \text{Const} \tag{9.40}$$

must hold along a minimal path.

The first integral is of limited value in the solution of optimal control problems because it is only one equation and, hence, can only replace one of the $n + n + m$ equations (9.30) through (9.33). On the other hand, if there is only one state, then the first integral represents the first integration of the Euler–Lagrange equations.

## 9.6   Example

To illustrate the application of the first differential necessary conditions, consider the following problem: Find the control $u(t)$ that minimizes the performance index

$$J = \int_{t_0}^{t_f} (u - x)^2 \, dt, \tag{9.41}$$

subject to the differential constraint

$$\dot{x} = u \tag{9.42}$$

and the prescribed boundary conditions

$$t_0 = 0, \quad x_0 = 1, \quad t_f = 1. \tag{9.43}$$

The control and the state are scalars. Note that there is no $\phi$ or $\beta$.

The first step is to form the endpoint function and the Hamiltonian as

$$G = \phi + \nu^T \beta = 0 \tag{9.44}$$

$$H = L + \lambda^T f = (u - x)^2 + \lambda u. \tag{9.45}$$

Then, Eqs.(9.32), (9.33), and (9.36) become

$$\dot{\lambda} = 2(u - x) \tag{9.46}$$

$$0 = 2(u - x) + \lambda \tag{9.47}$$

$$\lambda_f = 0. \tag{9.48}$$

The first two equations can be combined to yield

$$\dot{\lambda} = -\lambda \tag{9.49}$$

which can be integrated to obtain

$$\lambda = C\, e^{-t}, \tag{9.50}$$

where $C$ is a constant. Then, application of the natural boundary condition (9.48) gives $C = 0$ so that

$$\lambda = 0 \tag{9.51}$$

everywhere along the optimal path.

Next, Eq.(9.47) reduces to

$$u = x, \tag{9.52}$$

and Eq.(9.42) becomes

$$\dot{x} = x. \tag{9.53}$$

The solution of this equation subject to the prescribed initial conditions (9.43) is given by

$$x = e^t \tag{9.54}$$

so that the optimal control is

$$u = e^t. \tag{9.55}$$

Note that the first integral (9.40) exists but has not been used. Also, the solution $u = x$ makes the performance index an absolute minimum.

## 9.7   Acceleration Problem

Consider the straight-line motion of a car whose acceleration can be controlled ($\ddot{x} = u$). It is desired to transfer the vehicle from the origin at zero speed to $x_f = 1$ at $t_f = 1$ in such a way that the acceleration is not excessive. A formal statement of such a problem is the following: Find the control history $u(t)$ that minimizes the performance index

$$J = \frac{1}{2} \int_{t_0}^{t_f} u^2 dt, \tag{9.56}$$

subject to the differential constraints

$$\begin{aligned} \dot{x}_1 &= x_2 \\ \dot{x}_2 &= u \end{aligned} \tag{9.57}$$

and the prescribed boundary conditions

$$\begin{aligned} t_0 &= 0, \quad x_{1_0} = 0, \quad x_{2_0} = 0 \\ t_f &= 1, \quad \beta = x_{1_f} - 1 = 0. \end{aligned} \tag{9.58}$$

In the terminology of Section 9.4, the following functions are formed:

$$\begin{aligned} G &= \nu(x_{1_f} - 1) \tag{9.59} \\ H &= u^2/2 + \lambda_1 x_2 + \lambda_2 u. \tag{9.60} \end{aligned}$$

The Euler–Lagrange equations (9.31) through (9.33) can be written as

$$\begin{aligned} \dot{x}_1 &= x_2 \\ \dot{x}_2 &= u \\ \dot{\lambda}_1 &= -H_{x_1} = 0 \\ \dot{\lambda}_2 &= -H_{x_2} = -\lambda_1 \\ 0 &= H_u = u + \lambda_2 \end{aligned} \tag{9.61}$$

and can be solved to obtain

$$\begin{aligned} \lambda_1 &= C_1 \\ \lambda_2 &= -C_1 t + C_2 \\ u &= C_1 t - C_2 \\ x_2 &= C_1 t^2/2 - C_2 t + C_3 \\ x_1 &= C_1 t^3/6 - C_2 t^2/2 + C_3 t + C_4. \end{aligned} \tag{9.62}$$

The initial conditions (9.58) lead to $C_3 = C_4 = 0$, whereas the final conditions (9.58) and (9.36) require that

$$
\begin{aligned}
t_f &= 1 \\
\beta &= x_{1_f} - 1 = 0 \\
\lambda_{1_f} &= G_{x_{1_f}} = \nu \\
\lambda_{2_f} &= G_{x_{2_f}} = 0.
\end{aligned}
\tag{9.63}
$$

If the boundary conditions $t_f = 1$, $x_{1_f} = 1$, and $\lambda_{2_f} = 0$ are applied, it is seen that

$$
C_1 = -3, \quad C_2 = -3.
\tag{9.64}
$$

Hence, the optimal control is given by

$$
u = 3(1 - t)
\tag{9.65}
$$

while the states and the multipliers become

$$
\begin{aligned}
x_1 &= -t^3/2 + 3t^2/2 \\
x_2 &= -3t^2/2 + 3t \\
\lambda_1 &= -3 \\
\lambda_2 &= 3t - 3.
\end{aligned}
\tag{9.66}
$$

It remains to be established that the optimal control is a minimum. Furthermore, note that a first integral exists but that it is not used.

## 9.8 Navigation Problem

In Chapter 1, it is shown that the aircraft navigation problem can be formulated as a "fixed final time" problem. The simplest version of this problem is to find the velocity direction history $\theta(x)$ that minimizes the performance index

$$
J = t_f,
\tag{9.67}
$$

subject to the differential constraint

$$
\frac{dt}{dx} = \frac{1}{V \cos \theta},
\tag{9.68}
$$

the prescribed initial conditions

$$x_0 = 0, \quad t_0 = 0, \tag{9.69}$$

and the prescribed final condition

$$x_f = 1. \tag{9.70}$$

Note that the final value of the variable of integration is fixed, there are no prescribed final conditions, and the time is now a state.

With the endpoint function and the Hamiltonian defined as

$$G = t_f, \quad H = \frac{\lambda}{V \cos \theta}, \tag{9.71}$$

the Euler-Lagrange equations become

$$\frac{dt}{dx} = \frac{1}{V \cos \theta} \tag{9.72}$$

$$\frac{d\lambda}{dx} = -H_t = 0 \tag{9.73}$$

$$0 = H_\theta = \frac{\lambda \sin \theta}{V \cos^2 \theta}, \tag{9.74}$$

and the boundary conditions are given by

$$x_0 = 0, \quad t_0 = 0 \tag{9.75}$$

$$x_f = 1, \quad \lambda_f = G_{t_f} = 1. \tag{9.76}$$

Equation (9.73) says that $\lambda = $ Const. Since Eq.(9.76) indicates that $\lambda = 1$, the optimal control from Eq.(9.74) is $\sin \theta = 0$ or $\theta = 0$. The solution $\theta = \pi$ also exists, but it is not a minimizing solution. See Section 10.4. Finally, Eq.(9.72) can be integrated subject to the initial conditions (9.75) to obtain $t = x/V$, which evaluated at the final point gives $t_f = 1/V$ .

## 9.9    Minimum Distance on a Sphere

An interesting optimization problem is that of minimizing the distance on a sphere between a point O and a great circle (see Fig. 9.3). The coordinate

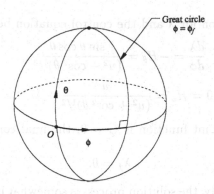

Figure 9.3: Minimum Distance on a Sphere

system is defined with the abscissa $\phi$ along the great circle passing through O which is perpendicular to the given great circle ($\phi = \phi_f$). The ordinate is $\theta$. For an arbitrary curve $\theta = \theta(\phi)$ through $\phi_0 = 0, \theta_0 = 0$, and $\phi_f = \phi_{f_s}$, the distance along the curve can be expressed as

$$J = \int_{\phi_0}^{\phi_f} [(d\theta/d\phi)^2 + \cos^2 \theta]^{1/2} \, d\phi. \tag{9.77}$$

In the optimal control format, the problem is to find the control history $u(\phi)$ that minimizes the performance index

$$J = \int_{\phi_0}^{\phi_f} (u^2 + \cos^2 \theta)^{1/2} \, d\phi, \tag{9.78}$$

subject to the differential constraint

$$\frac{d\theta}{d\phi} = u \tag{9.79}$$

and the prescribed boundary conditions

$$\phi_0 = 0, \quad \theta_0 = 0, \quad \phi_f = \phi_{f_s}. \tag{9.80}$$

Note that there is no condition imposed on $\theta_f$ (no $\beta$).

From the definition of the Hamiltonian, that is,

$$H = (u^2 + \cos^2 \theta)^{1/2} + \lambda u, \tag{9.81}$$

the differential equation for $\lambda$ and the control equation become

$$\frac{d\lambda}{d\phi} = -H_\theta = \frac{\sin\theta\cos\theta}{(u^2 + \cos^2\theta)^{1/2}} \tag{9.82}$$

$$0 = H_u = \frac{u}{(u^2 + \cos^2\theta)^{1/2}} + \lambda. \tag{9.83}$$

Then, since the endpoint function is $G = 0$, the final condition on $\lambda$ is given by

$$\lambda_f = 0. \tag{9.84}$$

Unfortunately, the solution process is somewhat involved. First, Eq.(9.89) can be solved for $u$ as

$$u = \left(\frac{\lambda^2 \cos^2\theta}{1 - \lambda^2}\right)^{1/2}, \tag{9.85}$$

where the plus sign has been chosen on the basis of symmetry (the $u < 0$ solution is the image of the $u > 0$ solution through the $\phi$ axis). Next, $\theta$ is chosen as the variable of integration, and the differential equation for $d\lambda/d\theta$, when combined with Eq. (9.85), can be integrated to give

$$\sqrt{1 - \lambda^2} = \frac{c}{\cos\theta}, \tag{9.86}$$

where $c$ is a constant of integration and must be $\geq 0$ for $-\pi/2 \leq \theta \leq \pi/2$. Hence, the control (9.85) can be rewritten as

$$u = \frac{1}{c}\cos\theta\sqrt{\cos^2\theta - c^2}. \tag{9.87}$$

Finally, the equation for $d\phi/d\theta$ can be integrated to obtain

$$\phi = \frac{1}{c}\arctan\frac{c\sin\theta}{(1 - c^2 - \sin^2\theta)^{1/2}}, \tag{9.88}$$

where the constant of integration is zero because of the initial conditions (9.80). The ordinate $\theta$ can then be expressed in terms of $\phi$ as

$$\sin\theta = \left[\frac{(1 - c^2)\tan^2 c\phi}{c^2 + \tan^2 c\phi}\right]^{1/2}. \tag{9.89}$$

To apply the natural boundary condition $\lambda_f = 0$, Eq.(9.86) is solved for $\lambda$, and the result is combined with Eq.(9.89) to obtain

$$\lambda^2 = (1 - c^2)\cos^2 c\phi. \tag{9.90}$$

Then, $\lambda_f = 0$ leads to

$$c = \pm 1 \quad \text{or} \quad c = \pi/2\phi_f. \tag{9.91}$$

The solution $c = -1$ violates Eq.(9.86), and $c = \pi/2\phi_f$ gives $1 - c^2 < 0$ for $\phi_f < \pi/2$, making $\theta$ and $\lambda$ imaginary. Hence, the only solution is $c = 1$ so that

$$\theta = 0, \quad \lambda = 0, \quad u = 0 \tag{9.92}$$

along the path. In other words, the optimal path is the great circle starting at $\phi_0 = \theta_0 = 0$ and ending perpendicular to the prescribed great circle.

Another version of this problem is to minimize the distance between two points. The problem statement is given by Eqs.(9.78) through (9.80) with the additional final condition

$$\theta_f = 0. \tag{9.93}$$

In other words, the $\phi\theta$ coordinate system is aligned so that the $\phi$-axis passes through the prescribed final point. With the exception of the natural boundary condition (9.84), Eqs.(9.81) through (9.89) are valid here. Then, application of the prescribed boundary condition (9.93) in Eq.(9.89) leads to $c = 1$ so that the optimal path is defined by Eqs.(9.92).

# Problems

The problems are stated as minimization problems or maximization problems although it is only possible at this point to determine the optimal solution. Remember that maximization problems are converted into minimization problems by multiplying the performance index by minus one (-1). In Chapter 10, two tests for a minimum are derived, and they are to be applied to these problems. The same is true for the Jacobi or conjugate point condition derived in Chapter 11.

9.1 Solve the example in Section 9.6 for the case where $x_f = 1$ in addition to the stated boundary conditions. Show that the optimal path is given by

$$x = \frac{e^t + e^{1-t}}{1 + e}.$$

9.2 Solve the example in Section 9.7 for the case where the final velocity must vanish in addition to the stated conditions, that is,

$$x_{2_f} = 0.$$

Show that the optimal control is given by

$$u = 6 - 12t \ .$$

9.3 Find the control $u(t)$ that minimizes the quadratic performance index

$$J = \frac{c}{2}x_f{}^2 + \int_{t_0}^{t_f} \frac{u^2}{2} \, dt,$$

where $c$ is a positive constant. The linear dynamical system is defined by

$$\dot{x} = u$$

and the prescribed boundary conditions are

$$t_0 = t_{0_s}, \quad x_0 = x_{0_s}, \quad t_f = t_{f_s}.$$

Show that the optimal control and state are given by

$$u = -\frac{cx_0}{1 + c(t_f - t_0)}, \quad x = x_0 - \frac{cx_0}{1 + c(t_f - t_0)}(t - t_0).$$

9.4 Solve Problem 9.3 for the case where $x_f$ is also prescribed. Show that the optimal control is a constant and that the optimal state is a straight line, that is,

$$u = \frac{x_f - x_0}{t_f - t_0}, \quad x = x_0 + \frac{x_f - x_0}{t_f - t_0}(t - t_0).$$

9.5 Minimize the performance index

$$J = \int_{t_0}^{t_f} (u^2 - x) \, dt,$$

subject to the differential constraint

$$\dot{x} = u$$

and the prescribed boundary conditions

$$t_0 = 0, \quad x_0 = 0, \quad t_f = 1.$$

Show that the optimal control and the optimal state are given by

$$u = -\frac{t}{2} + \frac{1}{2}, \quad x = -\frac{t^2}{4} + \frac{t}{2}.$$

9.6 Solve Problem 9.5 for the case where the final state is also prescribed, that is, $x_f = 1$. Show that the optimal state becomes

$$x = \frac{t}{4}(5 - t).$$

9.7 Consider the motion of a spring-mass system with spring constant $k$ and mass $m$ as shown in Fig. 9.4. Hamilton's principle states that the motion between fixed end points is such that the integral of the difference of kinetic and potential energy is an optimum. In optimal control ter-

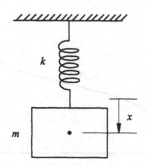

Figure 9.4: Spring-Mass System

minology, the problem is stated as follows: Optimize the performance index

$$J = \frac{1}{2} \int_{t_0}^{t_f} (u^2 - a^2 x^2) dt, \quad a^2 = k/m,$$

subject to the differential constraint

$$\dot{x} = u.$$

and the prescribed boundary conditions

$$t_0 = t_{0_s}, \quad x_0 = x_{0_s}, \quad t_f = t_{f_s}, \quad x_f = x_{f_s}.$$

Show that the optimal state has the form

$$x = c_1 \cos at + c_2 \sin at$$

and that the first integral is the energy of the system.

9.8 Solve Problem 9.7 for the case where $x_f$ is not prescribed. Show that the optimal control is given by

$$u = a x_0 \, \frac{\sin a(t_f - t)}{\cos a(t_f - t_0)},$$

9.9 Find the control history $u(x)$ such that the distance between two fixed points in a plane (Fig. 9.5) is an minimum, that is, minimize

$$J = \int_{x_0}^{x_f} (1 + u^2)^{1/2} dx,$$

subject to the differential constraint

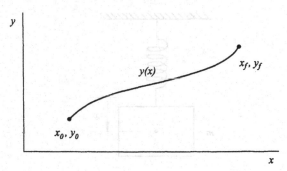

Figure 9.5: Minimum Distance Between Two Points

$$\frac{dy}{dx} = u$$

and the prescribed boundary conditions

$$x_0 = x_{0_s}, \quad y_0 = y_{0_s}, \quad x_f = x_{f_s}, \quad y_f = y_{f_s}.$$

Show that the optimal control is $u = \text{const}$ which leads to a straight line.

9.10 Solve Problem 9.9 for the case where $y_f$ is not specified. Show that the optimal control is $u = 0$.

9.11 The pressure drag of a body of revolution at hypersonic speeds (Fig. 9.6) is proportional to the integral

$$J = \int_{t_0}^{t_f} x\dot{x}^3 dt,$$

where $x(t)$ represents the shape of the body. If $\dot{x} = u$, find the control

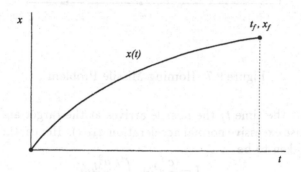

Figure 9.6: Projectile Shape of Minimum Pressure Drag

history that minimizes the drag subject to the boundary conditions

$$t_0 = x_0 = 0$$

$$t_f = x_f = 1.$$

Look at both the Euler–Lagrange equations and the first integral before solving a differential equation; then show that the optimal shape is given by

$$x = t^{3/4}.$$

9.12 Consider the problem of a constant speed missile intercepting a constant velocity target in a tail-chase scenario (Fig. 9.7) as discussed in more detail in Section 12.6. The angle $\theta$ between the missile velocity vector and that of the target is small. The vertical distance between the missile and the target is $y = \eta_T - \eta_M$, and the motion of the missile in the $y$-direction is governed by $\ddot{y} = -a_M$. It is desired to minimize the miss

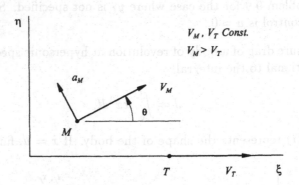

Figure 9.7: Homing Missile Problem

distance at the time $t_f$ the missile arrives at the target and at the same time not use excessive normal acceleration $a_M(t)$. Hence, the performance index is taken to be

$$J = \frac{c}{2}y_f^2 + \int_{t_0}^{t_f} \frac{a_M^2}{2} dt,$$

where $c$ is a positive constant. The equations of motion of the missile normal to the target velocity vector are given by

$$\dot{y} = v$$

$$\dot{v} = -a_M,$$

and the boundary conditions are

$$t_0 = t_{0_s}, \quad y_0 = y_{0_s}, \quad v_0 = v_{0_s}, \quad t_f = t_{f_s}.$$

Show that the optimal control is given by

$$a_M = \frac{c(t_f - t)}{(c/3)(t_f - t_0)^3 + 1} y_0 + \frac{c(t_f - t_0)(t_f - t)}{(c/3)(t_f - t_0)^3 + 1} v_0.$$

By adjusting the size of $c$, it is possible to change the size of $a_M$.

9.13 A bead is sliding down a wire under the action of gravity but without friction (Fig. 9.8). Find the path of the bead that minimizes the time

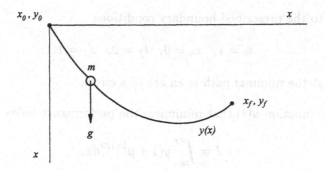

Figure 9.8: Brachistochrone Problem

required to move between fixed points in a plane. The problem is to minimize the performance index

$$J = \int_{x_0}^{x_f} [(1 + u^2)/y]^{1/2} dx,$$

where

$$\frac{dy}{dx} = u$$

subject to the prescribed boundary conditions

$$x_0 = 0, \quad y_0 = 0, \quad x_f = x_{f_s}, \quad y_f = y_{f_s}.$$

Show that the minimal path is given by the parametric equations

$$x = c(\theta - \sin\theta), \quad y = c(1 - \cos\theta),$$

where $c$ is a constant of integration.

9.14 Solve Problem 9.13 for the case where $y_f$ is not prescribed.

9.15 Find the control $u(t)$ that minimizes the performance index

$$J = \int_{t_0}^{t_f} [(1 + u^2)^{1/2}/t] dt,$$

where

$$\dot{x} = u$$

subject to the prescribed boundary conditions

$$t_0 = 1, \quad x_0 = 0, \quad t_f = 2, \quad x_f = 1.$$

Note that the minimal path is an arc of a circle.

9.16 Find the function $u(t)$ that minimizes the performance index

$$J = \int_{x_0}^{x_f} y(1 + u^2)^{1/2} dx,$$

where

$$\frac{dy}{dx} = u,$$

subject to the prescribed boundary conditions

$$x_0 = 0, \quad y_0 = y_{0_s}, \quad x_f = x_{f_s}, \quad y_f = y_{f_s}.$$

This performance index is the surface area of revolution generated by the curve $y(x)$ being rotated about the $x$-axis as shown in Fig. 9.9. Show

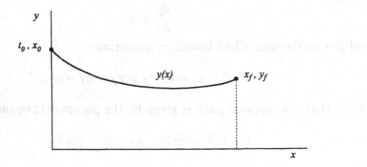

Figure 9.9: Minimum Surface Area of Revolution

that the optimal path is given by

$$y = b \cosh\frac{x - a}{b},$$

where $a$ and $b$ are constants of integration.

9.17 Minimize the performance index

$$J = \frac{1}{2}cx_f^2 + \frac{1}{2}\int_{t_0}^{t_f} u^2 dt,$$

subject to the differential constraint

$$\dot{x} = -ax + bu$$

and the prescribed boundary conditions

$$t_0 = t_{0_s}, \quad x_0 = x_{0_s}, \quad t_f = t_{f_s}.$$

The quantities $a$, $b$, and $c$ are positive constants.

9.18 The navigation problem with a fixed final point (Section 1.4.3) is stated as follows: Find the control history $\theta(x)$ that minimizes the performance index

$$J = t_f,$$

subject to the differential constraints

$$\frac{dt}{dx} = \frac{1}{V\cos\theta}$$

$$\frac{dy}{dx} = \frac{V\sin\theta + w}{V\cos\theta},$$

the prescribed initial conditions

$$x_0 = 0, \quad t_0 = 0, \quad y_0 = 0,$$

and the prescribed final conditions

$$x_f = 1, \quad y_f = 0.$$

Note that $x$ is the variable of integration and $t$ is a state. Show that the optimal control is given by

$$\sin\theta = -\frac{w}{V}$$

and that

$$t = \frac{x}{\sqrt{V^2 - w^2}}, \quad y = 0, \quad t_f = \frac{1}{\sqrt{V^2 - w^2}}.$$

9.19 The differential equation for $\lambda$ and its final condition can be written as

$$\dot{\lambda} = -f_x^T \lambda - L_x^T, \quad \lambda_f = \phi_{x_f}^T + \beta_{x_f}^T \nu.$$

(a) Show that if $\Phi(t, t_0)$ is the transition matrix that satisfies the equations $\dot{\Phi} = f_x \Phi$ and $\Phi_0 = I$, then $\Phi^{-T}$ satisfies the equations $\dot{\Phi}^{-T} = -f_x^T \Phi^{-T}$, $\Phi_0^{-T} = I$.

(b) Show that the solution of the linear $\lambda$ equation is given by

$$\lambda = \Phi^{-T} \Phi_f^T \beta_{x_f}^T \nu + \Phi^{-T} \Phi_f^T \phi_{x_f}^T + \Phi^{-T} \int_t^{t_f} \Phi^T L_x^T dt.$$

9.20 Another way to derive $H_u = 0$ for the optimal control problem with constrained final states is to assume

$$\delta u = d\alpha \, H_u^T$$

over $[t_0, t_f]$ where $d\alpha$ is a scalar. For this admissible comparison control, the first differential becomes

$$dJ' = d\alpha \int_{t_0}^{t_f} H_u H_u^T dt.$$

Hence, if $H_u \neq 0$, $\Delta J' \cong dJ'$ can be made positive or negative by changing the sign of $d\alpha$. What remains to be shown is that this choice of control variation can satisfy the prescribed final conditions. By using Eq.(7.15) with $\delta x_q = 0$, the definition $H_u = L_u + \lambda^T f_u$, and the expression for $\lambda$ derived in Problem 9.19, show that the boundary conditions can be satisfied by a judicious choice of $\nu$. Show that to solve for $\nu$ the controllability matrix must be invertible (Section 7.4).

9.21 In Section 9.3, adjoin the constraint $t_f = t_{f_s}$ by a scalar Lagrange multiplier $\mu$ and carry out the derivation of the first differential (9.9) assuming that $t_f$ is not fixed. Show that the coefficient of $dt_f$ can be made to vanish by choosing $\mu$ to satisfy

$$\mu + H_f = 0.$$

9.22 Consider the problem of minimizing the performance index

$$J = x_f^2,$$

subject to the differential constraint

$$\dot{x} = u$$

and the prescribed boundary conditions

$$t_0 = 0, \quad x_0 = 1, \quad t_f = 1,$$

Solve the problem by applying the Euler–Lagrange equations, the natural boundary condition, and the tests for a minimum. The solution can be explained by examine the problem. Note that any control history which gives $x_f = 0$ will give the lowest value of $J$. Hence, there is no optimal control.

# 10

# Fixed Final Time: Tests for a Minimum

## 10.1   Introduction

Before going on to the second differential, two tests for a minimal control
are derived. The Weierstrass condition is based on strong variations, which
means that control changes are large but that state changes are small. The
Weierstrass condition requires that the Hamiltonian be an absolute minimum
with respect to the control at every point of a minimal path. The Legendre–
Clebsch condition is obtained by reducing strong variations to weak variations.
It requires that the Hamiltonian be a local minimum with respect to the control
at every point of the minimal path.

It is important to apply some test for a minimum early in the solution process. For example, looking for the shape of a body of revolution that
minimizes the sonic boom on the ground is a reasonable endeavor. However,
in linearized supersonic aerodynamics, the analytical optimal shape is a maximum. There is no minimum sonic boom shape, and the Legendre–Clebsch
condition provides this information.

## 10.2 Weierstrass Condition

In Section 9.3, admissible comparison paths are generated by assuming that $\delta u$ and $\delta x$ are small over $[t_0, t_f]$. Such a system of variations is called weak variations. In this section, the optimal solution $u(t), x(t)$ is assumed to exist and satisfy conditions (9.31) through (9.36). Then, the following admissible comparison control is considered (Fig. 10.1):

$$t_0 \leq t \leq t_p, \quad u_* - u = 0$$

$$t_p \leq t \leq t_q, \quad u_* - u = \text{Const} \qquad (10.1)$$

$$t_q \leq t \leq t_f, \quad u_* - u \text{ such that } \beta_{x_f} \delta x_f = 0.$$

This is a large (strong) pulse over a small time interval $[t_p, t_q]$ followed by a small (weak) change in the control over $[t_q, t_f]$ that satisfies the final conditions. Even though the admissible comparison control does not completely lie in the neighborhood of the optimal control, the admissible comparison path does ($x_* - x$ is small). This class of variations is called *strong variations*.

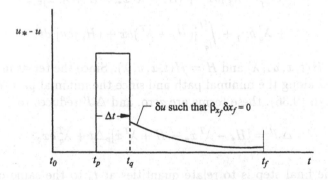

Figure 10.1: Strong Pulse, Scalar Control

The difference in the values of the augmented performance index (9.6) for the two controls is given by

$$\begin{aligned}
\Delta J' &= G(x_{*f}, \nu_*) - G(x_f, \nu) \\
&\quad + \int_{t_p}^{t_q} [H(t, x_*, u_*, \lambda_*) - \lambda_*^T \dot{x}_* - H(t, x, u, \lambda) + \lambda^T \dot{x}]\, dt \qquad (10.2) \\
&\quad + \int_{t_q}^{t_f} [H(t, x_*, u_*, \lambda_*) - \lambda_*^T \dot{x}_* - H(t, x, u, \lambda) + \lambda^T \dot{x}]\, dt.
\end{aligned}$$

Because all variations are not weak, the differential process is not valid, and the Taylor series approach must be used.

The small terms in $\Delta J'$ are expanded in a Taylor series, and only the first-order terms are kept. It is recalled that $t_q = t_p + \Delta t$ where $\Delta t$ is small and that over $[t_q, t_f]$ $u_* = u + \delta u$ and $x_* = x + \delta x$. Since the Lagrange multipliers $\lambda_*$ and $\nu_*$ are not required to satisfy any conditions, they are chosen to be $\lambda_* = \lambda$ and $\nu_* = \nu$. With these changes $\Delta J'$ becomes

$$\begin{aligned} \Delta J' \;=\; & G_{x_f}(x_f, \nu)\delta x_f \\ & + [H(t, x, u_*, \lambda) - \lambda^T \dot{x}_* - H(t, x, u, \lambda) + \lambda^T \dot{x}]_p \, \Delta t \\ & + \int_{t_q}^{t_f} [H_x(t, x, u, \lambda)\delta x + H_u(t, x, u, \lambda)\delta u - \lambda^T \delta \dot{x}] \, dt. \end{aligned} \tag{10.3}$$

Next, the term $\lambda^T \delta \dot{x}$ is integrated by parts so that Eq. (10.3) can be rewritten as

$$\begin{aligned} \Delta J' \;=\; & (G_{x_f} - \lambda_f^T)\delta x_f + [H_* - \lambda^T \dot{x}_* - H + \lambda^T \dot{x}]_p \, \Delta t \\ & + \lambda_q^T \delta x_q + \int_{t_q}^{t_f} [(H_x + \dot{\lambda}^T)\delta x + (H_u)\delta u] \, dt, \end{aligned} \tag{10.4}$$

where $H_* = H(t, x, u_*, \lambda)$ and $H = H(t, x, u, \lambda)$. Since the terms in parentheses are evaluated along the minimal path and since the minimal path satisfies Eqs. (9.31) through (9.36), these terms are zero, and $\Delta J'$ reduces to

$$\Delta J' = [H_* - \lambda^T \dot{x}_* - H + \lambda^T \dot{x}]_p \Delta t + \lambda_q^T \delta x_q. \tag{10.5}$$

The final step is to relate quantities at $t_q$ to the same quantities at $t_p$. Since $t_q = t_p + \Delta t$,

$$\lambda_q = \lambda_p + \dot{\lambda}_p \Delta t + \cdots \tag{10.6}$$

and

$$x_{*q} \;=\; x_{*p} + \dot{x}_{*p}\Delta t + \cdots \tag{10.7}$$

$$x_q \;=\; x_p + \dot{x}_p\Delta t + \cdots. \tag{10.8}$$

To first order, with $x_{*p} = x_p$,

$$\delta x_q = \text{fop}(x_{*q} - x_q) = (\dot{x}_* - \dot{x})_p \Delta t. \tag{10.9}$$

where fop denotes the first-order part. Substitution of these results into the expression for $\Delta J'$ leads to

$$\Delta J' = [H_* - H]_p \Delta t. \tag{10.10}$$

For $u(t), x(t)$ to be a minimum, $\Delta J'$ must be $> 0$ regardless of the choice of $u_*$. Since $\Delta t > 0$ and $t_p$ is arbitrary, the condition

$$H(t, x, u_*, \lambda) - H(t, x, u, \lambda) > 0 \tag{10.11}$$

must be satisfied at every point of the minimal path for all $u_* \neq u$. This condition is known as the *Weierstrass condition* and is a necessary condition for a strong relative minimum. In words, the Weierstrass condition states that the minimal control $u(t)$ must make the Hamiltonian an absolute minimum with respect to the control at every point of the minimal path $x(t)$.

## 10.3 Legendre–Clebsch Condition

Weak variations are a particular case of strong variations. Hence, if $u_* = u + \delta v$ where $\delta v$ is small, Eq. (10.11) can be expanded in a Taylor series to obtain

$$H(t, x, u + \delta v, \lambda) - H(t, x, u, \lambda) =$$
$$H_u(t, x, u, \lambda)\delta v + \frac{1}{2!}\delta v^T H_{uu}(t, x, u, \lambda)\delta v + \cdots . \tag{10.12}$$

Since $H_u = 0$ along the optimal path, the Weierstrass condition reduces to

$$\delta v^T H_{uu}(t, x, u, \lambda)\delta v > 0 \tag{10.13}$$

neglecting higher-order terms. The control variations $\delta v$ are arbitrary, so that the quadratic form (10.13) is positive definite if

$$H_{uu}(t, x, u, \lambda) > 0. \tag{10.14}$$

It is possible for $H_{uu}$ to vanish and a minimum still exist. Consider the case where $L = u^4, \dot{x} = u$. Hence, Eq. (10.13) is rewritten as

$$H_{uu}(t, x, u, \lambda) \geq 0. \tag{10.15}$$

This condition is known as the *Legendre–Clebsch condition*. It is a necessary condition for a minimum, and it must be satisfied at every point of the minimal path. Combined with $H_u = 0$, the Legendre–Clebsch condition requires that $H$ be a local minimum with respect to the control at every point of the minimal path.

If the Weierstrass condition is satisfied, the Legendre–Clebsch condition is satisfied. If the Legendre–Clebsch condition is satisfied, there is no guarantee that the Weierstrass condition is satisfied, and it must be checked independently.

## 10.4   Examples

For the example of Section 9.6, the Legendre-Clebsch condition is satisfied because

$$H_{uu} = 2. \tag{10.16}$$

To examine the Weierstrass condition, it is seen that

$$H(t, x, u_*, \lambda) - H(t, x, u, \lambda) = [(u_* - x)^2 + \lambda u_*] - [(u - x)^2 + \lambda u]. \tag{10.17}$$

Since $\lambda = -2(u - x)$, algebraic manipulations lead to

$$H_* - H = (u_* - u)^2 \tag{10.18}$$

which is $> 0$ at all points of the path for all $u_* \neq u$. Hence, the optimal path obtained in Section 9.6 satisfies the Legendre–Clebsch condition for a weak minimum and the Weierstrass condition for a strong minimum. These conditions are not sufficient for a minimum, but it is known that the optimal path is not a maximum.

For the car acceleration problem of Section 9.7, the Weierstrass condition (10.11) leads to

$$\left(\frac{u_*^2}{2} + \lambda_1 x_2 + \lambda_2 u_*\right) - \left(\frac{u^2}{2} + \lambda_1 x_2 + \lambda_2 u\right) > 0. \tag{10.19}$$

From Eq. (9.61), it is seen that $\lambda_2 = -u$ so that the Weierstrass condition becomes

$$\frac{u_*^2}{2} - u u_* + \frac{u^2}{2} = \frac{(u_* - u)^2}{2} > 0 \tag{10.20}$$

which is satisfied for all $u_* \neq u$ at every point of the optimal path. For the same problem, the Legendre–Clebsch condition is satisfied because

$$H_{uu} = 1 > 0. \tag{10.21}$$

For the aircraft navigation problem of Section 9.8, the Weierstrass condition is stated as

$$\frac{\lambda}{V \cos \theta_*} - \frac{\lambda}{V \cos \theta} > 0, \tag{10.22}$$

and since $\lambda$ and $V$ are positive and $\theta = 0$, it reduces to

$$\frac{1}{\cos \theta_*} - 1 > 0 \tag{10.23}$$

This condition is satisfied for all $-\pi/2 \leq \theta_* \leq \pi/2$ which is an implied constraint. The Legendre–Clebsch condition is stated as

$$H_{\theta\theta} = \frac{\lambda(1 + \sin^2 \theta)}{V \cos^3 \theta} = \frac{1}{V} > 0 \tag{10.24}$$

and is satisfied.

The Weierstrass condition for the problem of minimizing the distance on a sphere (Section 9.9) combined with Eq.(9.83) for $\lambda$ leads to

$$\sqrt{u_*^2 + \cos^2 \theta} - \frac{uu_*}{\sqrt{u^2 + \cos^2 \theta}} - \sqrt{u^2 + \cos^2 \theta} + \frac{u^2}{\sqrt{u^2 + \cos^2 \theta}} > 0. \tag{10.25}$$

If the radical is cleared out of the denominator, this condition requires that

$$\sqrt{(u_*^2 + \cos^2 \theta)(u^2 + cos^2\theta)} - (uu_* + cos^2\theta) > 0. \tag{10.26}$$

Finally, addition and subtraction of the term $2uu_* \cos^2 \theta$ under the radical and rearrangement yields

$$\sqrt{(uu_* + \cos^2 \theta)^2 + (u_* - u)^2 \cos^2 \theta} - (uu_* + \cos^2 \theta) > 0 \tag{10.27}$$

which is clearly satisfied for all $u_* \neq u$. Hence, the optimal solution of Section 9.9 satisfies the Weierstrass condition for a minimum.

The Legendre–Clebsch condition for this problem requires that

$$H_{uu} = \frac{\cos^2 \theta}{(u^2 + \cos^2 \theta)^{3/2}} \geq 0 \tag{10.28}$$

and is satisfied. Hence, the optimal path satisfies the Legendre–Clebsch condition for a weak minimum.

# Problems

10.1 Apply the Legendre–Clebsch and Weierstrass conditions to the minimum distance problem (Problem 9.9).

10.2 Apply the Legendre–Clebsch and Weierstrass conditions to the minimum pressure drag problem (Problem 9.11). This problem was responsible for the creation of the Weierstrass condition. It was noted that a shape with negative drag could be created by letting segments of the shape have a large negative slope. Hence, the Weierstrass condition is not satisfied. Once the constraint $\dot{x} \geq 0$ is added to the problem, the optimal shape satisfies the Weierstrass condition. See Chapter 17.

10.3 Apply the Legendre–Clebsch and Weierstrass conditions to the navigation problem (Problem 9.18) to show that

$$H_{\theta\theta} = \frac{V^2}{(V^2 - w^2)^{3/2}}$$

and

$$H_* - H = \frac{V[1 - \cos(\theta_* - \theta)]}{(V^2 - w^2)\cos\theta_*}.$$

Are they satisfied?

10.4 Find the control history $\gamma(t)$ that minimizes the performance index

$$J = \int_{t_0}^{t_f} \cos^2\gamma\, dt,$$

subject to the dynamics

$$\dot{x} = 2\cos\gamma$$

and the prescribed boundary conditions

$$t_0 = 0, \quad x_0 = 0, \quad t_f = 1, \quad x_f = 1.$$

Show that the problem yields two solutions but that the only solution that can satisfy the boundary conditions is $\gamma = \pi/6$. Show that the Legendre-Clebsch and Weierstrass conditions are satisfied.

10.5 Solve Problem 10.4 with no condition imposed on $x_f$. Show that the solution is $\gamma = 0$ but that the optimal control is now a maximum.

10.6 Solve Problem 10.5 as a maximization problem. Convert it to a minimization problem by multiplying $J$ by minus one (-1).

# 11

# Fixed Final Time: Second Differential

## 11.1 Introduction

In this chapter, the second differential is investigated to obtain the Legendre–Clebsch condition and the Jacobi or conjugate point condition. First, the second differential is obtained by taking the differential of the first differential. Second, the Legendre–Clebsch condition is derived from the second differential. Third, for the class of nonsingular problems ($H_{uu} > 0$) conditions are developed for the existence of a neighboring optimal path. Finally, these conditions are used to develop a sufficient condition for the existence of a weak relative minimum.

## 11.2 The Second Differential

The procedure followed in the derivation of the conjugate point condition involves the use of a neighboring optimal path from the perturbed initial point $t_0 = t_{0_s}$, $\delta x_0 = \delta x_{0_s}$ to the final constraint surface $t_f = t_{f_s}$, $\delta \beta = \beta_{x_f} \delta x_f = 0$ (Fig. 11.1). As a consequence, the first and second differentials must be derived for this case.

Taking the differential of the augmented performance index (9.6) and

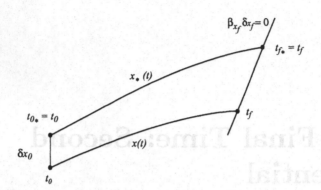

Figure 11.1: Admissible Comparison Path from $t_0$, $\delta x_0$ to $t_f$, $\delta\beta = 0$

performing the integration by parts leads to

$$dJ' = (G_{x_f} - \lambda_f^T)\delta x_f + \beta^T d\nu + \lambda_0^T \delta x_0$$
$$+ \int_{t_0}^{t_f} [(H_x + \dot\lambda^T)\delta x + (H_u)\delta u + (f^T - \dot x^T)\delta\lambda]dt, \tag{11.1}$$

where the substitutions $dx_f = \delta x_f$, $dx_0 = \delta x_0$, $G_\nu = \beta^T$, and $H_\lambda = f^T$ have been made. Actually, for a problem with fixed initial and final times, differentials of end coordinates become variations, that is, $d(\ ) = \delta(\ )$. This is also true for constants (see Section 6.7). Before taking the differential of Eq. (11.1), it is convenient to transpose it to obtain

$$dJ' = \delta x_f^T(G_{x_f}^T - \lambda_f) + d\nu^T\beta + \delta x_0^T\lambda_0$$
$$+ \int_{t_0}^{t_f} [\delta x^T(H_x^T + \dot\lambda) + \delta u^T H_u^T + \delta\lambda^T(f - \dot x)]dt. \tag{11.2}$$

Now, taking the differential and selectively replacing differentials by variations leads to

$$d^2 J' = \delta^2 x_f^T(G_{x_f}^T - \lambda_f) + \delta x_f^T(G_{x_f x_f}\delta x_f + G_{x_f\nu}d\nu - \delta\lambda_f)$$
$$+ d^2\nu^T\beta + d\nu^T\delta\beta + \delta^2 x_0^T\lambda_0 + \delta x_0^T\delta\lambda_0$$
$$+ \int_{t_0}^{t_f} [\delta^2 x^T(H_x^T + \dot\lambda) + \delta x^T(H_{xx}\delta x + H_{xu}\delta u \tag{11.3}$$
$$+ H_{x\lambda}\delta\lambda + \delta\dot\lambda) + \delta^2 u^T H_u^T + \delta u^T(H_{ux}\delta x + H_{uu}\delta u + H_{u\lambda}\delta\lambda)$$
$$+ \delta^2\lambda^T(f - \dot x) + \delta\lambda^T(f_x\delta x + f_u\delta u - \delta\dot x)]dt.$$

Because of the first differential conditions,

$$\beta = 0, \quad G_{x_f} - \lambda_f = 0, \quad f - \dot{x} = 0, \quad H_x^T + \dot{\lambda} = 0, \quad H_u^T = 0 \qquad (11.4)$$

and because $\delta x_0$ is fixed, the terms involving $\delta^2$ vanish. Also, because the differentials of the constraints must vanish, that is, Eqs.(9.12) through (9.14) must hold, the $d\nu$ terms vanish and the last $\delta\lambda$ term under the integral drops out. Finally, if the $\delta\lambda$ term is integrated by parts, the $\delta\lambda$ terms outside the integral cancel, and the remaining $\delta\lambda$ terms inside the integral cancel, leaving

$$d^2 J' = \delta x_f^T G_{x_f x_f} \delta x_f + \int_{t_0}^{t_f} (\delta x^T H_{xx} \delta x \qquad (11.5)$$
$$+ \delta x^T H_{xu} \delta u + \delta u^T H_{ux} \delta x + \delta u^T H_{uu} \delta u) dt.$$

If the integrand is written in a matrix form, the second differential becomes

$$d^2 J' = \delta x_f G_{x_f x_f} \delta x_f$$
$$+ \int_{t_0}^{t_f} [\delta x^T \quad \delta u^T] \begin{bmatrix} H_{xx} & H_{xu} \\ H_{ux} & H_{uu} \end{bmatrix} \begin{bmatrix} \delta x \\ \delta u \end{bmatrix} dt. \qquad (11.6)$$

It is emphasized that the $\delta x$ and the $\delta u$ are related by the differential equation

$$\frac{d}{dt} \delta x = f_x \delta x + f_u \delta u, \qquad (11.7)$$

the initial conditions

$$t_0 = t_{0_s}, \quad \delta x_0 = \delta x_{0_s}, \qquad (11.8)$$

and the final conditions

$$t_f = t_{f_s}, \quad \beta_{x_f} \delta x_f = 0. \qquad (11.9)$$

It is interesting to note that the terms involving $\delta^2$ drop out because they are multiplied by first differential conditions. This is true because of the order in which the manipulations are done. If the second differential had been taken before the integration by parts had been performed, then those terms would not have dropped out and would have been instrumental in obtaining the correct second differential. Also, if the differentials were being performed relative to an arbitrary path and not an optimal path, these terms would not drop out.

## 11.3    Legendre–Clebsch Condition

In Chapter 10 the Legendre–Clebsch condition is derived from the Weierstrass condition by reducing strong control variations to weak control variations. Here, it is shown that the Legendre–Clebsch condition can be derived from the second differential by using the admissible comparison control (9.22). The admissible comparison path is generated by the application of a small pulse $\delta u = \delta v$ starting at time $t_p$ and ending at time $t_q = t_p + \Delta t$. During this time, $\delta x$ goes from $\delta x_p = 0$ to $\delta x_q$. Then, at $t_q = t_p + \Delta t$, the control is changed to one that causes the system to go from $\delta x_q$ to the final conditions $t_f = t_{f_s}$, $\beta_{x_f} \delta x_f = 0$.

In Section 9.3.2, it is shown that

$$\delta x_q = (f_u)_p \, \Delta t \, \delta v. \tag{11.10}$$

Also, a control perturbation that goes from $t_q$, $\delta x_q$ to $t_f$, $\beta_{x_f} \delta x_f = 0$ is derived in Section 7.4 and is given by

$$\delta u = U(t)\delta x_q, \quad \delta x = X(t)\delta x_q. \tag{11.11}$$

Then, Eqs.(11.10) and (11.11) can be combined to give

$$\begin{aligned} \delta u &= U(t)\,(f_u)_p\,\delta v\,\Delta t \\ \delta x &= X(t)\,(f_u)_p\,\delta v\,\Delta t \end{aligned} \tag{11.12}$$

over $[t_q, t_f]$.

At this point, the second differential (11.5) is rewritten as

$$\begin{aligned} d^2 J' &= \int_{t_p}^{t_q} [\delta x^T H_{xx}\delta x + \delta x^T H_{xu}\delta u + \delta u^T H_{ux}\delta x + \delta u^T H_{uu}\delta u] \, dt \\ &\quad + \int_{t_q}^{t_f} [\delta x^T H_{xx}\delta x + \delta x^T H_{xu}\delta u + \delta u^T H_{ux}\delta x + \delta u^T H_{uu}\delta u] dt \\ &\quad + \delta x_f^T \, G_{x_f x_f} \, \delta x_f \end{aligned} \tag{11.13}$$

since $\delta u$ and $\delta x$ are zero over $[t_0, t_p]$. Next, the integral over $[t_p, t_q]$ can be expressed as

$$[\delta x^T H_{xx}\delta x + \delta x^T H_{xu}\delta u + \delta u^T H_{ux}\delta x + \delta u^T H_{uu}\delta u]_p \Delta t \tag{11.14}$$

or as
$$\delta v^T \ (H_{uu})_p \ \delta v \Delta t \tag{11.15}$$
since $\delta x_p = 0$. Because of Eqs.(11.12), the integral over $[t_q, t_f]$ in Eq.(11.13) and the final condition term are of higher order in $\Delta t$ than the first term, that is, Eq.(11.15). For the particular variation considered, the second differential reduces to
$$d^2 J' = \delta v^T (H_{uu})_p \ \delta v \ \Delta t \geq 0 \tag{11.16}$$
and, since $\delta v$ is arbitrary, requires that $(H_{uu})_p$ be positive semidefinite ($\geq 0$). Since the time $t_p$ is arbitrary, the condition
$$H_{uu} \ (t, x, u, \lambda) \geq 0 \tag{11.17}$$
must hold at every point of the optimal path. This is the Legendre–Clebsch condition which has been derived in Chapter 10 from the Weierstrass condition.

## 11.4 Neighboring Optimal Paths

The development of the Jacobi or conjugate point condition follows from the analysis of neighboring optimal paths that are paths which lie in the neighborhood of the optimal path and satisfy the first differential conditions. First, the neighboring optimal path from the neighboring initial point $t_0$, $\delta x_0 = \delta x_{0_s}$ to the final constraint manifold $t_f$, $\beta_{x_f} \delta x_f = 0$ is determined (Fig. 11.1). Then, an admissible comparison path is formed by allowing $\delta x_0$ to go to zero. In this development, the optimal control problem is assumed to be nonsingular, that is, the Legendre–Clebsch condition is assumed to be satisfied in the strengthened form $H_{uu} > 0$ so that its inverse exists. Problems where one or more of the eigenvalues of $H_{uu}$ are zero are called singular.

Taking the variation of Eqs.(9.31) through (9.33) and the differential of Eqs.(9.34) through (9.36) leads to the following equations for a neighboring optimal path from $t_0$, $\delta x_0$ to $t_f$, $\beta_{x_f} \delta x_f = 0$:
$$\delta \dot{x} = f_x \delta x + f_u \delta u \tag{11.18}$$
$$\delta \dot{\lambda} = -H_{xx} \delta x - H_{xu} \delta u - f_x{}^T \delta \lambda \tag{11.19}$$
$$0 = H_{ux} \delta x + H_{uu} \delta u + f_u{}^T \delta \lambda \tag{11.20}$$
$$t_0 = t_{0_s}, \quad \delta x_0 = \delta x_{0_s} \tag{11.21}$$
$$t_f = t_{f_s}, \quad \delta \beta = \beta_{x_f} \delta x_f = 0, \quad \delta \lambda_f = G_{x_f x_f} \delta x_f + \beta_{x_f}^T \ d\nu, \tag{11.22}$$

where use has been made of the relations

$$H_{x\lambda} = f_x^T, \quad H_{u\lambda} = f_u^T, \quad G_{x_f\nu} = \beta_{x_f}^T. \tag{11.23}$$

It is interesting to note that these equations are the same as those obtained from optimizing the second differential (11.6), subject to the differential constraints (11.7), the initial conditions (11.8), and final conditions (11.9), providing the Lagrange multipliers are written as $\delta\lambda$ and $d\nu$. This is called the *accessory optimization problem.*

Equations (11.18) through (11.22) form a linear two point boundary value problem. The third equation can be solved for $\delta u$ as follows:

$$\delta u = -H_{uu}^{-1}(H_{ux}\delta x + f_u^T\delta\lambda). \tag{11.24}$$

Substitution of $\delta u$ into the differential equations for $\delta x$ and $\delta\lambda$ leads to

$$\frac{d\delta x}{dt} = A\delta x - B\delta\lambda \tag{11.25}$$

$$\frac{d\delta\lambda}{dt} = -C\delta x - A^T\delta\lambda, \tag{11.26}$$

where

$$A = f_x - f_u H_{uu}^{-1} H_{ux} \tag{11.27}$$

$$B = f_u H_{uu}^{-1} f_u^T \tag{11.28}$$

$$C = H_{xx} - H_{xu} H_{uu}^{-1} H_{ux}. \tag{11.29}$$

Note that matrices $B$ and $C$ are symmetric.

One procedure for solving the linear two point boundary value problem is the sweep method. Another is the transition matrix method whose discussion is delayed until Chapter 12.

For the sweep method, the solution is assumed to have the form of the final conditions (11.22), that is,

$$\begin{aligned} \delta\lambda &= S(t)\delta x + R(t)d\nu \\ \delta\beta &= T(t)\delta x + Q(t)d\nu = 0 \end{aligned} \tag{11.30}$$

so that the final conditions are swept backward in time. Also, the final values of $S, R, T, Q$ are chosen to be

$$S_f = G_{x_f x_f}, \quad R_f = \beta_{x_f}^T, \quad T_f = \beta_{x_f}, \quad Q_f = 0 \tag{11.31}$$

to guarantee satisfaction of the final conditions. The next step is to find the differential equations for $S, R, T$, and $Q$, so that the differential equations (11.25) and (11.26) are satisfied. Differentiation of Eqs.(11.30) leads to

$$\frac{d\delta\lambda}{dt} = \dot{S}\delta x + S\frac{d\delta x}{dt} + \dot{R}d\nu$$
$$0 = \dot{T}\delta x + T\frac{d\delta x}{dt} + \dot{Q}d\nu. \tag{11.32}$$

Then, substitution of Eqs.(11.25), (11.26), and (11.30) gives

$$[\dot{S} - SBS + SA + A^T S + C]\delta x + [\dot{R} + (A^T - SB)R]d\nu = 0$$
$$[\dot{T} + T(A - BS^T)]\delta x + [\dot{Q} - TBR]d\nu = 0. \tag{11.33}$$

Hence, these equations are satisfied if $S, R, T$, and $Q$ are chosen to satisfy the differential equations

$$\dot{S} = SBS - SA - A^T S - C$$
$$\dot{R} = (SB - A^T)R$$
$$\dot{T} = T(BS - A)$$
$$\dot{Q} = TBR. \tag{11.34}$$

Note that the differential equation and boundary condition for $S^T$ are the same as those for $S$ since $B$, $C$, and $G_{x_f x_f}$ are symmetric. Hence, by the existence theorem, $S^T = S$ so that $S$ is symmetric. Then, the differential equation and boundary condition for $R^T$ and $T$ are identical so that $T = R^T$. Hence, the differential equations and boundary conditions for $S, R$, and $Q$ become

$$\dot{S} = SBS - SA - A^T S - C, \qquad S_f = G_{x_f x_f} \tag{11.35}$$

$$\dot{R} = (SB - A^T)R, \qquad R_f = \beta_{x_f}^T \tag{11.36}$$

$$\dot{Q} = R^T BR, \qquad Q_f = 0, \tag{11.37}$$

where the matrix $Q$ is symmetric.

Since $S(t)$, $R(t)$, and $Q(t)$ are now known, Eqs.(11.30) can be rewritten as

$$\delta\lambda = S\delta x + Rd\nu \tag{11.38}$$

$$\delta\beta = R^T\delta x + Qd\nu = 0 \tag{11.39}$$

and are valid over $[t_0, t_f]$. If $Q_0^{-1}$ is assumed to exist, Eq.(11.39) evaluated at $t_0$ leads to

$$d\nu = -Q_0^{-1} R_0^T \delta x_0 \qquad (11.40)$$

so that Eq.(11.38) evaluated at $t_0$ gives

$$\delta \lambda_0 = \bar{S}_0 \, \delta x_0, \qquad (11.41)$$

where

$$\bar{S} = S - R \, Q^{-1} \, R^T. \qquad (11.42)$$

Hence, given $t_0$ and $\delta x_0$, Eq.(11.41) gives $\delta \lambda_0$ so that Eqs.(11.25) and (11.26) can be integrated forward in time to obtain $\delta x(t)$ and $\delta \lambda(t)$. The control perturbation follows from Eq.(11.24). Another form of the control perturbation is obtained by combining Eqs.(11.24) and (11.38), that is,

$$\delta u = -H_{uu}^{-1}[(H_{ux} + f_u^T S)\delta x + f_u^T R \, d\nu]. \qquad (11.43)$$

In discussing neighboring optimal paths, three cases must be considered: (1) $\bar{S}$ finite over $[t_0, \, t_f)$, (2) $\bar{S}$ infinite at $t_0$, and (3) $\bar{S}$ infinite at some time within the interval. Note that the matrix $\bar{S}$ is infinite at $t_f$ because $Q_f = 0$. For the case where there are no prescribed final conditions (no $\beta$), $Q$ and $R$ do not exist, and $\bar{S} = S$.

(1) If $\bar{S}$ is finite over $[t_0, \, t_f)$, a given value of $\delta x_0$ leads to a value for $\delta \lambda_0$. Then, Eqs.(11.25) and (11.26) can be integrated forward in time to obtain $\delta x(t)$ and $\delta \lambda(t)$, and Eq.(11.24) gives $\delta u(t)$. These are the perturbations that define the neighboring optimal control and path from $t_0$, $\delta x_0$ to $t_f$, $\beta_{x_f} \delta x_f = 0$. Taking the initial state perturbation $\delta x_0$ to zero to form an admissible comparison path leads to $\delta \lambda_0 = 0$, and $\delta x = \delta \lambda = \delta u = 0$. In other words, if $\bar{S}$ is finite over $[t_0, \, t_f)$, there is no neighboring optimal path from $t_0$, $\delta x_0 = 0$ to $t_f$, $\beta_{x_f} \delta x_f = 0$ characterized by $\delta u \neq 0$ and $\delta x \neq 0$. As a consequence, no admissible comparison path exists that is a neighboring optimal path.

(2) If $\bar{S}$ is infinite at $t_0$, a finite $\delta x_0$ yields $\delta \lambda_0 = \infty$, meaning that no neighboring optimal path exists. On the other hand, taking $\delta x_0$ to zero in a special way allows $\delta \lambda_0$ to be finite so that a neighboring optimal path exists for which $\delta x$, $\delta \lambda$, and $\delta u$ are not zero. Also, this neighboring optimal path is an admissible comparison path. The time where $\bar{S}$ becomes infinite is called a *conjugate point* and is denoted by $t_{cp}$.

As an example, assume that $\bar{S}_0^{-1}$ exists and write $\delta x_0 = \bar{S}_0^{-1}\delta k$ where $\delta k$ is a constant $n$-vector. For this choice, $\delta\lambda_0 = \delta k$. If $t_0 > t_{cp}$, $\bar{S}_0^{-1}$ is finite so that $\delta x_0$ goes to zero as $\delta k$ goes to zero, and $\delta\lambda_0$ goes to zero. If $t_0 = t_{cp}$, $\bar{S}_0 = \infty$; $\bar{S}_0^{-1} = 0$; and $\delta x_0 = 0$ regardless of the choice of $\delta k$. However, $\delta\lambda_0$ has whatever value is given to $\delta k$, so there exist many nontrivial neighboring optimal paths that are admissible comparison paths.

(3) Finally, if $\bar{S}$ becomes infinite at time $t_{cp}$ within $(t_0, t_f)$, a neighboring optimal path can exist from that point to the final constraint manifold. An admissible comparison path can be formed by combining $\delta u = 0$ over $[t_0, t_{cp}]$ with the $\delta u$ for the neighboring optimal path over $[t_{cp}, t_f]$.

In summary, the following conclusions are valid and are based on the assumption that the strengthened Legendre–Clebsch condition ($H_{uu} > 0$) is satisfied:

(1) If there is no conjugate point or if $t_{cp} < t_0$, there is no neighboring optimal path from $t_0$, $\delta x_0 = 0$ to $t_f$, $\beta_{x_f}\delta x_f = 0$ such that $\delta u \neq 0$, $\delta x \neq 0$.

(2) If a conjugate point exists at $t_{cp} = t_0$, there are neighboring optimal paths with nonzero $\delta u$, $\delta x$.

(3) If a conjugate point exists within the interval $(t_0, t_f)$, a neighboring optimal path can be constructed by combining a $\delta u = 0$ subarc from $t_0$ to $t_{cp}$ with the $\delta u \neq 0$ neighboring optimal path from $t_{cp}$ to $t_f$.

For a neighboring optimal path, the integral appearing in the second differential (11.6) can be carried out explicitly. Consider the time interval $[t_1, t_2]$ over which Eq.(11.24) is assumed to hold along with the differential equations (11.18), (11.25), and (11.26). Since

$$F \triangleq H_{uu}^{-1}(H_{ux}\delta x + f_u^T\delta\lambda) + \delta u = 0, \qquad (11.44)$$

the form

$$F^T H_{uu} F = 0 \qquad (11.45)$$

can be expanded and combined with Eqs.(11.18) and (11.27) through (11.29) to obtain

$$[\delta x^T \delta u^T]\begin{bmatrix} H_{xx} & H_{xu} \\ H_{ux} & H_{uu} \end{bmatrix}\begin{bmatrix} \delta x \\ \delta u \end{bmatrix} = -\delta\lambda^T\delta\dot{x} - \delta\dot{x}^T\delta\lambda$$

$$-\delta x^T(-C\delta x - A^T\delta\lambda) + \delta\lambda^T(A\delta x - B\delta\lambda). \qquad (11.46)$$

Finally, use of Eqs.(11.25) and (11.26) leads to

$$[\delta x^T \delta u^T] \begin{bmatrix} H_{xx} & H_{xu} \\ H_{ux} & H_{uu} \end{bmatrix} \begin{bmatrix} \delta x \\ \delta u \end{bmatrix} = -\frac{d}{dt}(\delta x^T \delta \lambda) \qquad (11.47)$$

or

$$\int_{t_1}^{t_2} [\delta x^T \delta u^T] \begin{bmatrix} H_{xx} & H_{xu} \\ H_{ux} & H_{uu} \end{bmatrix} \begin{bmatrix} \delta x \\ \delta u \end{bmatrix} dt = -\delta x_2^T \delta \lambda_2 + \delta x_1^T \delta \lambda_1. \qquad (11.48)$$

## 11.5  Neighboring Optimal Paths on a Sphere

To illustrate these points, consider the problem of minimizing the distance on a sphere between a point and a great circle as discussed in Section 9.9. This problem has no prescribed final conditions so the conjugate point discussion is based on $S$ which is shown in Section 11.10 to be

$$S = -\tan(\phi_f - \phi). \qquad (11.49)$$

Note that a conjugate point exists at $\phi_f - \phi_0 = \pi/2$ which is at the north pole if the great circle is the equator. Consider Fig. 11.2 where $FC$ is the optimal path, $C$ is the final point located on the equator, and $B$ is the north pole. For any initial point $F$ between $B$ and $C$, $S$ is finite and no conjugate point exists. However, when $F$ is placed at $B$, $S$ becomes infinite, and a conjugate point exists at $F$. From the geometry of the situation, it is clear that a neighboring optimal path exists between $B$ and the great circle, for example, $BD$. In fact, there exist many neighboring optimal paths. If the initial point is placed at $A$, which is to the left of the conjugate point $B$, there is now a conjugate point within the interval $AC$. In addition, there is no optimal path from $A$ to the great circle that lies in the neighborhood of $AC$. An optimal path does, however, exist between $A$ and the great circle down the back side of the sphere.

To show that $\delta \lambda_0$ can be finite at a conjugate point, it is known that

$$\delta \lambda_0 = S_0 \delta x_0 = -\tan(\phi_f - \phi_0)\delta x_0. \qquad (11.50)$$

Then, choose $\delta x_0$ to be

$$\delta x_0 = -\frac{\delta k}{\tan(\phi_f - \phi_0)} \qquad (11.51)$$

so that

$$\delta \lambda_0 = S_0 \delta x_0 = \delta k. \qquad (11.52)$$

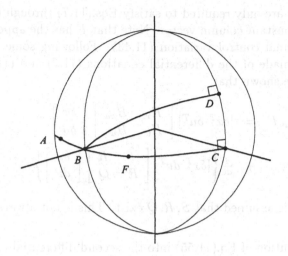

Figure 11.2: Conjugate Point on a Sphere

For $\phi_f - \phi_0 < \pi/2$, $\delta x_0$ can only be made zero by setting $\delta k = 0$, whereby $\delta \lambda_0 = 0$. However, for $\phi_f - \phi_0 = \pi/2$, that is, when the initial point is at the conjugate point, $\delta x_0 = 0$ regardless of the choice of $\delta k$, and $\delta \lambda_0 \neq 0$. Hence, there exist many neighboring optimal paths from the conjugate point that are admissible comparison paths.

## 11.6  Second Differential Condition

In this section, it is shown that, for the class of problems where $H_{uu} > 0$, a sufficient condition for a minimum ($d^2 J' > 0$) is

$$\bar{S} \text{ finite, } \quad t_0 \leq t < t_f. \tag{11.53}$$

This condition is equivalent to the nonexistence of a conjugate point, or if a conjugate point exists, it is outside the interval of integration.

Consider the form

$$F = H_{uu}^{-1}[(H_{ux} + f_u^T S)\delta x + f_u^T R d\nu] + \delta u, \tag{11.54}$$

where $\delta u$ and $\delta x$ are only required to satisfy Eqs.(11.7) through (11.9) and $d\nu$ is an arbitrary constant column vector. Note that $F$ has the appearance of the neighboring optimal control variation (11.43). Following some manipulation in which use is made of the differential equations (11.7) and (11.35) through (11.37), it can be shown that

$$F^T H_{uu} F = [\delta x^T \ \delta u^T] \begin{bmatrix} H_{xx} & H_{xu} \\ H_{ux} & H_{uu} \end{bmatrix} \begin{bmatrix} \delta x \\ \delta u \end{bmatrix} \tag{11.55}$$
$$+ \frac{d}{dt} \left\{ [\delta x^T \ d\nu^T] \begin{bmatrix} S & R \\ R^T & Q \end{bmatrix} \begin{bmatrix} \delta x \\ d\nu \end{bmatrix} \right\}.$$

At this point, it is assumed that $\dot{S}, \dot{R}, \dot{Q}$ exist. This is not always the case; see Section 11.11.

Substitution of Eq.(11.55) into the second differential (11.6), integration of the differential term and some rearrangement lead to

$$d^2 J' = -\delta x_f^T R_f d\nu - d\nu^T R_f^T \delta x_f + \delta x_0^T S_0 \delta x_0 + \delta x_0^T R_0 d\nu \tag{11.56}$$
$$+ d\nu^T R_0^T \delta x_0 + d\nu^T Q_0 d\nu + \int_{t_0}^{t_f} F^T H_{uu} F \ dt.$$

Note that the first two terms vanish because

$$R_f^T \delta x_f = \beta_{x_f} \delta x_f = \delta \beta = 0. \tag{11.57}$$

Next, choose $d\nu$ such that Eq.(11.40) is satisfied, that is,

$$d\nu = -Q_0^{-1} R_0^T \delta x_0. \tag{11.58}$$

Hence, Eq.(11.56) simplifies to

$$d^2 J' = \delta x_0^T \bar{S}_0 \delta x_0 + \int_{t_0}^{t_f} F^T H_{uu} F \ dt, \tag{11.59}$$

where $\bar{S} = S - RQ^{-1}R^T$ and $F$ is now given by

$$F = H_{uu}^{-1}[(H_{ux} + f_u^T S)\delta x - f_u^T R Q_0^{-1} R_0^T \delta x_0] + \delta u. \tag{11.60}$$

Assume that $\bar{S}$ is finite over $[t_0, \ t_f)$, and let $\delta x_0$ go to zero. Since $\bar{S}_0$ is finite, the $\delta x_0$ term vanishes. With $H_{uu} > 0$, the integrand is positive, unless an admissible comparison path ($\delta u \neq 0$, $\delta x \neq 0$) can be found that makes $F = 0$. From (11.60), it is seen that $F$ can vanish if

$$\delta u = -H_{uu}^{-1}(H_{ux} + f_u^T S)\delta x \qquad (11.61)$$

which is the control for a neighboring optimal path, since $d\nu = 0$ for $\delta x_0 = 0$. However, if $\bar{S}$ is finite, it has been shown in Section 11.4 that there is no neighboring optimal path with $\delta u \neq 0$, $\delta x \neq 0$ so that Eq.(11.61) cannot be satisfied. Then, since $d^2 J' > 0$ for arbitrary admissible comparison paths, the optimal path is a minimum.

If $\bar{S}$ becomes infinite at the initial time $t_0$ or, in other words, if there exists a conjugate point at $t_0$, the second differential can be made zero. First, since $\delta x_0 = 0$, $\delta x_0^T \bar{S}_0 \delta x_0 = \delta x_0^T \delta \lambda_0 = 0$ because $\delta \lambda_0$ can be finite. Second, there exists a nonzero control perturbation for which $F = 0$, that is, the control defined by Eq.(11.61). As a consequence, if a conjugate point occurs at $t_0$, there exists an admissible comparison path for which the second differential vanishes.

If the conjugate point occurs within the interval of integration, the optimal path is not a minimum. The minimal path is somewhere else. Consider the situation shown in Fig. 11.3 where $ABC$ represents an optimal path that satisfies $H_{uu} > 0$, $B$ is a conjugate point, and $BD$ represents a neighboring optimal path. Along $ABC$, $dJ' = 0$. Along $AB$, $d^2 J' = 0$ because $\delta u = \delta x = 0$,

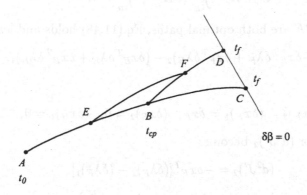

Figure 11.3: Effect of a Conjugate Point

and along $BC$, $d^2 J' = 0$ because $BD$ is a neighboring optimal path. Hence,

$$\Delta J' = J'_{ABD} - J'_{ABC} = 0 \qquad (11.62)$$

so that the performance index along $ABD$ equals that along $ABC$ which is supposed to be the minimum. Next, consider a point $E$ on the path $AB$ and a point $F$ on the path $BD$. The path that minimizes the performance index between $E$ and $F$ has a lower value of the performance index than does the path $EBF$ because forcing the path $EBF$ to pass through the point $B$ represents a constraint which increases the minimum value of the performance index. Since $J'_{AEFD} < J'_{ABD} = J'_{ABC}$, the second differential associated with the path $AEFD$ relative to the optimal path $ABC$ must be negative. As a consequence, if a conjugate point occurs within the interval of integration $(t_0 < t_{cp} < t_f)$, the optimal path is not a minimum.

This discussion can be made formal. Write the second differential (11.6) along paths $ABD$ (subscript 1) and $AEFD$ (subscript 2) as

$$
\begin{aligned}
(d^2 J')_1 &= \int_{t_B}^{t_F} (\ )_1 dt + \int_{t_F}^{t_D} (\ )_1 dt + \delta x_f{}^T (G_{x_f x_f})_1 \delta x_f \\
(d^2 J')_2 &= \int_{t_E}^{t_F} (\ )_2 dt + \int_{t_F}^{t_D} (\ )_2 dt + \delta x_f{}^T (G_{x_f x_f})_2 \delta x_f,
\end{aligned}
\tag{11.63}
$$

where $(\ )$ denotes the integrand. Next, $(d^2 J')_1 = 0$; $(\ )_2 = (\ )_1$ over $[t_F, t_D]$; and $(G_{x_f x_f})_2 = (G_{x_f x_f})_1$ so that subtraction of these expressions gives

$$
(d^2 J')_2 = \int_{t_E}^{t_F} (\ )_2 dt - \int_{t_B}^{t_F} (\ )_1 dt.
\tag{11.64}
$$

Since $EF$ and $BF$ are both optimal paths, Eq.(11.48) holds and leads to

$$
(d^2 J')_2 = (-\delta x_F{}^T \delta \lambda_F + \delta x_F{}^T \delta \lambda_E)_2 - (\delta x_F{}^T \delta \lambda_F + \delta x_B{}^T \delta \lambda_B)_1.
\tag{11.65}
$$

Then, with

$$
(\delta x_F)_2 = (\delta x_F)_1 = \delta x_F, \quad (\delta x_E)_2 = 0, \quad (\delta x_B)_1 = 0,
\tag{11.66}
$$

the expression for $(d^2 J')_2$ becomes

$$
(d^2 J')_2 = -\delta x_F{}^T [(\delta \lambda_F)_2 - (\delta \lambda_F)_1].
\tag{11.67}
$$

At this point, $t_F - t_B$ and $t_B - t_E$ are each taken to be $\Delta t$. Then, the right-hand side of Eq.(11.25) is expanded in a Taylor series about $t_F$, and Eq.(11.25) is integrated over $BF$ and $EF$ to obtain

$$
\begin{aligned}
(\delta x_F)_1 - (\delta x_B)_1 &= [A_F (\delta x_F)_1 - B_F (\delta \lambda_F)_1] \Delta t \\
(\delta x_F)_2 - (\delta x_E)_2 &= [A_F (\delta x_F)_2 - B_F (\delta \lambda_F)_2] 2\Delta t.
\end{aligned}
\tag{11.68}
\tag{11.69}
$$

In view of Eqs.(11.66), subtraction gives

$$A_F \delta x_F = B_F[2(\delta \lambda_F)_2 - (\delta \lambda_F)_1] \qquad (11.70)$$

so that, from (11.68),

$$\delta x_F = 2\Delta t B_F[(\delta \lambda_F)_2 - (\delta \lambda_F)_1], \qquad (11.71)$$

and Eq.(11.67) becomes

$$(d^2 J')_2 = -2\Delta t[(\delta \lambda_F^T)_2 - (\delta \lambda_F^T)_1]B_F[(\delta \lambda_F)_2 - (\delta \lambda_F)_1]. \qquad (11.72)$$

From the definition of $B$, that is, $B = f_u H_{uu}^{-1} f_u^T$, it is seen that $H_{uu} > 0$ only guarantees $B \geq 0$, because $f_u$ is an $n \times m$ matrix and $n$ is usually greater than $m$. In the next paragraph, it is shown that $(d^2 J')_2 < 0$.

Now, Eq.(11.20) is evaluated at $F$ on both paths to obtain

$$\begin{aligned}
(H_{ux})_F(\delta x_F)_2 + (H_{uu})_F(\delta u_F)_2 + (f_u^T)_F(\delta \lambda_F)_2 &= 0 \\
(H_{ux})_F(\delta x_F)_1 + (H_{uu})_F(\delta u_F)_1 + (f_u^T)_F(\delta \lambda_F)_1 &= 0.
\end{aligned} \qquad (11.73)$$

Since $(\delta x_F)_2 = (\delta x_F)_1$, subtraction gives

$$(f_u^T)_F[(\delta \lambda_F)_2 - (\delta \lambda_F)_1] = -(H_{uu})_F[(\delta u_F)_2 - (\delta u_F)_1] \qquad (11.74)$$

so that Eq.(11.72) can be rewritten as

$$(d^2 J')_2 = -2\Delta t[(\delta u_F^T)_2 - (\delta u_F^T)_1](H_{uu})_F[(\delta u_F)_2 - (\delta u_F)_1]. \qquad (11.75)$$

Hence, $H_{uu} > 0$ implies that $(d^2 J')_2 < 0$. This result proves that, if a conjugate point exists in the interval of integration, it is possible to find a comparison path which makes the second differential negative. This means that the optimal path is not a minimum.

In conclusion, if $H_{uu} > 0$ and if no conjugate point exists or there is no conjugate point in the interval of integration ($t_{cp} < t_0$), the optimal path is a minimum. If the conjugate point occurs at the initial point ($t_{cp} = t_0$), the second differential can be zero. Finally, if the conjugate point occurs within the interval of integration ($t_0 < t_{cp} < t_f$), the optimal path is not a minimum.

## 11.7  Example

For the example of Section 9.6, the optimal control has been shown in Section 10.4 to satisfy the strengthened Legendre–Clebsch condition $H_{uu} > 0$. The next step is to show that no conjugate point exists or, if it does, that it is not within the interval of integration $[t_0, t_f)$. For this problem,

$$f_x = 0, \quad f_u = 1, \quad H_{uu} = 2, \quad H_{ux} = -2, \quad H_{xx} = 2 \tag{11.76}$$

so that, from Eqs.(11.27) through (11.29),

$$A = 1, \quad B = \frac{1}{2}, \quad C = 0. \tag{11.77}$$

Since $G_{x_f x_f} = 0$ and there are no prescribed final conditions (no $\beta$ so that $R$ and $Q$ do not exist), the differential equation for $S$ and its boundary condition are given by

$$\dot{S} = \frac{S^2}{2} - 2S, \quad S_f = 0. \tag{11.78}$$

The differential equation has no forcing term and a zero boundary condition so the solution is $S = 0$ everywhere. Therefore, since $S$ is finite everywhere, there is no conjugate point. Since $H_{uu} > 0$ and there is no conjugate point, the optimal control derived in Section 9.6 is a minimum.

## 11.8  Acceleration Problem

For the car acceleration problem defined in Section 9.7, the optimal control has been shown in Section 10.4 to satisfy the strengthened Legendre–Clebsch condition $H_{uu} > 0$. The next step is to show that a conjugate point does not exist, or if it does, that it is outside the interval of integration $[t_0, t_f]$. For this problem,

$$f_x = \begin{bmatrix} 0 & 1 \\ 0 & 0 \end{bmatrix}, \quad f_u = \begin{bmatrix} 0 \\ 1 \end{bmatrix}, \quad H_{uu} = 1, \quad H_{ux} = [0 \ 0], \quad H_{xx} = \begin{bmatrix} 0 & 0 \\ 0 & 0 \end{bmatrix} \tag{11.79}$$

so that, from Eqs.(11.27) through (11.29),

$$A = \begin{bmatrix} 0 & 1 \\ 0 & 0 \end{bmatrix}, \quad B = \begin{bmatrix} 0 & 0 \\ 0 & 1 \end{bmatrix}, \quad C = \begin{bmatrix} 0 & 0 \\ 0 & 0 \end{bmatrix}. \tag{11.80}$$

Next,

$$G_{x_f x_f} = \begin{bmatrix} 0 & 0 \\ 0 & 0 \end{bmatrix}, \quad \beta_{x_f} = [1 \ 0] \tag{11.81}$$

so that the differential equations and boundary conditions for $S$, $R$, and $Q$ are given by

$$\dot{S} = SBS - SA - A^T S - C, \quad S_f = \begin{bmatrix} 0 & 0 \\ 0 & 0 \end{bmatrix}$$

$$\dot{R} = (SB - A^T)R, \qquad\qquad R_f = \begin{bmatrix} 1 \\ 0 \end{bmatrix} \tag{11.82}$$

$$\dot{Q} = R^T BR, \qquad\qquad\quad Q_f = 0.$$

Since the forcing term in the $S$ equation is $C = 0$ and the final condition is $S_f = 0$, the equation admits the solution $S = 0$ everywhere. Next, the equation for the components of $R$ are

$$\dot{R}_1 = 0, \qquad R_{1f} = 1$$
$$\dot{R}_2 = -R_1, \quad R_{2f} = 0 \tag{11.83}$$

and admit the solutions

$$R_1 = 1$$
$$R_2 = t_f - t. \tag{11.84}$$

Finally, the equation for $Q$ becomes

$$\dot{Q} = R_2{}^2 = (t_f - t)^2, \quad Q_f = 0 \tag{11.85}$$

and leads to

$$Q = -(1/3)(t_f - t)^3. \tag{11.86}$$

At this point, the components of $\bar{S} = S - RQ^{-1}R^T$ can be formed as follows:

$$\bar{S} = \begin{bmatrix} 0 & 0 \\ 0 & 0 \end{bmatrix} - \begin{bmatrix} 1 \\ t_f - t \end{bmatrix} \frac{(-3)}{(t_f - t)^3} [1 \quad t_f - t] \tag{11.87}$$

or

$$\bar{S} = \begin{bmatrix} \frac{3}{(t_f-t)^3} & \frac{3}{(t_f-t)^2} \\ \frac{3}{(t_f-t)^2} & \frac{3}{(t_f-t)} \end{bmatrix}. \tag{11.88}$$

Hence, each element of $\bar{S}$ is finite for $t < t_f$ so that there is no conjugate point.

In conclusion, since $H_{uu} > 0$ and there is no conjugate point, the optimal control of Section 8.6 is a minimum.

## 11.9   Navigation Problem

For the minimum time aircraft navigation problem defined in Section 9.8, the strengthened Legendre–Clebsch condition is satisfied (Section 10.4). To discuss the conjugate point condition, it is noted that there are no prescribed final conditions so that $R$ and $Q$ do not exist. Hence, the existence or nonexistence of a conjugate point depends on the characteristics of $S(x)$.

For this problem, $\theta = 0$ so that

$$f_t = 0, \quad f_\theta = 0, \quad H_{tt} = 0, \quad H_{t\theta} = 0, \quad H_{\theta\theta} = 1/V$$

$$A = f_t - f_\theta H_{\theta\theta}^{-1} H_{\theta t} = 0$$

$$B = f_\theta H_{\theta\theta}^{-1} f_\theta^T = 0 \tag{11.89}$$

$$C = H_{tt} - H_{t\theta} H_{\theta\theta}^{-1} H_{\theta t} = 0.$$

Hence, the differential equation for $S$ reduces to $dS/dx = 0$, and for the boundary condition $S_f = 0$, gives $S = 0$. Since S is finite, there is no conjugate point. Hence, $\theta(x)$ is a minimum because $H_{\theta\theta} > 0$ and there is no conjugate point.

## 11.10   Minimum Distance on a Sphere

In Section 9.9, the optimal path for the problem of minimizing the distance on a sphere between a point and a great circle has been obtained, and in Section 10.4, it has been shown to satisfy the strengthened Legendre–Clebsch condition. Here, the question of a conjugate point is investigated. Since there are no prescribed final conditions (no $\beta$), the discussion of conjugate points centers around the function $S(t)$ defined by Eq.(11.35).

Along the optimal path, $u = \theta = \lambda = 0$ so that

$$f_\theta = 0, \quad f_u = 1, \quad H_{\theta\theta} = -1, \quad H_{\theta u} = 0, \quad H_{u\theta} = 0, \quad H_{uu} = 1 \tag{11.90}$$

and

$$A = 0, \quad B = 1, \quad C = -1. \tag{11.91}$$

Then, the differential equation for $S$ and its boundary condition are given by

$$\frac{dS}{d\phi} = 1 + S^2, \quad S_f = 0. \tag{11.92}$$

This equation has the solution

$$S = -\tan(\phi_f - \phi) \tag{11.93}$$

and indicates that a conjugate point exists at $\phi_f - \phi_{cp} = \pi/2$. Hence, if the initial point is less than $\pi/2$ away from the prescribed great circle ($\phi_f - \phi_0 < \pi/2$), the optimal path is a minimum. If the initial point is at the north pole and the prescribed great circle is the equator ($\phi_f - \phi_0 = \pi/2$), a conjugate point exists at the initial point, and the optimal path is a minimum, but it is not unique. Finally, if $\phi_f - \phi_0 > \pi/2$, the conjugate point lies within the interval of integration, and the optimal path is not a minimum. In fact, the minimum distance path is the great circle arc going in the negative $\phi$ direction.

For this problem, Fig. 11.3 can be drawn explicitly and is shown in Fig. 11.4. Here, it is apparent that the distance along $EF$ is less than the distance along $EBF$.

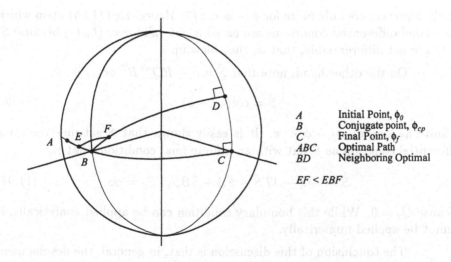

| | |
|---|---|
| $A$ | Initial Point, $\phi_0$ |
| $B$ | Conjugate point, $\phi_{cp}$ |
| $C$ | Final Point, $\phi_f$ |
| $ABC$ | Optimal Path |
| $BD$ | Neighboring Optimal |

$EF < EBF$

Figure 11.4: Conjugate Point on a Sphere

## 11.11   Minimum Distance Between Two Points on a Sphere

For the problem of minimizing the distance between two points on a sphere, the optimal solution is governed by $u = \theta = \lambda = 0$ so that Eqs.(11.90) and (11.91) are still valid. Hence, the differential equations for $S$, $R$, $Q$ and their boundary conditions are

$$\frac{dS}{d\phi} = 1 + S^2, \quad S_f = 0$$

$$\frac{dR}{d\phi} = SR, \quad R_f = 1 \tag{11.94}$$

$$\frac{dQ}{d\phi} = R^2, \quad Q_f = 0.$$

These equations admit the solutions

$$S = -\tan(\phi_f - \phi)$$
$$R = \sec(\phi_f - \phi) \tag{11.95}$$
$$Q = -\tan(\phi_f - \phi)$$

which, however, are only finite for $\phi_f - \phi < \pi/2$. Hence, Eq.(11.55) upon which the second differential conditions are based is not valid over $(t_0, t_f)$ because $S$, $R$, $Q$ are not differentiable, that is, they blow up.

On the other hand, note that $\bar{S} = S - RQ^{-1}R^T$ or

$$\bar{S} = \cot(\phi_f - \phi) \tag{11.96}$$

is finite as long as $\phi_f - \phi < \pi$. It is easily shown that $\bar{S}$ satisfies the same differential equation as $S$ but with an infinite final condition, that is,

$$\dot{\bar{S}} = -C - AT\bar{S} - \bar{S}A + \bar{S}B\bar{S}, \quad \bar{S}_f = \infty \tag{11.97}$$

because $Q_f = 0$. While this boundary condition can be applied analytically, it cannot be applied numerically.

The conclusion of this discussion is that, in general, the development of the second differential conditions cannot be based entirely on $S$, $R$, $Q$ or on $\bar{S}$. In the following paragraphs, the second differential conditions are developed on the assumption that $\bar{S}$ is finite (differentiable) over $(t_0, t_1]$ and $S$, $R$, $Q$ are finite over $[t_1, t_f)$. See Ref. WB.

Over $(t_0, t_1]$, consider the form

$$E = H_{uu}^{-1}[H_{ux} + f_u^T \bar{S}]\delta x + \delta u, \qquad (11.98)$$

where $\bar{S}$ satisfies the differential equation (11.97). It is easily shown that

$$E^T H_{uu} E = [\delta x^T \delta u^T] \begin{bmatrix} H_{xx} & H_{xu} \\ H_{ux} & H_{uu} \end{bmatrix} \begin{bmatrix} \delta x \\ \delta u \end{bmatrix} + \frac{d}{dt}(\delta x^T \bar{S}\delta x). \qquad (11.99)$$

Over $[t_1, t_f]$, Eqs.(11.54) and (11.55) are valid.

At this point, the integral in the second differential (11.6) is written as the sum of integrals over $[t_0, t_1]$ and $[t_1, t_f]$ and combined with Eqs.(11.99) and (11.55) to obtain

$$\begin{aligned} d^2 J' &= \delta x_f^T G_{x_f x_f} \delta x_f + \int_{t_0}^{t_1} E^T H_{uu} E dt - \left[\delta x^T \bar{S}\delta x\right]_{t_0}^{t_1} \\ &+ \int_{t_1}^{t_f} F^T H_{uu} F dt - \left[[\delta x^T d\nu] \begin{bmatrix} S & R \\ R^T & Q \end{bmatrix} \begin{bmatrix} \delta x \\ d\nu \end{bmatrix}\right]_{t_1}^{t_f}. \end{aligned} \qquad (11.100)$$

After $d\nu$ is chosen to be

$$d\nu = -Q_1^{-1} R_1^T \delta x_1 \qquad (11.101)$$

and Eq.(11.57) is recalled, the second differential becomes

$$d^2 J' = \delta x_0^T \bar{S}_0 \delta x_0 + \int_{t_0}^{t_1} E^T H_{uu} E dt + \int_{t_1}^{t_f} F^T H_{uu} F dt. \qquad (11.102)$$

By using this formula, all of the conclusions of Section 11.6 can be re-established.

What has been demonstrated here is that, if $S$, $R$, $Q$ blow up within the interval of integration, it is possible to form $\bar{S}$ to the right of the singularity and use it to prove minimality. This has already been done in Section 11.8 for a case where $S$, $R$, $Q$ did not blow up.

Equation (11.96) indicates that a conjugate point exists at $\phi_f - \phi_{cp} = \pi$, which corresponds to the case where the final point is the south pole and the conjugate point is the north pole. Hence, if $\phi_f - \phi_0 < \pi$, the optimal path is a minimum; if $\phi_f - \phi_0 = \pi$, the minimal path is not unique; and if $\phi_f - \phi_0 > \pi$, the optimal path is not a minimum. In the last case, the minimum is achieved by going in the negative $\phi$ direction.

It has been stated in this section that $\bar{S}$ satisfies the same differential equation as $S$, but a comparison of solutions (11.93) and (11.96) indicates that this is not obvious. The general solution of Eq.(11.97) for $\bar{S}$ is given by

$$\tan^{-1} \bar{S} = \phi + \text{Const.} \tag{11.103}$$

Then, since $Q_f$ approaches zero from the negative side,

$$\bar{S}_f = S_f - \frac{R_f^2}{Q_f} = \infty, \tag{11.104}$$

and Eq.(11.103) implies that

$$\text{Const} = \frac{\pi}{2} - \phi_f. \tag{11.105}$$

Hence, the expression for $\bar{S}$ becomes

$$\tan^{-1} \bar{S} = \phi - \phi_f + \frac{\pi}{2}. \tag{11.106}$$

Finally, with the trignometric identity

$$\cot^{-1} \bar{S} = \frac{\pi}{2} - \tan^{-1} \bar{S}, \tag{11.107}$$

Eq.(11.106) leads to Eq.(11.93).

# 11.12   Other Sufficient Conditions

Consider the transformation

$$\delta u = -H_{uu}{}^{-1} H_{ux} \delta x + \delta v \tag{11.108}$$

which introduces a new control $\delta v$ and requires that $H_{uu} > 0$. If $\delta u$ is substituted into the second differential (11.6), the following results:

$$d^2 J' = \delta x_f{}^T G_{x_f x_f} \delta x_f + \int_{t_0}^{t_f} (\delta x^T C \delta x + \delta v^T H_{uu} \delta v) dt \tag{11.109}$$

where $C = H_{xx} - H_{xu} H_{uu}{}^{-1} H_{ux}$ as defined previously. From this expression, it is seen that sufficient conditions for a minimum are

$$H_{uu} > 0, \quad G_{x_f x_f} \geq 0, \quad C \geq 0. \tag{11.110}$$

The new differential equation for the state differential becomes

$$\frac{d}{dt}\delta x = (f_x - f_u H_{uu}{}^{-1} H_{ux})\delta x + f_u \delta v$$
$$= A\delta x + f_u \delta v. \tag{11.111}$$

If these conditions are satisfied, it is not possible for a conjugate point to exist since $d^2 J' > 0$. However, it is possible for $G_{x_f x_f}$ to be negative and a minimum to exist (see Problem 11.6). Also, $C$ can be negative and a minimum still exist.

From Eq.(11.6), it is seen that

$$G_{x_f x_f} \geq 0, \quad \begin{bmatrix} H_{xx} & H_{xu} \\ H_{ux} & H_{uu} \end{bmatrix} > 0 \tag{11.112}$$

are also sufficient conditions for a minimum.

# Problems

11.1 For the case where there are no prescribed final conditions (no $\beta$), verify the sufficient condition for a minimum by rederiving the equations in Section 11.6 with no $\beta$.

11.2 Show that $\bar{S} = S - RQ^{-1}R^T$ satisfies the same differential equation as $S$.

11.3 Check Problem 9.7 for a conjugate point.

11.4 Show that the optimal path of Problem 9.9 satisfies the sufficient condition for a minimum.

11.5 Show that the optimal path of Problem 9.12 satisfies the sufficient condition for a minimum.

11.6 Find the control that minimizes the performance index

$$J = -\frac{1}{2}ax_f^2 + \frac{1}{2}\int_{t_0}^{t_f}(u^2 + x^2)dt,$$

where $a$ is a positive constant. The system is governed by the differential equation

$$\dot{x} = u$$

and the prescribed boundary conditions are given by

$$t_0 = t_{0_s}, \quad x_0 = x_{0_s}, \quad t_f = t_{f_s}.$$

Show that, if

$$a > \frac{1 + e^{-2(t_f - t_0)}}{1 - e^{-2(t_f - t_0)}},$$

a conjugate point exists in the interval of integration.

11.7 Continue the aircraft navigation problem whose optimal control has been derived in Problem 9.18 and which has been shown to satisfy the Legendre–Clebsch condition in Problem 10.3. Show that the sufficient condition for a minimum is satisfied. In particular, show that

$$H_{\theta\theta} = \frac{V^2}{(V^2 - w^2)^{3/2}} > 0$$

and that

$$S = \begin{bmatrix} 0 & 0 \\ 0 & 0 \end{bmatrix}, \quad R = \begin{bmatrix} 0 \\ 1 \end{bmatrix}, \quad Q = \frac{(V^2 - w^2)^{3/2}}{V^2}(x - 1).$$

Then, since

$$\bar{S} = \frac{V^2}{(V^2 - w^2)^{3/2}(1 - x)} \begin{bmatrix} 0 & 0 \\ 0 & 1 \end{bmatrix}$$

is finite, the solution $\sin\theta = -w/V$ is a minimum.

11.8 With regard to the aircraft navigation problem (see Chapter 1), find the velocity direction history that minimizes the control effort

$$J = \frac{1}{2} \int_{t_0}^{t_f} \theta^2 \, dt,$$

subject to the dynamics

$$\dot{x}_1 = V\cos\theta$$
$$\dot{x}_2 = V\sin\theta + w$$

and the prescribed boundary conditions

$$t_0 = 0, \quad x_{1_0} = 0, \quad x_{2_0} = 0$$
$$t_f = 1, \quad x_{1_f} = 1.$$

Show that the optimal control is given by

$$\theta = \arccos \frac{1}{V}$$

and that it satisfies the sufficient condition for a weak minimum.

11.9 Solve Problem 11.8 for the case where $\theta$ is assumed to be small, that is, $\sin \theta \cong \theta$ and $\cos \theta \cong 1$ in the statement of the problem.

11.10 Find the control $u(t)$ that minimizes the performance index

$$J = \frac{x_{1_f}^2 + x_{2_f}^2}{2} + \int_{t_0}^{t_f} \frac{u^2}{2} dt,$$

subject to the dynamics

$$\dot{x}_1 = u, \quad \dot{x}_2 = u$$

and the boundary conditions

$$t_0 = 0, \quad x_{1_0} = a, \quad x_{2_0} = b, \quad t_f = 1.$$

What is the controllability requirement for this problem?

11.11 Consider the problem of minimizing the quadratic performance index

$$J = \frac{1}{2} x_f^T T x_f + \frac{1}{2} \int_{t_0}^{t_f} (x^T U x + u^T W u) dt,$$

subject to the linear state equation

$$\dot{x} = Mx + Nu$$

and the prescribed boundary conditions

$$t_0 = t_{0_s}, \quad x_0 = x_{0_s}, \quad t_f = t_{f_s}.$$

In these equations, $T \geq 0$, $U \geq 0$, and $W > 0$ are symmetric constant matrices; $M$ and $N$ are constant matrices. By using the sweep method, show that the optimal control is given by

$$u = -W^{-1} N^T S x,$$

where $S$ satisfies the differential equation

$$\dot{S} = -U - M^T S - SM + SNW^{-1}N^T S, \quad S_f = T.$$

11.12 Consider the example of Section 9.6. Show that the second differential
(11.6) is given by

$$d^2 J' = 2 \int_{t_0}^{t_f} (\delta x - \delta u)^2 dt.$$

This means that the second differential is positive for an arbitrary $\delta u$
unless $\delta x = \delta u$. For this case show that the only solution of Eq.(11.7)
subject to $\delta x_0$ is $\delta x = 0$. Hence, there is no admissible comparison path
with $\delta u \neq 0$ for which the second differential can vanish, and the optimal
control is a minimum.

11.13 For the acceleration problem (Section 11.8) show that the second differential (11.6) is given by

$$d^2 J' = \int_{t_0}^{t_f} \delta u^2 dt.$$

Hence, the optimal control is a minimum because $d^2 J' > 0$ for all admissible comparison paths. It seems to be a good idea to check the second
differential before applying the sufficient condition.

# 12

# Fixed Final Time Guidance

## 12.1    Introduction

Guidance is the process by which an actual dynamical system is controlled to satisfy the final conditions while satisfying any path constraints imposed on the system. Examples of path constraints are control inequality constraints or state inequality constraints. Path constraints are usually difficult to handle in guidance so they are avoided by adding penalty terms to the performance index or by guiding relative to a reference or nominal path that satisfies the path constraints by some margin. Hence, many guidance problems fit into the format of the optimal control problem defined in Chapter 9, and this chapter can be considered as an application of the material in Chaps. 9 and 11.

To provide some physical feeling for guidance, consider the case of a constant speed homing missile attacking a constant velocity target. If the line of sight is defined as the line from the missile to the target, the function of the guidance system is to drive the line of sight rate to zero, putting the missile and target on a collision course. The guidance system itself is composed of a seeker that locks onto the target and provides the two components of line of sight rate, a guidance law which commands lateral accelerations proportional to the line of sight rates, actuators which deflect the missile fins to achieve the lateral accelerations, and lateral accelerometers which are used in a feedback loop to ensure that the commanded accelerations are achieved. Such a guidance system is analog in that measurements and control occur continuously.

A more advanced guidance system might contain a seeker, a computer, actuators, accelerometers, and even an inertial measuring unit that provides measurements of inertial position and velocity components. In the computer are algorithms for performing navigation, guidance, and control. The navigation algorithm processes measurements to determine missile and target states. The guidance algorithm uses the missile and target states to compute commanded missile accelerations. Finally, the control algorithm issues actuator commands and uses measurements to ensure that the proper accelerations are achieved. Because of the computer, such a guidance system would be digital. At a sample time, measurements are processed by the computer to obtain actuator inputs that are held constant over the sample period, the time elapsed between sample times. This is called a sample and hold control system.

Three kinds of guidance strategies are discussed in this chapter: optimal guidance, neighboring optimal guidance, and linear quadratic guidance. In optimal guidance, a new optimal control is computed to the final conditions continuously or at each sample time. In neighboring optimal guidance, it is assumed that the actual path lies in the neighborhood of the nominal optimal path so that the equations for the neighboring optimal control can be used. Finally, linear quadratic guidance has two applications. First, it can be used to create an optimal control law for a linear dynamical system with soft control bounds. Second, it can be used to keep a dynamical system in the neighborhood of a nominal path, not necessarily an optimum.

These guidance laws are derived within the framework of fixed final time. However, they can be applied in situations where the final time is free by predicting the final time at each sample point. Also, they can be applied to problems where the final conditions change between sample points.

Neighboring optimal guidance is first presented by using the sweep method of Chapter 11. Then, the transition matrix method is developed and related to the sweep method. Both methods are used to derive linear quadratic guidance, and the transition matrix method is used to solve a homing missile problem.

## 12.2  Optimal Guidance

A schematic of the operation of a dynamical system is shown in Fig. 12.1. The operation of the system begins at $t_0$ by using the optimal control developed for a model of the system. Because the model is not exact, the system does not follow the state corresponding to the model. At time $t_p$, the system has deviated somewhat from the intended path, and it is not going to hit the final conditions, which might be slightly different.

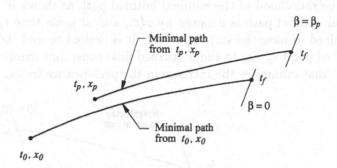

Figure 12.1: Optimal Guidance

In optimal guidance, a new optimal control is computed from the perturbed point $t_p, x_p$ to the perturbed final point $t_f, \beta = \beta_p$ and is applied at $t_p$. For optimal guidance it is not necessary to work in terms of perturbations from the nominal optimal path.

For the aircraft navigation problem considered in Section 9.8, the optimal control can be computed from any point $x_p, t_p, y_p$ to the line $x_f = x_{f_s}$ and is given by $\theta = 0$. Hence, the aircraft is to be steered directly toward the line at all times regardless of the position of the aircraft. Note that an estimate of the aircraft state is not needed for this guidance law.

If the aircraft is required to reach the point $x_f = x_{f_s}, y_f = y_{f_s}$, the optimal control is given by

$$\sin\theta = -\frac{\frac{w}{V} - \frac{y_f - y_p}{x_f - x_p}\sqrt{1 + \left(\frac{y_f - y_p}{x_f - x_p}\right)^2 - \left(\frac{w}{V}\right)^2}}{1 + \left(\frac{y_f - y_p}{x_f - x_p}\right)^2}. \tag{12.1}$$

To implement this guidance law, estimates of the aircraft position $(x_p, y_p)$, speed $(V)$, and the wind speed $(w)$ are needed at each sample point, as is the location of the final point.

## 12.3   Neighboring Optimal Guidance

In neighboring optimal guidance, the optimal path from $t_p, x_p$ is assumed to lie in the neighborhood of the nominal optimal path as shown in Fig. 12.2. The nominal optimal path is denoted by $x(t)$, and at some time $t_p$, the system is determined to have the state $x_p + \delta x_p$. It is desired to find the path from the perturbed point $t_p, \delta x_p$ to the perturbed final constraint manifold (represented by $\delta\beta_p$) that minimizes the increase in the performance index.

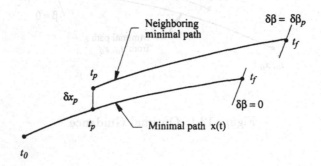

Figure 12.2: Neighboring Optimal Guidance

The total change in the augmented performance index experienced by operating along the perturbed path is given by

$$\Delta J' = dJ' + \frac{1}{2!}d^2 J' + \cdots. \tag{12.2}$$

Since the first variation does not completely vanish because $\delta x_p \neq 0$, the total change becomes

$$\Delta J' = \lambda_p^T \delta x_p + \frac{1}{2!}d^2 J'. \tag{12.3}$$

On the other hand, $\lambda_p$ is a known function of time along the minimal path, so that, since $\delta x_p$ is known, minimizing $\Delta J'$ is equivalent to minimizing the second differential $d^2 J'$. Hence, the problem is find the $\delta u(t)$ that minimizes the new performance index

$$\bar{J} = \frac{1}{2}\delta x_f^T G_{x_f x_f}\delta x_f + \delta\nu^T \delta\beta_p + \frac{1}{2}\int_{t_p}^{t_f} (\delta x^T H_{xx}\delta x$$
$$+ \delta x^T H_{xu}\delta u + \delta u^T H_{ux}\delta x + \delta u^T H_{uu}\delta u)\, dt, \tag{12.4}$$

subject to the differential constraints

$$\frac{d\,\delta x}{dt} = f_x \delta x + f_u \delta u, \tag{12.5}$$

the prescribed initial conditions

$$t_p = t_{p_s}, \quad \delta x_p = \delta x_{p_s}, \tag{12.6}$$

and the prescribed final conditions

$$t_f = t_{f_s}, \quad \beta_{x_f} \delta x_f = \delta \beta_p. \tag{12.7}$$

If the Lagrange multiplier functions are called $\delta\lambda$ and the constant Lagrange multipliers are called $d\nu$, the Euler–Lagrange equations and natural boundary conditions for this problem are the equations for a neighboring optimal path given by Eqs.(11.18) through (11.22). Hence, the control perturbation that minimizes the change in the performance index going from $t_p$, $\delta x_p$ to $t_f$, $\beta_{x_f}\delta x_f = \delta\beta_p$ generates the neighboring minimal path.

Given at $t_p$ a state perturbation $\delta x_p$ and a final constraint perturbation $\delta\beta_p$ from the nominal optimal path, Eq.(11.41) becomes

$$\delta\lambda_p = \bar{S}_p \delta x_p + R_p Q_p^{-1} \delta\beta_p \tag{12.8}$$

and the sampled data neighboring optimal control is given by

$$\delta u = -H_{uu}^{-1}[(H_{ux} + f_u^T S)\delta x - f_u^T R Q_p^{-1} R_p^T \delta x_p + f_u^T R Q_p^{-1} \delta\beta_p]. \tag{12.9}$$

Hence, given $t_p, \delta x_p$, and $\delta\beta_p$, $\delta\lambda_p$ is obtained from Eq.(12.8), and $\delta x$ follows from the integration of Eqs.(11.25) and (11.26).

The implementation of this guidance law can be simplified by computing the control perturbation at $t_p$ and holding it constant over the sample period. For a sample and hold control system, Eq.(12.9) becomes

$$\delta u_p = -C_x(t_p)\delta x_p - C_\beta(t_p)\delta\beta_p, \tag{12.10}$$

where

$$C_x(t) = H_{uu}^{-1}(H_{ux} + f_u^T \bar{S}), \quad C_\beta(t) = H_{uu}^{-1} f_u^T R Q^{-1}. \tag{12.11}$$

For a continuous control system, $t_p$ is replaced by $t$, and the control rule is given by

$$\delta u = -C_x(t)\delta x - C_\beta(t)\delta\beta. \tag{12.12}$$

Note that the gains $C_x$ and $C_\beta$ are evaluated on the nominal optimal path. Hence, they can be computed ahead of time and stored in the computer.

For the aircraft navigation problem to the line $x_f = 1$ discussed in Sections 9.8 and 11.9, $H_{\theta t} = f_\theta = 0$ . Hence, for $\delta\beta_p = 0$, either the sample and hold guidance law or the continuous guidance law gives $\delta\theta = 0$. This is consistent with the optimal guidance law $\theta = 0$.

The aircraft navigation problem to the point $x_f = 1, y_f = 0$ considered in Problems 9.18 and 11.7 leads to $(\delta\beta_p = 0)$

$$\delta\theta = -\frac{\delta y_p}{1 - x_p} \qquad (12.13)$$

for both the sample and hold and continuous control guidance laws. Equation (12.13) follows from Eq.(12.1) by setting $\theta = \theta_{opt} + \delta\theta$ and $y_p = y_{opt_p} + \delta y_p$ , where $y_{opt_p} = 0$ and by linearizing the result.

## 12.4   Transition Matrix Method

Formally, the sweep method leads to elegant results for neighboring optimal guidance. However, the equations are nonlinear, and the existence of an analytical solution is not obvious. On the other hand, the transition matrix method (Ref. Sp) leads to linear equations whose solution may be more apparent. The equations to be solved for a fixed final time neighboring optimal path are given in Chapter 11 to be

$$\begin{aligned}
\frac{d\delta x}{dt} &= A\delta x - B\delta\lambda \\
\frac{d\delta\lambda}{dt} &= -C\delta x - A^T\delta\lambda \\
t_p &= t_{p_s}, \quad \delta x_p = \delta x_{p_s} \\
t_f &= t_{f_s}, \quad \beta_{x_f}\delta x_f = \delta\beta_p \\
\delta\lambda_f &= G_{x_f x_f}\delta x_f + \beta_{x_f}^T d\nu,
\end{aligned} \qquad (12.14)$$

where $A$, $B$, $C$ are defined in Eqs.(11.18) through (11.22), and the control perturbation is given by

$$\delta u = -H_{uu}^{-1}H_{ux}\delta x - H_{uu}^{-1}f_u^T\delta\lambda. \qquad (12.15)$$

As stated in Section 11.4, Eqs.(12.14) represent a linear two point boundary value problem (TPBVP).

## 12.4.1   Solution of the Linear Equations

The differential equations in Eq.(12.14) can be rewritten in the form

$$\frac{d}{dt}\delta z = F\delta z, \tag{12.16}$$

where $\delta z = [\delta x^T \ \delta\lambda^T]^T$ and

$$F = \begin{bmatrix} A & -B \\ -C & -A^T \end{bmatrix}. \tag{12.17}$$

Equation (12.16) can be solved formally by introducing the transition matrix $\Phi(t, t_f)$ and assuming a solution of the form

$$\delta z(t) = \Phi(t, t_f)\delta z_f, \quad \Phi(t_f, t_f) = I, \tag{12.18}$$

which guarantees that the final conditions are satisfied. Differentiation of Eq.(12.18) and substitution into Eq.(12.16) shows that the differential equations are satisfied if the transition matrix is chosen to be

$$\dot{\Phi} = F\Phi, \quad \Phi_f = I. \tag{12.19}$$

This equation can be integrated to obtain $\Phi(t, t_f)$ once $A$, $B$, $C$ are known.

## 12.4.2   Symplectic Property of the Transition Matrix

The transition matrix for the linear two point boundary value problem satisfies the symplectic property

$$\Phi^T J\Phi = J. \tag{12.20}$$

where

$$J = \begin{bmatrix} 0 & I \\ -I & 0 \end{bmatrix}. \tag{12.21}$$

Here, $I$ is the $n \times n$ identity matrix. This property can be established by showing that $\Phi^T J\Phi = \text{Const}$ by differentiation and noting that $\Phi_f = I$.

One consequence of the symplectic property is a formula for $\Phi^{-1}$. Multiply Eq.(12.20) on the left by $J^{-1}$ and on the right by $\Phi^{-1}$ to obtain

$$J^{-1}\Phi^T J = \Phi^{-1}. \tag{12.22}$$

Since $J^T J = I$ and $J^T = J^{-1}$, meaning that $J$ is orthogonal, it is seen that

$$\Phi^{-1} = J^T \Phi^T J. \tag{12.23}$$

Next, $\Phi$ is written in the partitioned form

$$\Phi = \begin{bmatrix} \Phi_{11} & \Phi_{12} \\ \Phi_{21} & \Phi_{22} \end{bmatrix}, \tag{12.24}$$

where the partitions are of dimension $n \times n$, and Eq.(12.23) leads to

$$\Phi^{-1} = \begin{bmatrix} \Phi_{22}^T & -\Phi_{12}^T \\ -\Phi_{21}^T & \Phi_{11}^T \end{bmatrix}. \tag{12.25}$$

Hence, if both $\Phi$ and $\Phi^{-1}$ are needed it is only necessary to compute $\Phi$.

Another consequence of the symplectic property is an interesting identity. If Eq.(12.24) is substituted into Eq.(12.20), three relations result:

$$\Phi_{11}^T \Phi_{21} - \Phi_{21}^T \Phi_{11} = 0 \tag{12.26}$$

$$\Phi_{11}^T \Phi_{22} - \Phi_{21}^T \Phi_{12} = I \tag{12.27}$$

$$\Phi_{12}^T \Phi_{22} - \Phi_{22}^T \Phi_{12} = 0. \tag{12.28}$$

The first and third relations imply that $\Phi_{11}^T \Phi_{21}$ and $\Phi_{22}^T \Phi_{12}$ are symmetric. Next, Eq.(12.27) premultiplied by $\Phi_{11}^{-T}$ combined with Eq.(12.26) premultiplied by $\Phi_{11}^{-T}$ and postmultiplied by $\Phi_{11}^{-1}$ leads to

$$\Phi_{22} - \Phi_{21} \Phi_{11}^{-1} \Phi_{12} = \Phi_{11}^{-T} \tag{12.29}$$

which is to play an important role in future derivations.

## 12.4.3  Solution of the Linear TPBVP

The solution (12.18) of the linear differential equations can be written as

$$\begin{aligned} \delta x &= \Phi_{11} \delta x_f + \Phi_{12} \delta \lambda_f \\ \delta \lambda &= \Phi_{21} \delta x_f + \Phi_{22} \delta \lambda_f. \end{aligned} \tag{12.30}$$

Since the final time is fixed, the final condition $\delta \lambda_f$ is given by Eq.(12.14) to be

$$\delta \lambda_f = G_{x_f x_f} \delta x_f + \beta_{x_f}^T d\nu, \tag{12.31}$$

and Eqs.(12.30) become

$$\delta x = X\delta x_f + \Phi_{12}\beta_{x_f}^T d\nu \qquad (12.32)$$

$$\delta\lambda = \Lambda\delta x_f + \Phi_{22}\beta_{x_f}^T d\nu, \qquad (12.33)$$

where

$$X \triangleq \Phi_{11} + \Phi_{12}G_{x_f x_f}$$

$$\Lambda \triangleq \Phi_{21} + \Phi_{22}G_{x_f x_f}. \qquad (12.34)$$

The last boundary condition to apply is $\beta_{x_f}\delta x_f = \delta\beta_p$. If it is assumed that $X^{-1}$ exists, Eq.(12.32) can be rewritten as

$$X^{-1}\delta x = \delta x_f + X^{-1}\Phi_{12}\beta_{x_f}^T d\nu \qquad (12.35)$$

so that

$$\beta_{x_f}X^{-1}\delta x = \delta\beta_p + \beta_{x_f}X^{-1}\Phi_{12}\beta_{x_f}^T d\nu. \qquad (12.36)$$

Applied at $t_p$, this equation can be solved for $d\nu$ as

$$d\nu = \Gamma_p[\beta_{x_f}X_p^{-1}\delta x_p - \delta\beta_p], \qquad (12.37)$$

where

$$\Gamma_p \triangleq (\beta_{x_f}X_p^{-1}\Phi_{12_p}\beta_{x_f}^T)^{-1} \qquad (12.38)$$

With $\delta x_f$ obtained from Eq.(12.35), the following expressions for $\delta x$ and $\delta\lambda$ are obtained:

$$\delta x = [X - (\Phi_{12} - XX_p^{-1}\Phi_{12_p})\beta_{x_f}^T\Gamma_p\beta_{x_f}]X_p^{-1}\delta x_p$$
$$- (\Phi_{12} - XX_p^{-1}\Phi_{12_p})\beta_{x_f}^T\Gamma_p\delta\beta_p \qquad (12.39)$$

$$\delta\lambda = [\Lambda + (\Phi_{22} - \Lambda X_p^{-1}\Phi_{12_p})\beta_{x_f}^T\Gamma_p\beta_{x_f}]X_p^{-1}\delta x_p$$
$$- (\Phi_{22} - \Lambda X_p^{-1}\Phi_{12_p})\beta_{x_f}^T\Gamma_p\delta\beta_p. \qquad (12.40)$$

Finally, the sampled data neighboring optimal control perturbation becomes

$$\delta u = -C_x(t, t_p)\delta x_p - C_\beta(t, t_p)\delta\beta_p, \qquad (12.41)$$

where

$$C_x(t, t_p) = H_{uu}^{-1}H_{ux}[X - (\Phi_{12} - XX_p^{-1}\Phi_{12_p})\beta_{x_f}^T\Gamma_p\beta_{x_f}]X_p^{-1}$$
$$+ H_{uu}^{-1}f_u^T[\Lambda + (\Phi_{22} - \Lambda X_p^{-1}\Phi_{12_p})\beta_{x_f}^T\Gamma_p\beta_{x_f}]X_p^{-1} \qquad (12.42)$$

$$C_\beta(t, t_p) = - H_{uu}^{-1} H_{ux}(\Phi_{12} - XX_p^{-1}\Phi_{12_p})\beta_{x_f}^T \Gamma_p$$
$$- H_{uu}^{-1} f_u^T(\Phi_{22} - \Lambda X_p^{-1}\Phi_{12_p})\beta_{x_f}^T \Gamma_p. \tag{12.43}$$

For a sample and hold control system, Eq.(12.41) leads to

$$\delta u_p = -C_x(t_p)\delta x_p - C_\beta(t_p)\delta\beta_p, \tag{12.44}$$

where

$$C_x(t) = H_{uu}^{-1} H_{ux} + H_{uu}^{-1} f_u^T [\Lambda + X^{-T}\beta_{x_f}^T \Gamma\beta_{x_f}]X^{-1} \tag{12.45}$$
$$C_\beta(t) = -H_{uu}^{-1} f_u^T X^{-T}\beta_{x_f}^T \Gamma. \tag{12.46}$$

Then, for a continuous data control system, the control perturbation is given by

$$\delta u = -C_x(t)\delta x - C_\beta(t)\delta\beta. \tag{12.47}$$

It is interesting to note that $X$ and $\Lambda$ satisfy the same differential equations as $\delta x$ and $\delta\lambda$, that is,

$$\dot{X} = AX - B\Lambda$$
$$\dot{\Lambda} = -CX - A^T\Lambda, \tag{12.48}$$

but with different final conditions:

$$X_f = I$$
$$\Lambda_f = G_{x_f x_f}. \tag{12.49}$$

## 12.4.4   Relationship to the Sweep Method

In this section, the relationship of the transition matrix method to the sweep approach is developed. To do so, the transition matrix equations

$$\delta x = X\delta x_f + \Phi_{12}\beta_{x_f}^T d\nu$$
$$\delta\lambda = \Lambda\delta x_f + \Phi_{22}\beta_{x_f}^T d\nu \tag{12.50}$$

are rewritten in the sweep form. Since $\delta\beta = \beta_{x_f}\delta x_f$, Eqs.(12.50) can be rewritten as

$$\delta\lambda = \Lambda X^{-1}\delta x + (\Phi_{22} - \Lambda X^{-1}\Phi_{12})\beta_{x_f}^T d\nu$$
$$\delta\beta = \beta_{x_f}X^{-1}\delta x - \beta_{x_f}X^{-1}\Phi_{12}\beta_{x_f}^T d\nu. \tag{12.51}$$

At this point, it is noted that the matrix

$$\begin{bmatrix} X & \Phi_{12} \\ \Lambda & \Phi_{22} \end{bmatrix} = \begin{bmatrix} \Phi_{11} & \Phi_{12} \\ \Phi_{21} & \Phi_{22} \end{bmatrix} \begin{bmatrix} I & 0 \\ G_{x_f x_f} & I \end{bmatrix}. \tag{12.52}$$

is symplectic because its factors are symplectic (Problem 12.5). Hence, from the symplectic property

$$\Phi_{22} - \Lambda X^{-1} \Phi_{12} = X^{-T}, \tag{12.53}$$

it is seen that Eqs.(12.51) become

$$\begin{aligned} \delta\lambda &= \Lambda X^{-1} \delta x + X^{-T} \beta_{x_f}^T d\nu \\ \delta\beta &= \beta_{x_f} X^{-1} \delta x - \beta_{x_f} X^{-1} \Phi_{12} \beta_{x_f}^T d\nu. \end{aligned} \tag{12.54}$$

Comparison of Eqs.(12.54) with those for the sweep method, that is,

$$\begin{aligned} \delta\lambda &= S\delta x + Rd\nu & S_f = G_{x_f x_f}, & R_f - \beta_{u_f}^T \\ \delta\beta &= R^T \delta x + Qd\nu & Q_f = 0 \end{aligned} \tag{12.55}$$

shows that

$$S = \Lambda X^{-1}, \quad R = X^{-T} \beta_{x_f}^T, \quad Q = -\beta_{x_f} X^{-1} \Phi_{12} \beta_{x_f}^T. \tag{12.56}$$

Note that the boundary conditions match because $X_f = I$, $\Lambda_f = G_{x_f x_f}$ [Eqs.(12.49)]. Further verification comes from showing that $S$, $R$, and $Q$ as defined by Eqs.(12.56) satisfy the differential equations (9.35) through (9.37). This is done by differentiating Eqs.(12.56) and using Eqs. (12.9) and (12.48). The derivative $dX^{-1}/dt$ is also needed and is found from the identity $XX^{-1} = I$ to be

$$\frac{dX^{-1}}{dt} = -X^{-1} \dot{X} X^{-1}. \tag{12.57}$$

Finally, the symmetry of $\Lambda X^{-1}$ and $X^{-1} \Phi_{12}$ can be shown by applying the symmetry properties of a symplectic matrix, that is, Eqs.(12.26) and (12.28).

It is interesting to note that $S$ goes to infinity when $X$ becomes singular.

## 12.5   Linear Quadratic Guidance

Linear quadratic guidance can be used to control a system in the neighborhood of an optimal path or a nonoptimal path or to control a system defined by linear differential equations without having to introduce control bounds and compute switch times. Controlling relative to a nominal path is considered first. Then, the development of a controller for a linear system is demonstrated by solving a homing missile problem.

For guidance relative to a nominal path, the general statement of the linear quadratic problem is as follows: Find the control history $\delta u(t)$ that minimizes the performance index

$$
J = \tfrac{1}{2}\delta x_f^T T \delta x_f + \tfrac{1}{2}\int_{t_p}^{t_f} (\delta x^T U \delta x \tag{12.58}
$$
$$
+ \delta x^T V \delta u + \delta u^T V^T \delta x + \delta u^T W \delta u)\, dt,
$$

subject to the differential equations

$$
\frac{d\,\delta x}{dt} = M \delta x + N \delta u, \tag{12.59}
$$

to the prescribed initial conditions

$$
t_p = t_{p_s}, \quad \delta x_p = \delta x_{p_s}, \tag{12.60}
$$

and to the prescribed final conditions

$$
t_f = t_{f_s}, \quad D \delta x_f = \delta \beta_p. \tag{12.61}
$$

The dimensions of the matrices $\delta x$, $\delta u$, $T$, $U$, $V$, $W$, $M$, $N$, and $D$ are $n \times 1$, $m \times 1$, $n \times n$, $n \times n$, $n \times m$, $m \times m$, $n \times n$, $n \times m$, and $p \times n$, respectively. The matrices $T, D$ are constant while $U, V, W, M, N$ are constant or known functions of time. In the definition of the problem, the matrices $T$, $U$, and $W$ are symmetric; $T$ is positive semidefinite; and $W$ is positive definite. As shown later, if $U - VW^{-1}V^T$ is positive semidefinite, no conjugate point can exist.

If the matrices $T$, $U$, and so on are replaced by $G_{x_f x_f}$, $H_{xx}$, and so on along an optimal path, the linear quadratic problem is identical with the neighboring optimum problem. On the other hand, if the nominal path is an arbitrary path, then Eqs.(12.59) are the linearized system equations, and the quadratic performance index (12.58) is created to form an optimization problem. This process leads to unique control whose effect on the dynamical system can be modified by varying the matrices $T$, $U$, and so on. Finally, if the dynamical system is actually linear, the linear quadratic problem is formulated in terms of the actual states and not perturbations from a nominal path.

It is worth noting that the cross terms in the integrand of Eq.(12.58) can be eliminated by introducing a new control $\delta v$ which satisfies the relation $\delta u = -W^{-1}V^T\delta x + \delta v$. In terms of the new control, the equations for the linear quadratic problem become

$$J = \tfrac{1}{2}\delta x_f^T T\delta x_f + \tfrac{1}{2}\int_{t_p}^{t_f}[\delta x^T(U - VW^{-1}V^T)\delta x + \delta v^T W\delta v]dt$$
$$\tfrac{d\,\delta x}{dt} = (M - NW^{-1}V^T)\delta x + N\delta v. \tag{12.62}$$

Hence, if $T \geq 0$, $(U - VW^{-1}V^T) \geq 0$, and $W > 0$, a minimum exists because the performance index is positive. Also, the Legendre–Clebsch condition is satisfied, and no conjugate point exists because $J$ cannot be negative.

## 12.5.1  Sweep Solution

The minimal control for the general linear quadratic problem is obtained from the following sequence of equations:

$$
\begin{aligned}
A &= M - NW^{-1}V^T, \quad B = NW^{-1}N^T, \quad C = U - VW^{-1}V^T \\
\dot{S} &= SBS - SA - A^TS - C, \quad S_f = T \\
\dot{R} &= -(A^T - SB)R, \quad R_f = D^T \\
\dot{Q} &= R^TBR, \quad Q_f = 0 \\
\tfrac{d\delta x}{dt} &= A\delta x - B\delta\lambda, \quad \tfrac{d\delta\lambda}{dt} = -C\delta x - A^T\delta\lambda \\
t_p &= t_{p_s}, \quad \delta x_p = \delta x_{p_s}, \quad \delta\lambda_p = (S - RQ^{-1}R^T)_p\delta x_p \\
\delta u &= -W^{-1}[(V^T + N^TS)\delta x - N^T RQ_p^{-1}R_p^T\delta x_p + N^T RQ_p^{-1}\delta\beta_p].
\end{aligned}
\tag{12.63}
$$

Note that $S - RQ^{-1}R^T$ can be computed beforehand for a sequence of times and stored in a computer. Then, given a $t_p$ and $\delta x_p$, $(S - RQ^{-1}R)_p$ can be computed by interpolation, and $\delta\lambda_p$ computed. For a time-dependent system, $\delta\dot{x}$ and $\delta\dot{\lambda}$ must be integrated to obtain $\delta x(t)$ and $\delta\lambda(t)$ and ultimately $\delta u(t)$. This control law is called a *sampled data feedback control law*. For a *sample and hold system*, only $\delta x_p$ is needed to compute $\delta u_p$.

If the sampling is performed continuously, that is, if $\delta x(t)$ is known, the control law becomes

$$\delta u = -C_x(t)\delta x(t) - C_\beta(t)\delta\beta, \tag{12.64}$$

where the gains are given by

$$\begin{aligned}
C_x(t) &= W^{-1}[V^T + N^T(S - RQ^{-1}R^T)] \\
C_\beta(t) &= W^{-1}N^TRQ^{-1}.
\end{aligned} \tag{12.65}$$

This is called the *continuous data feedback control law*. Note that the gains can be computed offline and stored for future use.

If there are no prescribed final conditions (no $D$), the terms $R$ and $Q$ do not exist and are deleted from the above equations.

## 12.5.2 Transition Matrix Solution

In terms of the transition matrix (Ref. Sp), the *sampled data feedback control law* is obtained from Eqs.(12.41) through (12.43), that is,

$$\delta u = -C_x(t, t_p)\delta x_p - C_\beta(t, t_p)\delta\beta_p, \tag{12.66}$$

where

$$\begin{aligned}
C_x(t, t_p) &= W^{-1}V^T[X - (\Phi_{12} - XX_p^{-1}\Phi_{12_p})D^T\Gamma_pD]X_p^{-1} \\
&\quad + W^{-1}N^T[\Lambda + (\Phi_{22} - \Lambda X_p^{-1}\Phi_{12_p})D^T\Gamma_pD]X_p^{-1} \\
C_\beta(t, t_p) &= -W^{-1}V^T(\Phi_{12} - XX_p^{-1}\Phi_{12_p})D^T\Gamma_p \\
&\quad - W^{-1}N^T(\Phi_{22} - \Lambda X_p^{-1}\Phi_{12_p})D^T\Gamma_p
\end{aligned} \tag{12.67}$$

and where

$$\Gamma_p = (DX_p^{-1}\Phi_{12_p}D^T)^{-1}. \tag{12.68}$$

Then, the *continuous data feedback control law* becomes

$$\delta u = -C_x(t)\delta x(t) - C_\beta(t)\delta\beta, \tag{12.69}$$

where

$$\begin{aligned}
C_x(t) &= W^{-1}V^T + W^{-1}N^T[\Lambda + X^{-T}D^T\Gamma D]X^{-1} \\
C_\beta(t) &= W^{-1}N^TX^{-T}\Gamma.
\end{aligned} \tag{12.70}$$

If there are no prescribed final conditions (no $D$), the terms involving $D$ are omitted. Hence, the *sampled data feedback law* becomes

$$\delta u = -C_x(t, t_p)\delta x_p, \tag{12.71}$$

where

$$C_x(t, t_p) = W^{-1}[V^T X + N^T \Lambda]X_p^{-1}. \tag{12.72}$$

In turn, the *continuous data feedback control law* becomes

$$\delta u = -C_x(t)\delta x, \tag{12.73}$$

where

$$C_x(t) = W^{-1}[V^T + N^T \Lambda X^{-1}]. \tag{12.74}$$

Note that only $X$ and $\Lambda$ are now needed.

## 12.6  Homing Missile Problem

Consider the problem of a constant speed missile intercepting a constant speed target moving in a straight line (Fig. 12.3). Assume that the target is moving along the $\xi$-axis but that the missile is launched somewhat off the $\xi$-axis. Hence, the problem is how to control the missile about the $\xi$-axis so that intercept occurs when the missile gets to the target. The missile is controlled by varying its normal acceleration.

In relative coordinates,

$$x = \xi_T - \xi_M, \quad y = \eta_T - \eta_M, \tag{12.75}$$

the kinematic equations of motion are given by

$$
\begin{aligned}
\dot{x} &= \dot{\xi}_T - \dot{\xi}_M = V_T - V_M \cos\theta \\
\dot{y} &= \dot{\eta}_T - \dot{\eta}_M = -V_M \sin\theta \\
\dot{\theta} &= \frac{a_M}{V_M},
\end{aligned}
\tag{12.76}
$$

where $\xi$, $\eta$ are absolute coordinates, $V_T$ is the constant speed of the target, $V_M > V_T$ is the constant speed of the missile, $\theta$ is the orientation of the missile

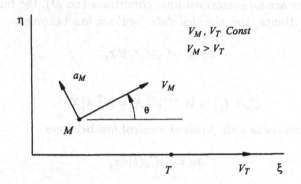

Figure 12.3: Intercept Geometry

velocity vector to the $\xi$-axis, and $a_M$ is the missile normal acceleration. For small angles, the equations of motion become

$$
\begin{aligned}
\dot{x} &= V_T - V_M \\
\dot{y} &= -V_M\theta \\
\dot{\theta} &= \frac{a_M}{V_M}.
\end{aligned}
\tag{12.77}
$$

Note that the $x$ equation can be integrated to give

$$x = (V_T - V_M)t + \text{Const}, \tag{12.78}$$

which, when the integration constant is evaluated at the current time $t_p$, leads to

$$x = x_p + (V_T - V_M)(t - t_p). \tag{12.79}$$

The intercept is over when $x = x_f = 0$ so that the time to intercept or time-to-go is given by

$$t_f - t_p = \frac{x_p}{V_M - V_T} = \frac{x_p}{V_c}, \tag{12.80}$$

where $V_c$ is the closing speed.

Next, the angle $\theta$ can be eliminated from the $y$ equation to obtain

$$\ddot{y} = -a_M \tag{12.81}$$

which can be rewritten as a system of first-order equations as

$$\dot{y} = v$$
$$\dot{v} = -a_M. \tag{12.82}$$

Hence, the system equations are linear in the state and the control. The initial conditions on the state are

$$t_p = t_{p_s}, \quad y_p = y_{p_s}, \quad v_p = v_{p_s} \tag{12.83}$$

while the final condition is

$$t_f = t_{f_s} \tag{12.84}$$

as predicted by Eq.(12.80).

The homing missile problem could be formulated as the minimization of the miss distance, that is, $J = (1/2)y_f^2$. However, the optimal control would be $a_M = \infty$ to rotate $V_M$ to the intercept direction followed by $a_M = 0$. Since the missile normal acceleration cannot physically be infinite, it would also be necessary to put a bound on $a_M$. Doing so would require the compution of the switch time between the $a_M = a_{M\max}$ and $a_M = 0$ subarcs. To place a bound on $a_M$ without doing so directly, a penalty on normal acceleration is added to the performance index. Hence, the performance index for this problem is taken to be

$$J = \frac{1}{2}cy_f^2 + \frac{1}{2}\int_{t_p}^{t_f} a_M^2 dt, \tag{12.85}$$

where $c$ is a positive constant weight that can be used to adjust the largest normal accleration.

In summary, the problem of minimizing the miss distance can be stated as follows: Find the control history $a_M(t)$ that minimizes the performance index (12.85) subject to the differential constraints (12.82), the prescribed initial conditions (12.83), and the prescribed final condition (12.84). This optimal control problem fits into the category of a linear quadratic problem so those results can be applied here. Since the dynamics are actually linear, the state and the control of the linear quadratic problem are not perturbations but are the actual state and control. Hence, in the equations of the linear quadratic problem, $\delta x$, $\delta u$ are replaced by $x$, $u$.

In the terminology of the linear quadratic problem defined by Eqs. (12.58) through (12.61), the coefficient matrices are given by

$$x = \begin{bmatrix} y & v \end{bmatrix}^T, \quad u = \begin{bmatrix} a_M \end{bmatrix}$$

$$T = \begin{bmatrix} c & 0 \\ 0 & 0 \end{bmatrix}, \quad U = \begin{bmatrix} 0 & 0 \\ 0 & 0 \end{bmatrix}, \quad V = \begin{bmatrix} 0 \\ 0 \end{bmatrix}, \quad W = \begin{bmatrix} 1 \end{bmatrix} \quad (12.86)$$

$$M = \begin{bmatrix} 0 & 1 \\ 0 & 0 \end{bmatrix}, \quad N = \begin{bmatrix} 0 \\ -1 \end{bmatrix},$$

where there is no $D$ matrix. Hence, the gains for the sampled data feedback control law defined in Eq.(12.72) become

$$C_x(t, t_p) = W^{-1} N^T \Lambda X_p^{-1} \quad (12.87)$$

or

$$C_x(t, t_p) = \begin{bmatrix} 0 & -1 \end{bmatrix} \begin{bmatrix} \Lambda_{11} & \Lambda_{12} \\ \Lambda_{21} & \Lambda_{22} \end{bmatrix} \begin{bmatrix} X_{11_p} & X_{12_p} \\ X_{21_p} & X_{22_p} \end{bmatrix}^{-1}. \quad (12.88)$$

In this equation, $X(t)$ and $\Lambda(t)$ are obtained from Eqs. (12.48) and (12.49) with $A, B, C$ defined by the first of Eqs. (12.63). For this problem it is seen that

$$\begin{aligned}
\dot{X}_{11} &= X_{21}, & X_{11_f} &= 1 \\
\dot{X}_{12} &= X_{22}, & X_{12_f} &= 0 \\
\dot{X}_{21} &= -\Lambda_{21}, & X_{21_f} &= 0 \\
\dot{X}_{22} &= -\Lambda_{22}, & X_{22_f} &= 1 \\
\dot{\Lambda}_{11} &= 0, & \Lambda_{11_f} &= c \\
\dot{\Lambda}_{12} &= 0, & \Lambda_{12_f} &= 0 \\
\dot{\Lambda}_{21} &= -\Lambda_{11}, & \Lambda_{21_f} &= 0 \\
\dot{\Lambda}_{22} &= -\Lambda_{12}, & \Lambda_{22_f} &= 0.
\end{aligned} \quad (12.89)$$

The solution process begins with the $\Lambda$s and moves to the $X$s resulting in

$$\begin{aligned}
X_{11} &= -\tfrac{c}{6}(t_f - t)^3 + 1 \\
X_{12} &= -(t_f - t) \\
X_{21} &= \tfrac{c}{2}(t_f - t)^2 \\
X_{22} &= 1 \\
\Lambda_{11} &= c \\
\Lambda_{12} &= 0 \\
\Lambda_{21} &= c(t_f - t) \\
\Lambda_{22} &= 0.
\end{aligned} \quad (12.90)$$

Substitution of these equations into Eqs.(12.88) yields the following gains:

$$C_y(t, t_p) = -\frac{3c(t_f - t)}{c\Delta t^3 + 3}, \quad C_v(t, t_p) = -\frac{3c(t_f - t)\Delta t}{c\Delta t^3 + 3}, \tag{12.91}$$

where $\Delta t = t_f - t_p$, and the sampled data feedback guidance law is

$$a_M = -C_y(t, t_p)y_p - C_v(t, t_p)v_p. \tag{12.92}$$

For the continuous data case, $t_p$ is replaced by $t$ so that the guidance law becomes

$$a_M = -C_y y - C_v v, \tag{12.93}$$

where the gains are given by

$$C_y = -\frac{3c(t_f - t)}{c(t_f - t)^3 + 3}, \quad C_v = -\frac{3c(t_f - t)^2}{c(t_f - t)^3 + 3}. \tag{12.94}$$

For the case where $c \to \infty$, that is, for $y_f \to 0$, the guidance law can be written as

$$a_M = \frac{3}{(t_f - t)^2}y + \frac{3}{t_f - t}v. \tag{12.95}$$

From Fig. 12.4, it is seen that the line-of-sight angle $\sigma$ is for small angles defined as

$$\sigma = \frac{y}{R}, \tag{12.96}$$

where R is the range from the missile to the target. If $V_c$ denotes the closing

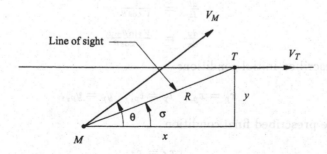

Figure 12.4: Definition of Line-of-Sight Angle $\sigma$

speed, the range is given by

$$R = (V_M - V_T)(t_f - t) = V_c(t_f - t). \tag{12.97}$$

Differentiation of Eq.(12.96) leads to

$$\dot{\sigma} = \frac{R\dot{y} - y\dot{R}}{R^2} = \frac{1}{V_c}\left[\frac{y}{(t_f - t)^2} + \frac{v}{(t_f - t)}\right]. \tag{12.98}$$

Hence, Eq.(12.95) can be written in the form

$$a_M = 3V_c\dot{\sigma} \tag{12.99}$$

which is the well-known proportional navigation guidance law. Here, $\dot{\sigma}$ is the line-of-sight rate; $V_c$ is the closing speed; and the numerical coefficient 3 is called the navigation constant.

# Problems

12.1 Optimal guidance for the navigation problem where only $x_f$ is prescribed is to find the control history that minimizes the time to reach the line $x_f = x_{f_s}$ from a fixed sample point. The optimal control problem can be stated as follows: Find the control history $\theta(t)$ that minimizes the final time

$$J = t_f,$$

subject to the dynamics

$$\frac{dt}{dx} = \frac{1}{V \cos \theta}$$

$$\frac{dy}{dx} = \frac{V \sin \theta + w}{V \cos \theta},$$

the prescribed initial conditions

$$x_p = x_{p_s}, \quad t_p = t_{p_s}, \quad y_p = y_{p_s},$$

and the prescribed final condition

$$x_f = x_{f_s}.$$

Show that the optimal control is $\theta = 0$ which is independent of the prescribed sample conditions.

12.2 For the navigation problem (Problem 12.1), show that the optimal control from $x_p, t_p, y_p$ to the point $x_f, y_f$ is given by Eq.(12.1).

12.3 For Problem 12.1, show that the neighboring optimal control is given by $\delta\theta = 0$.

12.4 For Problem 12.2 show that the neighboring optimal control is given by Eq.(12.13).

12.5 Show that, if two square matrices are symplectic, their product is symplectic.

12.6 Show that $X$ and $\Lambda$ satisfy the differential equations (12.48) and the boundary conditions (12.49).

12.7 Using the symmetry properties associated with symplectic matrices, show that $\Lambda X^{-1}$ and $X^{-1}\Phi_{12}$ are symmetric so that $S$ and $Q$ defined by Eqs. (12.56) are symmetric.

12.8 Solve the following linear quadratic problem using the transition matrix approach:

$$J = \frac{1}{2}cx_f^2 + \int_{t_p}^{t_f} \frac{1}{2}u^2 dt$$

$$c = \text{Const} > 0$$

$$\dot{x} = u$$

$$t_p = t_{p_s}, \quad x_p = x_{p_s}, \quad t_f = t_{f_s}$$

12.9 Solve the following problem as a linear quadratic problem using the transition matrix approach:

$$J = \int_{t_p}^{t_f} \frac{u^2}{2} dt$$

$$\dot{x} = u$$

$$t_p = t_{p_s}, \quad x_p = x_{p_s}, \quad t_f = t_{f_s}, \quad x_f = 0.$$

Compare the control law with that from Problem 12.8 when $c$ becomes infinite.

12.10 Set up the missile intercept problem (Section 12.6) as a linear quadratic problem with $c = 0$ but with a final condition $y_f = 0$. Solve the problem using the transition matrix approach.

12.11 Set up the car acceleration problem (Section 9.7) as a linear quadratic problem using the transition matrix approach.

12.12 It is desired to develop an optimal guidance law for the airplane navigation problem (Chapter 1) with the equations of motion written as

$$\dot{x} = V \cos \theta$$

$$\dot{y} = V \sin \theta + w$$

and the sample and final conditions

$$t_p = t_{p_s}, \quad x_p = x_{p_s}, \quad y_p = y_{p_s}, \quad x_f = x_{f_s}.$$

If the steering angle $\theta$ is assumed to be small, the equations of motion become

$$\dot{x} = V$$

$$\dot{y} = V\theta + w,$$

the first of which can be solved for the final time. To minimize the miss distance at the final point $y_f$, the performance index is chosen to be

$$J = \frac{1}{2}cy_f^2 + \frac{1}{2}\int_{t_p}^{t_f} \theta^2 dt,$$

where $c$ is a positive weight. Show that the optimal control is given by

$$\theta = -\frac{cy_p + cw(t_f - t_p)}{1/V + cV(t_f - t_p)}.$$

For $c \to \infty$ to make $y_f = 0$, the control becomes

$$\theta = -\frac{y_p + w(t_f - t_p)}{V(t_f - t_p)}.$$

Note that this is not a linear quadratic problem.

# 13

# Free Final Time

## 13.1   Introduction

In this chapter, the optimal control problem is extended to allow the final time to be free, and the conditions for a minimum are developed anew. Allowing the final time to be free gives rise to a natural boundary condition for determining the optimal final time. Next, the Weierstrass condition and the Legendre–Clebsch condition are presented. Then, the conjugate point condition is developed for nonsingular optimal control problems ($H_{uu} > 0$). Finally, several example problems are solved.

## 13.2   First Differential Conditions

The optimal control problem being considered is to find the control history $u(t)$ that minimizes the performance index

$$J = \phi(t_f, x_f) + \int_{t_0}^{t_f} L(t, x, u)dt, \qquad (13.1)$$

subject to the differential constraints

$$\dot{x} = f(t, x, u), \qquad (13.2)$$

to the prescribed initial conditions

$$t_0 = t_{0_s}, \quad x_0 = x_{0_s}, \qquad (13.3)$$

and to the prescribed final conditions

$$\psi(t_f, x_f) = 0. \tag{13.4}$$

Here, $\psi$ is a $(p+1) \times 1$ vector, where $0 \le p \le n$; there must be at least one final condition that draws the optimal path to the final point. The final time is now free, and its optimal value is to be determined.

The conditions to be satisfied by a minimal path are derived from the first and second differentials of the augmented performance index

$$J' = G(t_f, x_f, \nu) + \int_{t_0}^{t_f} [H(t, x, u, \lambda) - \lambda^T \dot{x}] dt, \tag{13.5}$$

where the endpoint function $G$ and the Hamiltonian $H$ are defined as

$$G = \phi(t_f, x_f) + \nu^T \psi(t_f, x_f) \tag{13.6}$$

$$H = L(t, x, u) + \lambda^T f(t, x, u). \tag{13.7}$$

The differential is taken of the augmented performance index (13.5) to give

$$dJ' = G_{t_f} dt_f + G_{x_f} dx_f + G_\nu d\nu + [(H - \lambda^T \dot{x}) dt]_{t_0}^{t_f}$$
$$+ \int_{t_0}^{t_f} [H_x \delta x + H_u \delta u + (H_\lambda - \dot{x}^T) \delta \lambda - \lambda^T \delta \dot{x}] dt, \tag{13.8}$$

where the coefficients of $d\nu$ and $\delta\lambda$ are zero because $G_\nu = \psi$ and $H_\lambda = f^T$. Next, the last term in the integrand is integrated by parts, and the differential $dx_f$ is related to $\delta x_f$ and $dt_f$ through the formula (Section 6.7)

$$dx_f = \delta x_f + \dot{x}_f dt_f. \tag{13.9}$$

The first differential becomes

$$dJ' = (G_{t_f} + L_f + G_{x_f} f_f) dt_f + (G_{x_f} - \lambda_f^T) \delta x_f$$
$$+ \int_{t_f}^{t_0} [(H_x + \dot{\lambda}^T) \delta x + H_u \delta u] dt. \tag{13.10}$$

Following the approach of Section 9.3, choose Lagrange multipliers ($\lambda$s and one $\nu$) such that

$$\dot{\lambda} = -H_x^T, \quad \lambda_f = G_{x_f}^T, \quad G_{t_f} + L_f + G_{x_f} f_f = 0. \tag{13.11}$$

Then, the first differential reduces to

$$dJ' = \int_{t_0}^{t_f} H_u dt. \qquad (13.12)$$

Finally, the proof that the vanishing of the first differential requires $H_u = 0$ is the same as that in Section 9.3 if the assumption of controllability for free-final time is made.

In summary, the first differential necessary conditions for a minimum are the Euler–Lagrange equations

$$\dot{x} = f(t, x, u) \qquad (13.13)$$

$$\dot{\lambda} = -H_x^T(t, x, u, \lambda) \qquad (13.14)$$

$$0 = H_u^T(t, x, u, \lambda) \qquad (13.15)$$

and the prescribed and natural boundary conditions

$$t_0 = t_{0_s}, \quad x_0 = x_{0_s} \qquad (13.16)$$

$$\psi(t_f, x_f) = 0 \qquad (13.17)$$

$$G_{t_f}(t_f, x_f, \nu) + L(t_f, x_f, u_f) + G_{x_f}(t_f, x_f, \nu) f(t_f, x_f, u_f) = 0 \qquad (13.18)$$

$$\lambda_f = G_{x_f}^T(t_f, x_f, \nu). \qquad (13.19)$$

Equation (13.15) determines the optimal control so that Eqs.(13.13) and (13.14) give the corresponding state and Lagrange multiplier histories. For $n$ states, the solutions of these equations involve $2n+2$ unknowns, $2n$ constants of integration plus the initial and final times. Part of these unknowns is determined from the $n+1$ initial conditions (13.16). The $p+2+n$ final conditions (13.17) through (13.19) determine the constant multipliers $\nu$, the final time $t_f$, and the final state $x_f$. If $\nu$ is eliminated, the resulting equations provide $n+1$ boundary conditions for determining the remaining unknowns.

By combining Eqs.(13.18) and (13.19), the more familiar form of Eq.(13.18) is given by

$$H_f = -G_{t_f}(t_f, x_f, \nu). \qquad (13.20)$$

Note that the first integral $H = \text{Const}$ is valid here, if it exists, because it comes from Eqs.(13.13) through (13.15) which are the same as those for the fixed final time problem. See Section 9.5.

## 13.3    Tests for a Minimum

The development of the Weierstrass condition for the free final time problem
is similar to that for the fixed final time problem. Here, Eq.(10.3) has an
additional term involving $dt_f$ which vanishes on the minimal path. Hence, the
Weierstrass condition

$$H(t, x, u_*, \lambda) - H(t, x, u, \lambda) > 0 \qquad (13.21)$$

for all $u_* \neq u$, and the Legendre–Clebsch condition

$$H_{uu}(t, x, u, \lambda) \geq 0 \qquad (13.22)$$

must be satisfied at every point of a minimal path.

## 13.4    Second Differential

The procedure followed in the derivation of the second differential conditions is
to use the equations of a neighboring optimal path (Fig. 13.1) from a perturbed
initial point $\delta x_0 = \delta x_{0_*}$ to the final constraint surface $d\psi = 0$ to write the second
differential as a perfect square. If $\delta x_0$ is not set to zero in the derivation of the
first differential (13.10), the result is

$$
\begin{aligned}
dJ' \;=\; & \Omega dt_f + (G_{x_f} - \lambda_f^T)\delta x_f + \psi^T d\nu + \lambda_0^T \delta x_0 \\
& + \int_{t_0}^{t_f} [(H_x + \dot\lambda^T)\delta x + (H_u)\delta u + (f^T - \dot x^T)\delta\lambda]dt,
\end{aligned}
\qquad (13.23)
$$

where

$$\Omega(t_f, x_f, \nu, u_f) \triangleq G_{t_f} + L_f + G_{x_f} f_f. \qquad (13.24)$$

The next step is to take the differential of the transpose of $dJ'$ using
the first differential conditions (13.13) through (13.19). The second differential

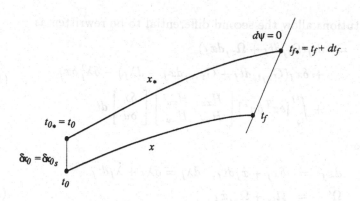

Figure 13.1: Neighboring Optimal Path

is given by

$$
\begin{aligned}
d^2 J' \;=\; & dt_f(\Omega_{t_f}dt_f + \Omega_{x_f}dx_f + \Omega_\nu d\nu + \Omega_{u_f}du_f) \\[4pt]
& + \delta x_f^T(G_{x_f t_f}dt_f + G_{x_f x_f}dx_f + \psi_{x_f}^T d\nu - d\lambda_f) \\[4pt]
& + \delta\lambda_0^T \delta x_0 + \int_{t_0}^{t_f}[\delta x^T(H_{xx}\delta x + H_{xu}\delta u \\[4pt]
& + f_x^T\delta\lambda + \delta\dot\lambda) + \delta u^T(H_{ux}\delta x + H_{uu}\delta u + f_u^T\delta\lambda) \\[4pt]
& + \delta\lambda^T(f_x\delta x + f_u\delta u - \delta\dot x)]\, dt.
\end{aligned}
\tag{13.25}
$$

At this point, the term $\delta x^T\,\delta\dot\lambda$ is integrated by parts, and use is made of the the relations

$$
\begin{aligned}
d\psi &= \psi' dt_f + \psi_{x_f}\delta x_f = 0 \\
\psi' &= \psi_{t_f} + \psi_{x_f}\dot x_f \\
\delta\dot x &= f_x\delta x + f_u\delta u \\
\Omega_\nu &= \psi'^T, \quad \Omega_{u_f} = 0.
\end{aligned}
\tag{13.26}
$$

These substitutions allow the second differential to be rewritten as

$$
\begin{aligned}
d^2 J' &= dt_f(\Omega_{t_f} dt_f + \Omega_{x_f} dx_f) \\
&\quad + \delta x_f(G_{x_f t_f} dt_f + G_{x_f x_f} dx_f - d\lambda_f) - \delta \lambda_f^T \delta x_f \\
&\quad + \int_{t_0}^{t_f} [\delta x^T \, \delta u^T] \begin{bmatrix} H_{xx} & H_{xu} \\ H_{ux} & H_{uu} \end{bmatrix} \begin{bmatrix} \delta x \\ \delta u \end{bmatrix} dt.
\end{aligned}
\tag{13.27}
$$

Finally, since

$$
\begin{aligned}
dx_f &= \delta x_f + \dot{x}_f dt_f, \quad d\lambda_f = \delta \lambda_f + \dot{\lambda}_f dt_f \\
\Omega' &= \Omega_{t_f} + \Omega_{x_f} \dot{x}_f \\
\Omega_{x_f} &= (L_f)_{x_f} + G_{x_f}(f_f)_{x_f} + f_f^T G_{x_f x_f} + G_{t_f x_f},
\end{aligned}
\tag{13.28}
$$

the second differential assumes the elegant form

$$
\begin{aligned}
d^2 J' &= [\delta x_f^T \, dt_f] \begin{bmatrix} G_{x_f x_f} & \Omega_{x_f}^T \\ \Omega_{x_f} & \Omega' \end{bmatrix} \begin{bmatrix} \delta x_f \\ dt_f \end{bmatrix} \\
&\quad + \int_{t_0}^{t_f} [\delta x^T \, \delta u^T] \begin{bmatrix} H_{xx} & H_{xu} \\ H_{ux} & H_{uu} \end{bmatrix} \begin{bmatrix} \delta x \\ \delta u \end{bmatrix} dt.
\end{aligned}
\tag{13.29}
$$

Note that for the first time the changes in the multipliers do not all drop out. While it is true that $\delta \lambda$ can be set equal to zero (not taken), $d\lambda_f = \dot{\lambda}_f dt_f$ and must be included.

## 13.5  Neighboring Optimal Paths

The differential equations for a neighboring optimal path are obtained by taking the variation of the first differential conditions (13.13) through (13.15), that is,

$$
\begin{aligned}
\delta \dot{x} &= f_x \delta x + f_u \delta u \\
\delta \dot{\lambda} &= -H_{xx} \delta x - H_{xu} \delta u - f_x^T \delta \lambda \\
0 &= H_{ux} \delta x + H_{uu} \delta u + f_u^T \delta \lambda.
\end{aligned}
\tag{13.30}
$$

If it is assumed that the optimal control problem is nonsingular ($H_{uu} > 0$), the control perturbation can be written as

$$
\delta u = -H_{uu}^{-1}(H_{ux} \delta x + f_u^T \delta \lambda)
\tag{13.31}
$$

so that the differential equations for $\delta x$ and $\delta \lambda$ become

$$\begin{aligned} \delta \dot{x} &= A\delta x - B\delta \lambda \\ \delta \dot{\lambda} &= -C\delta x - A^T \delta \lambda, \end{aligned} \tag{13.32}$$

where

$$\begin{aligned} A &= f_x - f_u H_{uu}^{-1} H_{ux} \\ B &= f_u H_{uu}^{-1} f_u^T \\ C &= H_{xx} - H_{xu} H_{uu}^{-1} H_{ux}. \end{aligned} \tag{13.33}$$

The boundary conditions for the differential equations are obtained by taking the differential of the initial conditions (13.16) and the final conditions (13.17) through (13.19). The varied initial conditions are given by

$$t_0 = t_{0_s}, \quad \delta x_0 = \delta x_{0_s}. \tag{13.34}$$

After the final conditions are written as

$$\lambda_f = G_{x_f}^T, \quad \psi = 0, \quad \Omega = G_{t_f} + L_f + G_{x_f} f_f = 0, \tag{13.35}$$

taking the differential and converting to the variation leads to

$$\delta \lambda_f = G_{x_f x_f} \delta x_f + \psi_{x_f}^T d\nu + \Omega_{x_f}^T dt_f \tag{13.36}$$

$$d\psi = \psi_{x_f} \delta x_f + \psi' dt_f \tag{13.37}$$

$$d\Omega = \Omega_{x_f} \delta x_f + \psi^T d\nu + \Omega' dt_f, \tag{13.38}$$

where $d\psi = 0$ and $d\Omega = 0$.

At this point, there are several ways to proceed. Two of the approaches involve converting the free final time problem into something that looks like a fixed final time problem. First, Bryson and Ho (Ref. BrH) solve Eq. (13.38) for $dt_f$ and eliminate it from Eqs.(13.36) and (13.37) to get something that looks like Eqs.(11.38) and (11.39). The development of the neighboring optimal path and the sufficient condition for a minimum follows the procedure of Chapter 11. Since this approach requires that $\Omega' \neq 0$, Wood and Bryson (Ref. WB) suggest that $d\psi$ be partitioned to isolate one element that can be solved for $dt_f$. Then, $dt_f$ is eliminated from the remaining equations as was done above. This approach leads to fewer differential equations, but it requires some preprocessing of the problem.

The third approach (Ref. BrH) is to extend the sweep approach of Chapter 11 by assuming that

$$
\begin{bmatrix} \delta\lambda \\ d\psi \\ d\Omega \end{bmatrix} = \begin{bmatrix} S & R & m \\ R^T & Q & n \\ m^T & n^T & \alpha \end{bmatrix} \begin{bmatrix} \delta x \\ d\nu \\ dt_f \end{bmatrix}, \tag{13.39}
$$

where

$$
\begin{aligned}
S_f &= G_{x_f x_f}, & R_f &= \psi_{x_f}^T, & m_f &= \Omega_{x_f}^T, \\
Q_f &= 0, & n_f &= \psi', & \alpha_f &= \Omega'.
\end{aligned} \tag{13.40}
$$

To obtain the differential equations for $S$, $R$, $Q$, $m$, $n$, and $\alpha$, Eqs. (13.39) are differentiated and combined with Eqs.(13.32). The results are

$$
\begin{aligned}
\dot{S} &= -C - A^T S - SA + SBS \\
\dot{R} &= (SB - A^T)R \\
\dot{m} &= (SB - A^T)m \\
\dot{Q} &= R^T BR \\
\dot{n} &= R^T Bm \\
\dot{\alpha} &= m^T Bm.
\end{aligned} \tag{13.41}
$$

Hence, the solutions of Eqs.(13.41) subject to the boundary conditions (13.40) lead to the functions $S(t)$, $R(t)$, $m(t)$, $Q(t)$, $n(t)$, and $\alpha(t)$, so that Eqs.(13.39) can be used along the optimal path.

The equations for $d\psi = 0$ and $d\Omega = 0$ can be evaluated at $t_0$ to obtain

$$
\begin{aligned}
R_0^T \delta x_0 + Q_0 d\nu + n_0 dt_f &= 0 \\
m_0^T \delta x_0 + n_0^T d\nu + \alpha_0 dt_f &= 0.
\end{aligned} \tag{13.42}
$$

At this point, Bryson and Ho solve $d\Omega = 0$ for $dt_f$ and eliminate it from the remaining equations. This approach requires that $\alpha_0 \neq 0$. However, there are optimal control problems where $\alpha = 0$. Another approach is that of Chapter 11. Solve $d\psi = 0$ for $d\nu$ and eliminate it from the remaining equations. This requires that $Q_0^{-1}$ exist, and there are optimal control problems in which $Q = 0$ or $\det Q = 0$. The only direct way to develop a theory that can handle these particular cases is to solve Eqs.(13.42) simultaneously for $d\nu$ and $dt_f$ as

$$
\begin{bmatrix} d\nu \\ dt_f \end{bmatrix} = -V_0^{-1} U_0^T \delta x_0, \tag{13.43}
$$

where

$$U = [R \ m], \quad V = \begin{bmatrix} Q & n \\ n^T & \alpha \end{bmatrix}. \tag{13.44}$$

Then, the expression for $\delta\lambda_0$ can be obtained from Eq.(13.39) as

$$\delta\lambda_0 = \hat{S}_0 \delta x_0, \tag{13.45}$$

where

$$\hat{S} = S - UV^{-1}U^T. \tag{13.46}$$

Given $t_0$ and $\delta x_0$, Eq.(13.45) yields $\delta\lambda_0$ so that Eqs.(13.32) can be solved for $\delta x(t)$ and $\delta\lambda(t)$. Then, Eq. (13.31) leads to the control perturbation $\delta u(t)$ for the neighboring optimal path.

In the development of the second differential conditions, a different equation for $\delta u$ is useful. The expression for $\delta\lambda$, Eq.(13.39), can be rewritten as

$$\delta\lambda = S\delta x + U \, dp, \tag{13.47}$$

where

$$dp = \begin{bmatrix} d\nu \\ dt_f \end{bmatrix}. \tag{13.48}$$

Hence, the control perturbation is given by

$$\delta u = -H_{uu}^{-1}[(H_{ux} + f_u^T S)\delta x + f_u^T U \, dp]. \tag{13.49}$$

While the equations for a neighboring optimal path have been presented from $t_0$ to $t_f$, they could have been developed from $t_p$ to $t_f$ where $t_p > t_0$. Hence, for a neighboring optimal path to exist, the sweep variables must be finite over $(t_0, t_f)$. In particular, $V$ must be invertible and $V^{-1}U^T$ and $\hat{S} = S - UV^{-1}U^T$ must be finite over the interval $(t_0, t_f)$. Also, all of the comments made in Section 11.4 about $\bar{S}$ are valid here for $\hat{S}$. Finally, $\hat{S}$ satisfies the same differential equation as $S$, that is,

$$\dot{\hat{S}} = -C - A^T\hat{S} - \hat{S}A + \hat{S}B\hat{S}. \tag{13.50}$$

This can be established by differentiating Eq.(13.46) and using Eqs.(13.41) and (13.44).

If the sweep variables blow up within the interval of integration, it is possible to switch to $\hat{S}$ before the blowup occurs, that is, at a time greater than the blowup time.

Finally, for problems where there is only one final condition ($p = 1$ or scalar Q), $V_f^{-1}$ may be finite so that $\hat{S}_f^{-1}$ is finite. Then, the discussion of neighboring optimal paths can be based on $\hat{S}$ over the entire interval of integration.

## 13.6   Second Differential Conditions

To write the integrand of the second differential (13.29) as a positive quantity, it can be shown that

$$
\begin{aligned}
[\delta x^T \delta u^T] \begin{bmatrix} H_{xx} & H_{xu} \\ H_{ux} & H_{uu} \end{bmatrix} \begin{bmatrix} \delta x \\ \delta u \end{bmatrix} = \\
F^T H_{uu} F - \frac{d}{dt}\left\{ [\delta x^T dp^T] \begin{bmatrix} S & U \\ U^T & V \end{bmatrix} \begin{bmatrix} \delta x \\ dp \end{bmatrix} \right\},
\end{aligned}
\tag{13.51}
$$

where $F$ has the form of the neighboring optimal control variation (13.49), that is,

$$
F = H_{uu}^{-1}[(H_{ux} + f_u^T S)\delta x + f_u^T U dp] + \delta u. \tag{13.52}
$$

The procedure for showing this is to substitute Eq.(13.52) into $F^T H_{uu} F$ and show that Eq.(13.51) results. The differential equations for $U$ and $V$ are given by

$$
\begin{aligned}
\dot{U} &= (SB - A^T)U \\
\dot{V} &= U^T BU.
\end{aligned}
\tag{13.53}
$$

Note that to write Eq.(13.51) the sweep variables must be differentiable over $(t_0, t_f)$, that is, they must be finite over $(t_0, t_f)$.

With Eqs.(13.40) and (13.51), the second differential (13.29) can be rewritten as

$$
\begin{aligned}
d^2 J' &= [\delta x_f \, dt_f] \begin{bmatrix} S_f & m_f \\ m_f^T & \alpha_f \end{bmatrix} \begin{bmatrix} \delta x_f \\ dt_f \end{bmatrix} + \int_{t_0}^{t_f} F^T H_{uu} F dt \\
&\quad - \left[ \delta x^T S \delta x + \delta x^T U dp + dp^T U^T \delta x + dp^T V dp \right]_{t_0}^{t_f}.
\end{aligned}
\tag{13.54}
$$

First, if the definitions of $U$ and $V$ are employed, some of the final point terms cancel out and the rest vanish because $d\psi = 0$. Next, if $dp$ is written as

$dp = -V_0^{-1}U_0^T\delta x_0$ [see Eq.(13.43)] the initial point terms become $\delta x_0^T\hat{S}_0\delta x_0$. Hence, the second differential can be rewritten as

$$d^2 J' = \delta x_0^T \hat{S}_0 \delta x_0 + \int_{t_0}^{t_f} F^T H_{uu} F\, dt. \qquad (13.55)$$

For an admissible comparison path, that is, a path which lies in the neighborhood of the optimal path and satisfies all the constraints, $\delta x_0 = 0$. If $\hat{S}_0$ is finite over $[t_0, t_f)$, the initial point term in Eq.(13.55) vanishes, and the second differential reduces to

$$d^2 J' = \int_{t_0}^{t_f} F^T H_{uu} F\, dt. \qquad (13.56)$$

Since $H_{uu} > 0$, the second differential is positive for arbitrary variations $\delta u$ unless a $\delta u$ exists for which $F$ can vanish. From the definition (13.52) of $F$, this can happen only if the admissible comparison path is a neighboring optimal path. However, Eq.(13.45) shows that, for $\hat{S}$ finite and $\delta x_0 = 0$, $\delta\lambda_0 = 0$ and that the only solution of Eqs.(13.32) is $\delta x = \delta\lambda = 0$. This implies that $\delta u = 0$ or that there is no admissible comparison path that is a neighboring optimal path. Therefore, the optimal path is a minimum. On the other hand, if $\hat{S}_0$ is infinite, $\delta x_0 = 0$ can lead to a finite $\delta\lambda_0$ and an admissible path which is a neighboring optimal path. Then, the optimal path may not be a minimum. When $\hat{S}$ becomes infinite at some point within the interval of integration, the optimal path is not a minimum. The time at which $\hat{S}$ becomes infinite, if it does, is called a *conjugate point* and is denoted by $t_{cp}$.

In summary, for the free final time optimal control problem with $H_{uu} > 0$, the sufficient condition ($d^2 J' > 0$) for a minimum is that

$$\hat{S} \text{ finite},\quad t_0 \le t < t_f. \qquad (13.57)$$

Formally, this is the same sufficient condition as that for the fixed final time problem. The difference is that $\hat{S}$ is used to determine the conjugate point, if it exists, instead of $\bar{S}$. In terms of the conjugate point $t_{cp}$, this condition is stated as

$$t_{cp} \text{ nonexistent or } t_0 > t_{cp}. \qquad (13.58)$$

If the sweep variables blow up within the interval of integration (Ref. WB), it is possible to develop the second differential conditions by switching to $\hat{S}$ before the blowup occurs. The proof is similar to that contained in Section 11.11. If there is just one final condition ($p = 1$), the derivation of the second differential conditions can be based entirely on $\hat{S}$.

## 13.7    Example

Consider a scalar system whose velocity can be controlled ($\dot{x} = u$), and find the control $u(t)$ that minimizes the time to drive the system from the origin ($t_0 = 0$, $x_0 = 0$) to the final state $x_f = 1$. As the problem is stated so far, the minimal control is $u = \infty$. However, since an infinite control cannot exist physically, the control must be bounded. One way of doing this is to impose the constraint $u \leq u_{max}$, and this approach is considered in a later chapter. Another approach is to add a penalty to the performance index that is proportional to the integral of $u^2$. The constant weight $k > 0$ can then be adjusted to keep $u \leq u_{max}$.

The formal statement of the problem is to find the control history $u(t)$ that minimizes the performance index

$$J = t_f + k \int_{t_0}^{t_f} u^2 \, dt, \qquad (13.59)$$

subject to the system dynamics

$$\dot{x} = u, \qquad (13.60)$$

the prescribed initial conditions

$$t_0 = 0, \quad x_0 = 0, \qquad (13.61)$$

and the prescribed final condition

$$x_f = 1. \qquad (13.62)$$

For this problem, $\phi = t_f$, $L = ku^2$, $f = u$, and $\psi = x_f - 1$, so that the endpoint function and the Hamiltonian become

$$G = t_f + \nu(x_f - 1) \qquad (13.63)$$
$$H = ku^2 + \lambda u. \qquad (13.64)$$

Equations (13.14) and (13.15) lead to

$$\lambda = 0 \qquad (13.65)$$
$$0 = 2ku + \lambda \qquad (13.66)$$

which imply that $\lambda$ is a constant and that $u$ is given by

$$u = -\frac{\lambda}{2k} \tag{13.67}$$

and is also constant. Next, the natural boundary conditions (13.18) and (13.19) become

$$H_f = -1 \tag{13.68}$$
$$\lambda_f = \nu, \tag{13.69}$$

where the latter defines $\nu$ and the former combined with Eqs. (13.64) and (13.66) gives

$$u = \pm\frac{1}{\sqrt{k}}. \tag{13.70}$$

Here is a case where the first differential conditions indicate the possible existence of multiple solutions that can lead to an optimal solution composed of multiple subarcs. While these solutions are discussed in the chapter on control discontinuities, it is assumed here that the plus solution is the valid one. This choice can be substantiated physically by recognizing that a minimum time solution would not have any parts in which the system is going backward ($\dot{x} < 0$). Hence, the optimal control is taken to be

$$u = \frac{1}{\sqrt{k}}. \tag{13.71}$$

Note that increasing $k$ decreases $u$.

Integration of the state equation (13.60) subject to the boundary conditions (13.61) and (13.62) leads to

$$x = \frac{t}{\sqrt{k}}, \quad t_f = \sqrt{k}. \tag{13.72}$$

Also, the Lagrange multiplier $\lambda$ is obtained from Eq. (13.66) as

$$\lambda = -2\sqrt{k}. \tag{13.73}$$

For the solution to be a minimum, the Weierstrass condition requires that

$$ku_*^2 + \lambda u_* - ku^2 - \lambda u > 0 \tag{13.74}$$

for all $u_* \neq u$. Since $\lambda = -2ku$ and $k > 0$, it reduces to

$$(u_* - u)^2 > 0 \tag{13.75}$$

and is satisfied for all $u_* \neq u$. Hence, the control $u = 1/\sqrt{k}$ satisfies the Weierstrass necessary condition for a strong minimum. Also,

$$H_{uu} = 2k > 0 \tag{13.76}$$

so that the Legendre–Clebsch condition is satisfied.

Since the strengthened Legendre–Clebsch condition is satisfied, the next step is to investigate the conjugate point condition. It is observed that

$$f_x = 0, \quad f_u = 1, \quad H_{xx} = 0, \quad H_{xu} = 0, \quad H_{uu} = 2k \tag{13.77}$$

so that

$$A = 0, \quad B = \frac{1}{2k}, \quad C = 0. \tag{13.78}$$

Next, the solution of the sweep equations (13.41) subject to the boundary conditions (13.40) is performed. To obtain the boundary conditions, it is noted that

$$\psi = x_f - 1$$
$$\Omega = 1 + ku^2 + \nu u \tag{13.79}$$

leading to

$$S_f = 0, \quad R_f = 1, \quad m_f = 0, \quad Q_f = 0, \quad n_f = \frac{1}{\sqrt{k}}, \quad \alpha_f = 0. \tag{13.80}$$

Then, the sweep variables become

$$S = 0, \quad R = 1, \quad m = 0, \quad Q = \frac{t - \sqrt{k}}{2k}, \quad n = \frac{1}{\sqrt{k}}, \quad \alpha = 0 \tag{13.81}$$

so that

$$U = [1 \ 0], \quad V = \begin{bmatrix} \frac{t - \sqrt{k}}{2k} & \frac{1}{\sqrt{k}} \\ \frac{1}{\sqrt{k}} & 0 \end{bmatrix} \tag{13.82}$$

and

$$V^{-1} = \begin{bmatrix} 0 & \sqrt{k} \\ \sqrt{k} & -\frac{t - \sqrt{k}}{2} \end{bmatrix}. \tag{13.83}$$

Finally,
$$\hat{S} = S - UV^{-1}U^T = 0 \qquad (13.84)$$
and indicates that no conjugate point exists so that the optimal control (13.71) is a minimum.

The particular form (13.79) of the final conditions shows that Eqs.(13.37) and (13.38) can be solved for $d\nu$ and $dt_f$. The end result is that $\hat{S}_f = 0$ so that the solution of Eq.(13.50) is $\hat{S} = 0$, and it is not necessary to use the sweep variables.

It is interesting to note that the second differential (13.29) written specifically for this problem is given by
$$d^2 J' = \int_{t_0}^{t_f} \delta u^T H_{uu} \delta u \, dt. \qquad (13.85)$$
Hence, $H_{uu} > 0$ directly implies that $d^2 J' > 0$. It is comforting to see that the more complex analysis gives the same result.

## 13.8  Distance Problem

An interesting problem is that of finding the path which minimizes the distance to the parabola shown in Fig. 13.2. Formally, the problem statement is to find

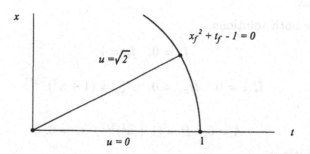

Figure 13.2: Minimum Distance to a Parabola

the control $u(t)$ that minimizes the performance index
$$J = \int_{t_0}^{t_f} \sqrt{1 + u^2} \, dt, \qquad (13.86)$$

subject to the differential constraint

$$\dot{x} = u, \tag{13.87}$$

the prescribed initial conditions

$$t_0 = 0, \quad x_0 = 0, \tag{13.88}$$

and the prescribed final condition

$$x_f^2 + t_f - 1 = 0. \tag{13.89}$$

For this problem,

$$H = \sqrt{1 + u^2} + \lambda u, \quad G = \nu(x_f^2 + t_f - 1), \tag{13.90}$$

and the first differential conditions (13.13) through (13.19) show that two optimal solutions exist:

$$\text{(a)} \quad u = 0, \quad x = 0, \quad t_f = 1, \quad \nu = -1 \tag{13.91}$$

$$\text{(b)} \quad u = \sqrt{2}, \quad x = \sqrt{2}\,t, \quad t_f = \frac{1}{2}, \quad \nu = \frac{-1}{\sqrt{3}}. \tag{13.92}$$

This problem is solved as a parameter optimization problem in Chapters 2 and 3, and it is shown that solution (a) is a local maximum and solution (b) is a local minimum. Here, it is interesting to see what the second differential conditions say.

For both solutions,

$$f_x = 0, \quad f_u = 1$$

$$H_{xx} = 0, \quad H_{xu} = 0, \quad H_{uu} = (1 + u^2)^{-3/2} \tag{13.93}$$

so that

$$A = 0, \quad B = (1 + u^2)^{3/2}, \quad C = 0. \tag{13.94}$$

For both solutions,

$$H_{uu} > 0. \tag{13.95}$$

For solution (a), the sweep variables are obtained from Eqs.(13.40) and (13.41) and are given by

$$S = \frac{2}{1 - 2t}, \quad R = 0, \quad m = 0, \quad Q = 0, \quad n = 1, \quad \alpha = 0. \tag{13.96}$$

Because

$$\hat{S} = \frac{2}{1 - 2t} \qquad (13.97)$$

and blows up at $t = 1/2$, there is a conjugate point within the interval of integration. Hence, solution (a) is not a minimum.

For solution (b), the sweep variables are given by

$$S = \frac{-1}{\sqrt{3}(3t-1)}, \quad R = \frac{1}{\sqrt{2}(3t-1)}, \quad m = \frac{-\sqrt{2}}{\sqrt{3}(3t-1)},$$

$$Q = \frac{3\sqrt{3}(2t-1)}{2(3t-1)}, \quad n = \frac{3t}{3t-1}, \quad \alpha = \frac{-2}{\sqrt{3}(3t-1)}. \qquad (13.98)$$

Since the sweep variables blow up at $t = 1/3$, it is not possible to determine the existence of a conjugate point with them alone. It is possible to calculate $\hat{S}$ to the right of the blowup and switch to $\hat{S}$ at that point. However, it is shown in the next paragraph that $\hat{S}_f$ is finite so that $\hat{S}$ can be used over the entire interval of integration.

For both solutions, Eqs.(13.40) give

$$S_f = 2\nu, \quad R_f = 2x_f, \quad m_f = 2\nu u$$

$$Q_f = 0, \quad n_f = 1 + 2x_f u, \quad \alpha_f = 2\nu u^2 \qquad (13.99)$$

so that

$$\hat{S}_f = 2\nu \left[ 1 - \frac{4u x_f (1 + u x_f)}{(1 + 2x_f u)^2} \right]. \qquad (13.100)$$

For solution (a), Eq.(13.100) becomes

$$\hat{S}_f = -2, \qquad (13.101)$$

and the differential equation for $\hat{S}$ gives

$$\hat{S} = \frac{2}{1 - 2t}. \qquad (13.102)$$

Note that $\hat{S}$ becomes infinite at $t = 1/2$ so that a conjugate point exists in the interval of integration. Again, solution (a) is not a minimum.

For solution (b), Eq.(13.96) becomes

$$\hat{S}_f = \frac{-2}{9\sqrt{3}} \qquad (13.103)$$

so that

$$\hat{S} = \frac{-1}{3\sqrt{3}(1+t)}. \tag{13.104}$$

Hence, since $\hat{S}$ is finite over $[t_0, t_f)$, solution (b) is a minimum.

To prove that solution (a) is a maximum, a minus sign must be put in front of the integral sign in the performance index (13.86) and the problem resolved.

## 13.9   Navigation Problem

With time as the variable of integration, the aircraft navigation problem defined in Chapter 1 is to find the velocity direction history $\theta(t)$ that minimizes the performance index

$$J = t_f, \tag{13.105}$$

subject to the differential contraints

$$\begin{aligned} \dot{x}_1 &= V\cos\theta \\ \dot{x}_2 &= V\sin\theta + w, \end{aligned} \tag{13.106}$$

and the prescribed boundary conditions

$$\begin{aligned} t_0 = 0, \quad x_{1_0} = 0, \quad x_{2_0} = 0 \\ x_{1_f} = 1, \quad x_{2_f} = 0 \end{aligned} \tag{13.107}$$

For this problem, the endpoint function and the Hamiltonian are given by

$$\begin{aligned} G &= t_f + \nu_1(x_{1_f} - 1) + \nu_2(x_{2_f}) \\ H &= \lambda_1(V\cos\theta) + \lambda_2(V\sin\theta + w). \end{aligned} \tag{13.108}$$

Application of the Euler–Lagrange equations leads to

$$\begin{aligned} \dot{\lambda}_1 &= -H_{x_1} &= 0 \\ \dot{\lambda}_2 &= -H_{x_2} &= 0 \\ 0 &= H_\theta &= -\lambda_1 V\sin\theta + \lambda_2 V\cos\theta, \end{aligned} \tag{13.109}$$

where the first two equations imply that

$$\lambda_1 = \text{Const}, \quad \lambda_2 = \text{Const}. \tag{13.110}$$

Then, the equation for $\theta$ gives

$$- \lambda_1 \sin \theta + \lambda_2 \cos \theta = 0 \qquad (13.111)$$

which, since $\lambda_1$ and $\lambda_2$ are constant, implies that $\theta$ is constant.

Since the optimal control is a constant, the system equations (13.106) can be integrated subject to the prescribed initial conditions to obtain

$$\begin{aligned} x_1 &= V \cos \theta \, t \\ x_2 &= (V \sin \theta + w)t. \end{aligned} \qquad (13.112)$$

Use of the prescribed final conditions $x_{1_f} = 1$ and $x_{2_f} = 0$ leads to

$$t_f = \frac{1}{V \cos \theta}, \quad \sin \theta = -\frac{w}{V}. \qquad (13.113)$$

Then, from the triangle in Fig. 13.3, it is seen that

$$\cos \theta = \frac{\sqrt{V^2 - w^2}}{V}. \qquad (13.114)$$

Hence, the optimal control and the final time can be written as

$$\theta = -\arcsin(w/V), \quad t_f = \frac{1}{\sqrt{V^2 - w^2}}. \qquad (13.115)$$

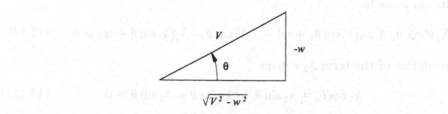

Figure 13.3: Control Triangle

The natural boundary conditions for this problem are given by

$$H_f = -1, \quad \lambda_{1_f} = \nu_1, \quad \lambda_{2_f} = \nu_2, \qquad (13.116)$$

where the last two equations determine the values of $\nu_1$ and $\nu_2$. The first condition combined with the optimal control leads to

$$\lambda_1 = -\frac{1}{\sqrt{V^2 - w^2}}. \tag{13.117}$$

The remaining multiplier comes from Eq.(13.111) and is given by

$$\lambda_2 = \frac{w}{V^2 - w^2}. \tag{13.118}$$

A summary of the first differential results is as follows:

$$
\begin{aligned}
\theta &= -\arcsin(w/V) & t_0 &= 0 & t_f &= (V^2 - w^2)^{-1/2} \\
x_1 &= \sqrt{V^2 - w^2}\,t & x_{1_0} &= 0 & x_{1_f} &= 1 \\
x_2 &= 0 & x_{2_0} &= 0 & x_{2_f} &= 0 \\
\lambda_1 &= -(V^2 - w^2)^{-1/2} & &= \lambda_{1_0} & &= \lambda_{1_f} \\
\lambda_2 &= w/(V^2 - w^2) & &= \lambda_{2_0} & &= \lambda_{2_f}.
\end{aligned}
\tag{13.119}
$$

Note that the aircraft is pointing upwind but that it is moving directly toward the final point. An interesting check of the results can be made for the case where $w \ll V$. Here, $\theta \cong 0$, $x_1 \cong Vt$, and $x_2 \cong 0$, which seem reasonable.

To test the minimality of the solution, application of the Weierstrass condition leads to

$$\lambda_1 V \cos\theta_* + \lambda_2(V \sin\theta_* + w) - \lambda_1 V \cos\theta - \lambda_2(V \sin\theta + w) > 0. \tag{13.120}$$

Cancellation of the term $\lambda_2 w$ gives

$$\lambda_1 \cos\theta_* + \lambda_2 \sin\theta_* - \lambda_1 \cos\theta - \lambda_2 \sin\theta > 0 \tag{13.121}$$

since $V \neq 0$. Next, the optimality condition (13.111) or

$$\lambda_2 = \lambda_1 \tan\theta \tag{13.122}$$

can be used to eliminate $\lambda_2$ so that, after some manipulation, the Weierstrass condition becomes

$$\frac{\lambda_1[\cos(\theta - \theta_*) - 1]}{\cos\theta} > 0. \tag{13.123}$$

Finally, since $\lambda_1 < 0$, $\cos\theta > 0$ and $-1 \le \cos(\theta - \theta_*) \le 1$, the Weierstrass condition is satisfied.

The application of the Legendre–Clebsch condition to this problem starts with

$$H_{\theta\theta} = -\lambda_1 V \cos\theta - \lambda_2 V \sin\theta. \tag{13.124}$$

Then, if the values for $\lambda_1$, $\lambda_2$, and $\theta$ from Eq.(13.119) are substituted,

$$H_{\theta\theta} = \frac{V^2}{V^2 - w^2} > 0. \tag{13.125}$$

It is positive as long as the wind speed $w$ is less than the aircraft speed $V$, as it should be.

As an aside, consider a different approach for obtaining the optimal control. If Eq.(13.111) is solved for the control as

$$\tan\theta = \frac{\lambda_2}{\lambda_1}, \tag{13.126}$$

the triangle shown in Fig. 13.4 can be solved for $\sin\theta$ and $\cos\theta$ as

$$\sin\theta = \frac{\lambda_2}{\sqrt{\lambda_1^2 + \lambda_2^2}}, \quad \cos\theta = \frac{\lambda_1}{\sqrt{\lambda_1^2 + \lambda_2^2}}. \tag{13.127}$$

Substitution of these expressions into (13.124) leads to

$$H_{\theta\theta} = -V\sqrt{\lambda_1^2 + \lambda_2^2} \tag{13.128}$$

so that the Legendre–Clebsch condition is not satisfied. In trying to find out

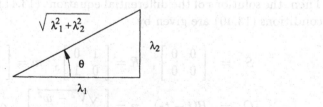

Figure 13.4: Control Triangle

what is wrong, it is noted that Eq.(13.126) can also be written as

$$\tan\theta = \frac{-\lambda_2}{-\lambda_1} \tag{13.129}$$

and gives

$$\sin\theta = \frac{-\lambda_2}{\sqrt{\lambda_1^2 + \lambda_2^2}}, \quad \cos\theta = \frac{-\lambda_1}{\sqrt{\lambda_1^2 + \lambda_2^2}}. \tag{13.130}$$

Then, the Legendre–Clebsch condition becomes

$$H_{\theta\theta} = V\sqrt{\lambda_1^2 + \lambda_2^2} \tag{13.131}$$

and is satisfied. Hence, the correct sine and cosine of the optimal control in terms of the Lagrange multipliers are given by Eq.(13.130).

To check the conjugate point condition, it is seen that

$$f_x = \begin{bmatrix} 0 & 0 \\ 0 & 0 \end{bmatrix}, \quad f_\theta = \begin{bmatrix} w \\ \sqrt{V^2 - w^2} \end{bmatrix}$$

$$\tag{13.132}$$

$$H_{xx} = \begin{bmatrix} 0 & 0 \\ 0 & 0 \end{bmatrix}, \quad H_{x\theta} = \begin{bmatrix} 0 \\ 0 \end{bmatrix}, \quad H_{\theta\theta} = \frac{V^2}{V^2 - w^2}$$

so that

$$A = \begin{bmatrix} 0 & 0 \\ 0 & 0 \end{bmatrix}, \quad C = \begin{bmatrix} 0 & 0 \\ 0 & 0 \end{bmatrix}$$

$$\tag{13.133}$$

$$B = \frac{V^2 - w^2}{V^2} \begin{bmatrix} w^2 & w\sqrt{V^2 - w^2} \\ w\sqrt{V^2 - w^2} & V^2 - w^2 \end{bmatrix}.$$

Then, the solutions of the differential equations (13.41) subject to the boundary conditions (13.40) are given by

$$S = \begin{bmatrix} 0 & 0 \\ 0 & 0 \end{bmatrix}, \quad R = \begin{bmatrix} 1 & 0 \\ 0 & 1 \end{bmatrix}, \quad m = \begin{bmatrix} 0 \\ 0 \end{bmatrix},$$

$$\tag{13.134}$$

$$Q = B(t - t_f), \quad n = \begin{bmatrix} \sqrt{V^2 - w^2} \\ 0 \end{bmatrix}, \quad \alpha = 0.$$

Note that det $Q = 0$ and $\alpha = 0$ for this problem.

From the definitions (13.44) of $U$ and $V$, it is seen that

$$U = \begin{bmatrix} 1 & 0 & 0 \\ 0 & 1 & 0 \end{bmatrix}, \quad V = \begin{bmatrix} Q_{11} & Q_{12} & n_1 \\ Q_{21} & Q_{22} & 0 \\ n_1 & 0 & 0 \end{bmatrix}, \tag{13.135}$$

where the components of $V$ are the nonzero elements of $Q$ and $n$. Next, it can be shown that

$$V^{-1} = \frac{V^2}{(V^2 - w^2)^3(t_f - t)} \begin{bmatrix} 0 & 0 & -n_1 Q_{22} \\ 0 & -n_1^2 & n_1 Q_{21} \\ -n_1 Q_{22} & n_1 Q_{12} & Q_{11} Q_{22} - Q_{12} Q_{21} \end{bmatrix} \tag{13.136}$$

so that

$$\hat{S} = \frac{V^2}{(V^2 - w^2)^2(t_f - t)} \begin{bmatrix} 0 & 0 \\ 0 & 1 \end{bmatrix}. \tag{13.137}$$

Since $H_{\theta\theta} > 0$ and $\hat{S}$ is finite (no conjugate point), the optimal control is a minimum.

Note that it is possible to form $\hat{S}$ from the sweep variables and use it throughout the interval of integration.

# Problems

13.1 Derive the Weierstrass condition for the free final time problem following the approach of Chapter 10.

13.2 Find the control $u(t)$ that minimizes the performance index

$$J = \frac{t_f^2}{2} + \int_{t_0}^{t_f} \frac{u^2}{2} dt$$

for the dynamical system

$$\dot{x} = u$$

and the prescribed boundary conditions

$$t_0 = 0, \quad x_0 = 0, \quad x_f = 1.$$

Show that the optimal control $u = \sqrt[3]{2}$ is a relative minimum. Also, show that it satisfies the Weierstrass condition for a strong relative minimum.

13.3  Consider the problem of minimizing the performance index

$$J = \int_{t_0}^{t_f} (u - x)^2 dt,$$

subject to the differential constraint

$$\dot{x} = u$$

and the prescribed boundary conditions

$$t_0 = 0, \quad x_0 = 1, \quad x_f = 2.$$

Find the optimal control, and show that it is a relative minimum.

13.4  Formulate the problem of Section 13.8 as a maximization problem and show that the solution $u = 0$ is a maximum.

13.5  Consider the problem of minimizing the distance from the origin to the straight line $x_f = 1 - t_f$. Applying both the first and second differential conditions, show that the optimal control $u = 1$ is a minimum.

13.6  (a) Solve the aircraft navigation problem (Section 13.9) for the case where the aircraft is only required to reach the line $x_{1_f} = 1$. Show that the optimal control is $\theta = 0$, and derive expressions for the states and Lagrange multipliers. Note that the aircraft is to be pointed directly at the line at all times. (b) Apply the Weierstrass condition and the Legendre–Clebsch condition. (c) Finally, apply the sufficient conditions of Section 13.6, and show that the optimal control is a minimum. In other words, show that $H_{\theta\theta} > 0$ and

$$\hat{S} = \begin{bmatrix} 0 & 0 \\ 0 & 0 \end{bmatrix},$$

meaning that there is no conjugate point. Note that $Q = 0$ for this problem.

13.7  Find the control history $u(t)$ that minimizes the distance from a point $(t_0, 0)$ on the positive $t$-axis to the circle $t_f^2 + x_f^2 = r^2$, where $0 \le t_0 \le r$. Show that the minimal control is $u = 0$ for all $t_0 > 0$ and that a conjugate point exists at $t_0 = 0$. Explain what is different for the cases $t_0 > 0$ and $t_0 = 0$.

13.8 Consider the problem of minimizing the distance from a point on the positive x-axis to the ellipse $(x_f/a)^2 + (y_f/b)^2 = 1$ where $a \geq b$ and $0 \leq x_0 \leq a$. At what value of $x_0$ does the minimal control switch from $u = 0$ to

$$u = \left[\frac{(a^2 - b^2)^2 - a^2 x_0^2}{b^2 x_0^2}\right]^{1/2}?$$

13.9 A form of the brachistochrone problem is that of minimizing the time $(J = t_f)$ to drive the system

$$\dot{x} = v \cos u, \quad \dot{y} = v \sin u, \quad \dot{v} = g \sin u$$

from the initial state

$$t_0 = 0, \quad x_0 = 0, \quad y_0 = 0, \quad v_0 = 0$$

to the final state

$$x_f = 1.$$

An illustration of the problem is shown in Fig 13.5 where $m$ is the mass of the particle, $g$ is the constant acceleration of gravity, $v$ is the velocity of the particle, and $u$ is the angle between the velocity vector and the $x$-axis. Applying just the first differential conditions, show that the optimal steering law is

$$u = \frac{\pi}{2} - \frac{\sqrt{\pi}}{2}t.$$

What do the Legendre–Clebsch and Weierstrass conditions say about the nature (maximum, minimum) of the optimal control?

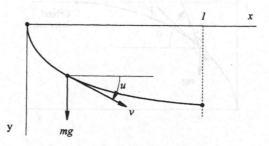

Figure 13.5: Brachistochrone Problem

13.10 A problem that has often been used to test the performance of a numerical optimization technique is the lunar launch problem illustrated in Fig 13.6. Here, the orientation $\theta$ of the thrust vector is to be controlled so that the rocket achieves orbit in minimum time assuming that the thrust acceleration $a = T/m$ of the rocket and the acceleration of gravity $g$ of the moon are constant. Formally, the problem is stated as follows: Find the thrust angle $\theta(t)$ that minimizes the performance index

$$J = t_f,$$

subject to the state differential equations

$$\dot{x} = u, \quad \dot{y} = v, \quad \dot{u} = a\cos\theta, \quad \dot{v} = a\sin\theta - g,$$

the prescribed initial conditions

$$t_0 = 0, \quad x_0 = 0, \quad y_0 = 0, \quad u_0 = 0, \quad v_0 = 0,$$

and the prescribed final conditions

$$y_f = 50,000 \text{ ft}, \quad u_f = 5,444 \text{ ft/s}, \quad v_f = 0 \text{ ft/s}.$$

The quantities $u$ and $v$ denote the horizontal and vertical components of the velocity $V$. Typical values of unspecified parameters are $a = 20.8 \text{ ft/s}^2$ and $g = 5.32 \text{ ft/s}^2$. Using first differential conditions and the Legendre–Clebsch condition, carry out the solution of the problem to the point where a system of nonlinear algebraic equations must be solved numerically.

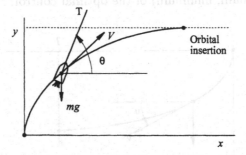

Figure 13.6: Lunar Launch Problem

# 14

# Parameters

## 14.1 Introduction

Here, the optimal control problem is extended to include parameters whose optimal values are to be determined. The simplest example of such a problem is the free final time problem in which the time is normalized by the final time. The final time then appears in the optimal control problem as a parameter. Another example is the case where the dynamical system and its operation are being optimized simultaneously. Consider the minimization of the launch mass of a multistage missile that is to hit a fixed final point in a given time. In addition to the optimization of the trajectory, it is desired to optimize the missile design with respect to parameters such as the stage payload masses, the stage burn times, and so on.

  After the problem is defined and controllability is discussed, first differential and second differential conditions are derived for the optimal control problem involving parameters. As an illustration of the application of the theory, the navigation problem is reformulated as a fixed final time problem and solved.

## 14.2 Problem Formulation

The free final time problem can be converted into a fixed final time problem by normalizing the time by the final time, that is, by introducing the nondimensional time

$$\tau = \frac{t - t_0}{t_f - t_0},$$                                           (14.1)

where

$$\tau_0 = 0, \quad \tau_f = 1.$$                                               (14.2)

Then, the optimal control problem of Chapter 13 can be restated as follows:
Find the control history $u(\tau)$ and the parameter $t_f$ which minimize the performance index

$$J = \phi(x_f, t_f) + \int_{\tau_0}^{\tau_f} (t_f - t_0) L(t_0 + (t_f - t_0)\tau, x, u) d\tau,$$   (14.3)

subject to the system dynamics

$$\frac{dx}{d\tau} = (t_f - t_0) f(t_0 + (t_f - t_0)\tau, x, u),$$               (14.4)

to the prescribed initial conditions

$$\tau_0 = 0, \quad x_0 = x_{0_s},$$                                            (14.5)

and to the prescribed final conditions

$$\tau_f = 1, \quad \beta(x_f, t_f) = 0.$$                                       (14.6)

Hence, while the optimal control problem now has a fixed final time, the quantity $t_f$ appears as a parameter whose optimal value is to be determined.

The fixed final time problem (Chapters 9 and 11) including a parameter can be stated as follows: Find the control history $u(t)$ and parameter $a$ that minimize the performance index

$$J = \phi(x_f, a) + \int_{t_0}^{t_f} L(t, x, u, a) dt,$$                         (14.7)

subject to the system dynamics

$$\dot{x} = f(t, x, u, a),$$                                                    (14.8)

to the prescribed initial conditions

$$t_0 = t_{0_s}, \quad x_0 = x_{0_s},$$                                         (14.9)

and to the prescribed final conditions

$$t_f = t_{f_s}, \quad \beta(x_f, a) = 0. \tag{14.10}$$

where $a$ is a $p \times 1$ vector, $\beta$ is an $s \times 1$ vector, and $0 \le s \le n + r$. Also, these final conditions must be a proper set. An example is

$$\beta \triangleq \begin{bmatrix} g(x_f) \\ h(a) \end{bmatrix} = 0, \tag{14.11}$$

where $g$ is a $p \times 1$ vector, $h$ is a $q \times 1$ vector, $p \le n$, and $q \le r$.

The fixed final time problem with parameters contains the converted free final time problem if $t$ is repaced by $\tau$, $t_0$, by 0, $t_{f_s}$ by 1, and the functions $L$ and $f$ are properly defined.

In taking variations, the variation of a constant equals the differential of the constant (Section 6.7). The convention followed here is that variations of constants such as $a$ are written as differentials.

## 14.3 Controllability

Some discussion of controllability is in order. For the dynamics (14.8) and the final condition (14.10), controllability is concerned with guaranteeing that a solution of the linearized dynamics

$$\frac{d}{dt}\delta x = f_x \delta x + f_u \delta u + f_a da \tag{14.12}$$

starting from

$$t_q = t_{q_s}, \quad \delta x_q = \delta x_{q_s} \tag{14.13}$$

satisfies the linearized final conditions

$$t_f = t_{f_s}, \quad \beta_{x_f}\delta x_f + \beta_a da = 0. \tag{14.14}$$

It is observed that the parameter $a$ can be converted into a state by adding the state equation $\dot{a} = 0$ or linearized state equation $dda/dt = 0$. Then, since the state perturbation at $t_q$ is prescribed [see Eq. (14.13)], the parameter perturbation $da$ is fixed. Hence, controllability for this problem requires that $Q^{-1}$ as defined in Section 7.4 must exist.

## 14.4    First Differential Conditions

In the following development, the functional dependence of the variables is stated with the parameter $a$ written last. This is done to make the relationship of this problem to the fixed final time problem clear. In other words, if there is no $a$, the equations reduce to those of Chapters 9 and 11.

Adjoining the constraints to the performance index by Lagrange multipliers leads to the augmented performance index

$$J' = G(x_f, \nu, a) + \int_{t_0}^{t_f} [H(t, x, u, \lambda, a) - \lambda^T \dot{x}] dt, \qquad (14.15)$$

where the endpoint function and the Hamiltonian are defined as

$$G = \phi + \nu^T \beta, \quad H = L + \lambda^T f \qquad (14.16)$$

Then, taking the differential of $J'$ and performing the integration by parts gives

$$dJ' = (G_{x_f} - \lambda_f^T)\delta x_f + \beta^T d\nu + \left(G_a + \int_{t_0}^{t_f} H_a dt\right) da + \lambda_0 \delta x_0$$
$$+ \int_{t_0}^{t_f} [(H_x + \dot{\lambda}^T)\delta x + H_u \delta u + (f^T - \dot{x}^T)\delta\lambda] dt, \qquad (14.17)$$

where $\beta, \delta x_0$, and $f^T - \dot{x}^T$ are all zero.

Since the parameter $a$ and control $u$ are independent, consider the class of admissible comparison paths where there is no change in the parameter ($da = 0$). The problem reduces to the fixed final time problem (Chapter 9) for which $H_x + \dot{\lambda}^T = 0, H_u^T = 0,$ and $G_{x_f} - \lambda_f^T = 0$. Then, for admissible comparison paths where $da \neq 0$, the first differential becomes

$$dJ' = \left(G_a + \int_{t_0}^{t_f} H_a dt\right) da. \qquad (14.18)$$

At this point, $\nu$s are chosen so that the coefficients of the dependent $da$s vanish. Then, the coefficients of the independent $da$s must vanish because they can be plus or minus and the first differential must vanish. The final result is that

$$G_a + \int_{t_0}^{t_f} H_a dt = 0. \qquad (14.19)$$

In summary, the first differential necessary conditions for a minimum are given by

$$\dot{x} = f$$

$$\dot{\lambda} = -H_x^T$$

$$0 = H_u$$

$$G_a + \int_{t_0}^{t_f} H_a dt = 0 \qquad (14.20)$$

$$t_0 = t_{0_s}, \quad x_0 = x_{0_s}$$

$$t_f = t_{f_s}, \quad \beta = 0$$

$$\lambda_f = G_{x_f}^T.$$

If the optimal control problem does not explicitly contain the time, the first integral

$$H = \text{Const} \qquad (14.21)$$

exists. Also, with the realization that the parameters behave like states, the Weierstrass condition

$$H(t, x, u_*, \lambda, a) - H(t, x, u, \lambda, a) \geq 0 \qquad (14.22)$$

and the Legendre–Clebsch condition

$$H_{uu}(t, x, u, \lambda, a) \geq 0 \qquad (14.23)$$

must hold.

## 14.5  Second Differential Conditions

Taking the differential of the transpose of the first differential (14.17), integrating by parts, and accounting for Eqs.(14.12) and (14.13) leads to the second differential

$$d^2 J' = [\delta x_f^T da] \begin{bmatrix} G_{x_f x_f} & G_{x_f a} \\ G_{a x_f} & G_{aa} \end{bmatrix} \begin{bmatrix} \delta x_f \\ da \end{bmatrix}$$

$$+ \int_{t_0}^{t_f} [\delta x^T \, \delta u^T \, da^T] \begin{bmatrix} H_{xx} & H_{xu} & H_{xa} \\ H_{ux} & H_{uu} & H_{ua} \\ H_{ax} & H_{au} & H_{aa} \end{bmatrix} \begin{bmatrix} \delta x \\ \delta u \\ da \end{bmatrix} dt. \qquad (14.24)$$

To write the second differential as a perfect square, the concept of the neighboring optimal path is employed. First, the fourth of Eqs.(14.20) is converted into a differential equation and boundary conditions. By introducing the function $\mu(t)$ defined by

$$\dot{\mu} = -H_a^T, \quad \mu_0 = 0, \tag{14.25}$$

it becomes

$$\mu_f = G_a^T. \tag{14.26}$$

Then, the complete set of equations defing the optimal path is given by

$$
\begin{aligned}
\dot{x} &= f(t, x, u, a) \\
\dot{\lambda} &= -H_x^T(t, x, u, \lambda, a) \\
\dot{\mu} &= -H_a^T(t, x, u, \lambda, a) \\
0 &= H_u^T(t, x, u, \lambda, a) \\
t_0 &= t_{0_s}, \quad x_0 = x_{0_s}, \quad \mu_0 = 0 \\
t_f &= t_{f_s}, \quad \beta(x_f, a) = 0 \\
\lambda_f &= G_{x_f}^T(x_f, \nu, a), \quad \mu_f = G_a(x_f, \nu, a).
\end{aligned}
\tag{14.27}
$$

The equations governing the neighboring optimal path from the perturbed initial point to the perturbed final constraint manifold are obtained by taking the variation of the path equations. For the class of problems where $H_{uu} > 0$, the varied $H_u^T$ equation can be solved for $\delta u$ as

$$\delta u = -H_{uu}^{-1}(H_{ux}\delta x + f_u^T\delta\lambda + H_{ua}da). \tag{14.28}$$

Then, the varied equations for $x$, $\lambda$ and $\mu$ can be written as

$$
\begin{aligned}
\frac{d}{dt}\delta x &= \phantom{-}A\delta x - B\delta\lambda + Dda \\
\frac{d}{dt}\delta\lambda &= -C\delta x - A^T\delta\lambda - Eda \\
\frac{d}{dt}\delta\mu &= -E^T\delta x - D^T\delta\lambda - Fda,
\end{aligned}
\tag{14.29}
$$

where

$$
\begin{aligned}
A &= f_x - f_u H_{uu}^{-1} H_{ux}, & B &= f_u H_{uu}^{-1} f_u^T \\
C &= H_{xx} - H_{xu} H_{uu}^{-1} H_{ux}, & D &= f_a - f_u H_{uu}^{-1} H_{ua} \\
E &= H_{xa} - H_{xu} H_{uu}^{-1} H_{ua}, & F &= H_{aa} - H_{au} H_{uu}^{-1} H_{ua}.
\end{aligned}
\tag{14.30}
$$

Next, taking the differential of the final conditions and converting to variations leads to

$$
\begin{aligned}
\delta\lambda_f &= G_{x_f x_f}\delta x_f + \beta_{x_f}^T \delta\nu + G_{x_f a}da \\
\delta\beta &= \beta_{x_f}\delta x_f + \beta_a da \\
\delta\mu_f &= G_{ax_f}\delta x_f + \beta_a^T d\nu + G_{aa}da,
\end{aligned}
\tag{14.31}
$$

where $\delta\beta = 0$. These equations motivate the definition of the sweep variables; in other words, it is assumed that

$$
\delta\lambda = S\delta x + Rd\nu + mda \tag{14.32}
$$

$$
0 = R^T\delta x + Qd\nu + nda \tag{14.33}
$$

$$
\delta\mu = m^T\delta x + n^T d\nu + \alpha da, \tag{14.34}
$$

where the sweep variables have the same names as those in previous chapters, but they are not the same variables. Substitution of Eqs.(14.32) through (14.34) into Eqs.(14.29) leads to the following differential equations for the sweep variables:

$$
\begin{aligned}
\dot{S} &= -C - A^T S - SA + SBS \\
\dot{R} &= (SB - A^T)R \\
\dot{m} &= (SB - A^T)m - SD - E \\
\dot{Q} &= R^T BR \\
\dot{n} &= R^T(Bm - D) \\
\dot{\alpha} &= m^T(Bm - D) - D^T m - F.
\end{aligned}
\tag{14.35}
$$

Finally, evaluation of Eqs.(14.32) through (14.34) at the final point and comparison with Eqs.(14.31) gives the following final conditions for the sweep variables:

$$
\begin{aligned}
S_f &= G_{x_f x_f}, & R_f &= \beta_{x_f}^T, & m_f &= G_{x_f a} \\
Q_f &= 0, & n_f &= \beta_a, & \alpha_f &= G_{aa}.
\end{aligned}
\tag{14.36}
$$

Equations (14.33) and (14.34) applied at the initial point can be solved simultaneously for $d\nu$ and $da$, that is,

$$
\begin{bmatrix} d\nu \\ da \end{bmatrix} = -V_0^{-1}U_0^T\delta x_0,
\tag{14.37}
$$

where $\delta\mu_0 = 0$ and

$$
U = [R\ m], \qquad V = \begin{bmatrix} Q & n \\ n^T & \alpha \end{bmatrix}.
\tag{14.38}
$$

Finally, $\delta\lambda_0$ is obtained from Eq.(14.32) as

$$\delta\lambda_0 = (S_0 - U_0 V_0^{-1} U_0^T)\delta x_0. \tag{14.39}$$

If $\delta x_0$ is known, $\delta\lambda_0$ and $da$ are obtained from Eqs.(14.39) and (14.37). Then, $\delta x, \delta\lambda$, and $\delta\mu$, where $\delta\mu_0 = 0$, follow from Eq.(14.29), and Eq.(14.28) gives $\delta u$. These results determine the neighboring optimal path from the perturbed initial point to the perturbed final constraint manifold.

At this point, the control perturbation (14.28) is combined with Eq. (14.32) to obtain

$$\delta u = -H_{uu}^{-1}[(H_{ux} + f_u^T S)\delta x + f_u^T R d\nu + (H_{ua} + f_u^T m)da]. \tag{14.40}$$

Then, if F is defined as

$$F = H_{uu}^{-1}[(H_{ux} + f_u^T S)\delta x + f_u^T R d\nu + (H_{ua} + f_u^T m)da] + \delta u, \tag{14.41}$$

it can be shown with considerable manipulation that the integrand of the second differential (14.24) can be written as

$$F^T H_{uu} F - \frac{d}{dt}\left\{ [\delta x^T \, d\nu^T \, da^T] \begin{bmatrix} S & R & m \\ R^T & Q & n \\ m^T & n^T & \alpha \end{bmatrix} \begin{bmatrix} \delta x \\ d\nu \\ da \end{bmatrix} \right\}. \tag{14.42}$$

As a result, the second differential combined with Eqs.(14.42) and (14.37) becomes

$$\delta^2 J' = \delta x_0^T \hat{S}_0 \delta x_0 + \int_{t_0}^{t_f} F^T H_{uu} F dt, \tag{14.43}$$

where

$$\hat{S} = S - UV^{-1}U^T. \tag{14.44}$$

All of the comments made in Section 13.6 relative to $\bar{S}$ are valid here relative to $\hat{S}$. Hence, for the class of nonsingular problems in which $H_{uu} > 0$, the sufficient condition for a minimum is

$$\hat{S} \text{ finite}, \quad t_0 \geq t < t_f. \tag{14.45}$$

If the conjugate point is defined as the time $t_{cp}$ where $\hat{S}$ becomes infinite, the sufficient condition is equivalent to no conjugate point or to $t_{cp} < t_0$.

## 14.6 Navigation Problem

To illustrate the solution of a free final time problem as a fixed final time problem, the navigation problem defined in Section 13.9 is reconsidered. If the transformation $\tau = t/t_f$ is introduced, the optimal control problem is to find the control $\theta(\tau)$ and parameter $t_f$ that minimize the performance index

$$J = t_f, \tag{14.46}$$

subject to the differential contraints

$$\begin{aligned}
\frac{dx_1}{d\tau} &= t_f V \cos\theta \\
\frac{dx_2}{d\tau} &= t_f(V \sin\theta + w)
\end{aligned} \tag{14.47}$$

and the prescribed boundary conditions

$$\begin{aligned}
\tau_0 &= 0, \quad x_{1_0} = 0, \quad x_{2_0} = 0 \\
\tau_f &= 1, \quad x_{1_f} = 1, \quad x_{2_f} = 0.
\end{aligned} \tag{14.48}$$

In applying the results of Sections 14.4 and 14.5, $t$ is $\tau$ and $a$ is $t_f$, a scalar.

Since

$$\begin{aligned}
G &= t_f + \nu_1(x_{1_f} - 1) + \nu_2 x_{2_f} \\
H &= \lambda_1 t_f V \cos\theta + \lambda_2 t_f(V \sin\theta + w),
\end{aligned} \tag{14.49}$$

the first differential conditions (14.20) become

$$\begin{aligned}
&\frac{dx_1}{d\tau} = t_f V \cos\theta \\
&\frac{dx_2}{d\tau} = t_f(V \sin\theta + w) \\
&\frac{d\lambda_1}{d\tau} = 0 \\
&\frac{d\lambda_2}{d\tau} = 0 \\
&0 = -\lambda_1 t_f V \sin\theta + \lambda_2 t_f V \cos\theta \\
&1 + \int_0^1 [\lambda_1 V \cos\theta + \lambda_2(V \sin\theta + w)]d\tau = 0 \\
&\tau_0 = 0, \quad x_{1_0} = 0, \quad x_{2_0} = 0 \\
&\tau_f = 1, \quad x_{1_f} = 1, \quad x_{2_f} = 0 \\
&\lambda_{1_f} = \nu_1, \quad \lambda_{2_f} = \nu_2.
\end{aligned} \tag{14.50}$$

Since $\lambda_1$ and $\lambda_2$ are constant, $\theta$ is constant. Then, integration of the state equations and application of the initial and final conditions leads to

$$\sin\theta = -\frac{w}{V}, \quad \cos\theta = \frac{\sqrt{V^2 - w^2}}{V} \tag{14.51}$$

and

$$t_f = \frac{1}{\sqrt{V^2 - w^2}}. \tag{14.52}$$

Finally, the Lagrange multipliers are given by

$$\lambda_1 = -\frac{1}{\sqrt{V^2 - w^2}}, \quad \lambda_2 = \frac{w}{V^2 - w^2}. \tag{14.53}$$

To apply the second differential conditions, it is noted that along the optimal path

$$f_x = \begin{bmatrix} 0 & 0 \\ 0 & 0 \end{bmatrix}, \quad f_\theta = \begin{bmatrix} \frac{w}{\sqrt{V^2 - w^2}} \\ 1 \end{bmatrix}, \quad f_{t_f} = \begin{bmatrix} \sqrt{V^2 - w^2} \\ 0 \end{bmatrix}$$

$$H_{xx} = \begin{bmatrix} 0 & 0 \\ 0 & 0 \end{bmatrix}, \quad H_{x\theta} = \begin{bmatrix} 0 \\ 0 \end{bmatrix}, \quad H_{\theta\theta} = \frac{V^2}{(V^2 - w^2)^{3/2}}$$

$$\tag{14.54}$$

$$H_{xt_f} = \begin{bmatrix} 0 \\ 0 \end{bmatrix}, \quad H_{\theta t_f} = 0, \quad H_{t_f t_f} = 0$$

$$G_{x_f x_f} = \begin{bmatrix} 0 & 0 \\ 0 & 0 \end{bmatrix}, \quad G_{x_f t_f} = \begin{bmatrix} 0 \\ 0 \end{bmatrix}, \quad G_{t_f t_f} = 0.$$

Since $H_{\theta\theta} > 0$, the results of Section 14.5 apply. From Eqs.(14.30), it is seen that

$$A = \begin{bmatrix} 0 & 0 \\ 0 & 0 \end{bmatrix}, \quad C = \begin{bmatrix} 0 & 0 \\ 0 & 0 \end{bmatrix}$$

$$B = \frac{\sqrt{V^2 - w^2}}{V^2} \begin{bmatrix} w^2 & w\sqrt{V^2 - w^2} \\ w\sqrt{V^2 - w^2} & V^2 - w^2 \end{bmatrix} \tag{14.55}$$

$$D = \begin{bmatrix} \sqrt{V^2 - w^2} \\ 0 \end{bmatrix}, \quad E = \begin{bmatrix} 0 \\ 0 \end{bmatrix}, \quad F = 0.$$

Then, the solutions of the differential equations (14.35) subject to the boundary conditions (14.36) are given by

$$S = \begin{bmatrix} 0 & 0 \\ 0 & 0 \end{bmatrix}, \quad R = \begin{bmatrix} 1 & 0 \\ 0 & 1 \end{bmatrix}, \quad m = \begin{bmatrix} 0 \\ 0 \end{bmatrix},$$

$$Q = B(\tau - 1), \quad n = \begin{bmatrix} \sqrt{v^2 - w^2} \\ 0 \end{bmatrix} (1 - \tau), \quad \alpha = 0. \tag{14.56}$$

To evaluate $\hat{S}$, Eq.(14.44), it is necessary to form

$$U = \begin{bmatrix} 1 & 0 & 0 \\ 0 & 1 & 0 \end{bmatrix}, \quad V = \begin{bmatrix} Q_{11} & Q_{12} & n_1 \\ Q_{21} & Q_{22} & 0 \\ n_1 & 0 & 0 \end{bmatrix} \tag{14.57}$$

and obtain

$$V^{-1} = \frac{1}{B_{22}\sqrt{V^2 - w^2}(1 - \tau)} \begin{bmatrix} 0 & 0 & B_{22} \\ 0 & -\sqrt{V^2 - w^2} & -B_{21} \\ B_{22} & -B_{12} & 0 \end{bmatrix}. \tag{14.58}$$

Finally, $\hat{S}$ is given by

$$\hat{S} = \frac{1}{B_{22}(1 - \tau)} \begin{bmatrix} 0 & 0 \\ 0 & 1 \end{bmatrix} \tag{14.59}$$

and is finite over $0 \le \tau < 1$. Hence, the optimal control is a minimum. These results agree with those of Chapter 13, as they should.

# Problems

14.1 Derive the Weierstrass condition for the fixed final time problem with parameters following the approach of Chapter 10.

14.2 Formulate and solve the navigation problem as a fixed final time problem for the case where no final condition is imposed on $x_2$.

# 15

# Free Initial Time and States

## 15.1 Introduction

The next logical extension of the optimal control problem could be to the case where the control is discontinuous. However, the development of the second differential requires notions that are related to problems with free initial states and time. Hence, this problem is considered first; the optimal control is still assumed to be continuous.

## 15.2 Problem Statement

The optimal control problem under consideration is to find the control $u(t)$ that minimizes the performance index

$$J = \phi(t_f, x_f) + \int_{t_0}^{t_f} L(t, x, u)dt, \tag{15.1}$$

subject to the system dynamics

$$\dot{x} = f(t, x, u), \tag{15.2}$$

the prescribed initial conditions

$$\theta(t_0, x_0) = 0, \tag{15.3}$$

and the prescribed final conditions

$$\psi(t_f, x_f) = 0. \tag{15.4}$$

Recall that $\phi$ and $L$ are scalars, $x$ and $f$ are $n \times 1$ vectors, $u$ is an $m \times 1$ vector, and $\psi$ is a $(p+1) \times 1$ vector. The dimension of the initial point constraint is taken to be $(q+1) \times 1$, where $q \le n$. This problem is illustrated in Fig. 15.1.

The augmented performance index for this problem is given by

$$J' = G(t_f, x_f, \nu, t_0, x_0, \xi) + \int_{t_0}^{t_f} [H(t, x, u, \lambda) - \lambda^T \dot{x}] dt, \tag{15.5}$$

where the endpoint function and the Hamiltonian are defined as

$$\begin{aligned} G &= \phi(t_f, x_f) + \nu^T \psi(t_f, x_f) + \xi^T \theta(t_0, x_0) \\ H &= L(t, x, u) + \lambda^T f(t, x, u). \end{aligned} \tag{15.6}$$

The $(q+1) \times 1$ vector $\xi$ is the constant Lagrange multiplier associated with the initial point constraints.

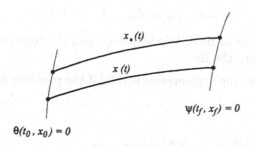

$\theta(t_0, x_0) = 0$

Figure 15.1: Free Initial States and Time

## 15.3 First Differential Conditions

Taking the differential of the augmented performance index (15.5), performing the integration by parts, and relating $\delta$ to $d$ outside the integral leads to the following expression for the first differential:

$$dJ' = \quad (G_{t_f} + H_f)dt_f + (G_{x_f} - \lambda_f^T)dx_f + \psi d\nu$$

$$+ \quad (G_{t_0} - H_0)dt_0 + (G_{x_0} + \lambda_0^T)dx_0 + \theta d\xi \qquad (15.7)$$

$$+ \quad \int_{t_0}^{t_f} [(H_x + \dot{\lambda}^T)\delta x + (H_u)\delta u + (f^T - \dot{x}^T)\delta \lambda]dt.$$

First, the coefficients of $d\nu, d\xi$, and $\delta\lambda$ vanish. Second, the coefficients of the $dt_f, dx_f, dt_0$, and $dx_0$ are made zero by the choice of $\lambda_f, \lambda_0, \nu$, and $\xi$. Third, the class of admissible comparison paths for which $dt_0$ and $dx_0$ vanish is considered. This reduces the problem to the free final time problem of Chapter 13 and leads to $H_u = 0$. As a consequence, the first differential conditions are given by the equations

$$\dot{x} = f(t, x, u)$$

$$\dot{\lambda} = -H_x^T(t, x, u, \lambda) \qquad (15.8)$$

$$0 = H_u^T(t, x, u, \lambda)$$

and the boundary conditions

$$\psi = 0, \quad H_f = -G_{t_f}, \quad \lambda_f = G_{x_f}^T$$

$$\theta = 0, \quad H_0 = G_{t_0}, \quad \lambda_0 = -G_{x_0}^T. \qquad (15.9)$$

Should the initial time or final time be prescribed, the corresponding condition on $H$ is dropped from the list.

Once again, the first integral is valid if the problem does not explicitly contain the time.

## 15.4  Tests for a Minimum

Both the Weierstrass condition,

$$H(t, x, u_*, \lambda) - H(t, x, u, \lambda) > 0 \qquad (15.10)$$

for all $u_* \neq u$ at every point of the minimal path, and the Legendre–Clebsch condition,

$$H_{uu}(t, x, u, \lambda) \geq 0 \qquad (15.11)$$

at every point of a minimal path, are local conditions. Hence, they must be satisfied at each point of the optimal path for it to be a minimum.

## 15.5   Second Differential Conditions

In the development of the second differential, it is convenient to express $dx_f$ in terms of $\delta x_f$ in the first differential (15.7) to obtain

$$
\begin{aligned}
dJ' = \quad & \Omega dt_f + (G_{x_f} - \lambda_f^T)\delta x_f + \psi d\nu \\
+ \quad & \Gamma dt_0 + (G_{x_0} + \lambda_0^T)\delta x_0 + \theta d\xi \\
+ \quad & \int_{t_0}^{t_f} [(H_x + \dot{\lambda}^T)\delta x + (H_u)\delta u + (f^T - \dot{x}^T)\delta\lambda]dt,
\end{aligned}
\tag{15.12}
$$

where

$$
\begin{aligned}
\Omega(t_f, x_f, u_f, \nu) &= G_{t_f}(t_f, x_f, \nu) + L(t_f, x_f, u_f) \\
&\quad + G_{x_f}(t_f, x_f, \nu)f(t_f, x_f, u_f) \\
\Gamma(t_0, x_0, u_0, \xi) &= G_{t_0}(t_0, x_0, \xi) - L(t_0, x_0, u_0) \\
&\quad - G_{x_0}(t_0, x_0, \xi)f(t_0, x_0, u_0).
\end{aligned}
\tag{15.13}
$$

The development of the second differential for this problem follows that of the free final time problem (Section 13.4). The differential of the transpose of $dJ'$ is taken; integration by parts is performed; and differentials are related to variations. Note that, because the initial conditions and final conditions in $G$ are separated, there are no cross derivatives such as $G_{t_0 t_f}$. The final result is given by

$$
\begin{aligned}
d^2 J' = \quad & [\delta x_f^T \ \ dt_f]
\begin{bmatrix} G_{x_f x_f} & \Omega_{x_f}^T \\ \Omega_{x_f} & \Omega' \end{bmatrix}
\begin{bmatrix} \delta x_f \\ dt_f \end{bmatrix} \\
+ \quad & [\delta x_0^T \ \ dt_0]
\begin{bmatrix} G_{x_0 x_0} & \Gamma_{x_0}^T \\ \Gamma_{x_0} & \Gamma' \end{bmatrix}
\begin{bmatrix} \delta x_0 \\ dt_0 \end{bmatrix} \\
+ \quad & \int_{t_0}^{t_f} [\delta x^T \ \ \delta u^T]
\begin{bmatrix} H_{xx} & H_{xu} \\ H_{ux} & H_{uu} \end{bmatrix}
\begin{bmatrix} \delta x \\ \delta u \end{bmatrix} dt,
\end{aligned}
\tag{15.14}
$$

where

$$
\Gamma' \triangleq \Gamma_{t_0} + \Gamma_{x_0}\dot{x}_0, \quad \Omega' \triangleq \Omega_{t_f} + \Omega_{x_f}\dot{x}_f.
\tag{15.15}
$$

The next step is to introduce the neighboring optimal path relative to the final point as in Sections 13.5 and 13.6, so that the second differential becomes

$$d^2 J' = [\delta x_0^T \quad dt_0] \begin{bmatrix} G_{x_0 x_0} + \hat{S}_0 & \Gamma_{x_0}^T \\ \Gamma_{x_0} & \Gamma' \end{bmatrix} \begin{bmatrix} \delta x_0 \\ dt_0 \end{bmatrix}$$

$$+ \int_{t_0}^{t_f} F^T H_{uu} F \, dt. \tag{15.16}$$

Necessary conditions are obtained by considering particular admissible comparison paths. Two classes of comparison paths are used: that which leaves the initial point unchanged and that which varies the initial point.

For the class of admissible paths for which $\delta x_0 = dt_0 = 0$ (see Fig. 15.2), the second differential reduces to the integral term, that is,

$$d^2 J' = \int_{t_0}^{t_f} F^T H_{uu} F \, dt \tag{15.17}$$

and, since $H_{uu} > 0$, is positive unless a conjugate point exists. If $t_{cp} = t_0$, there exist admissible comparison paths (neighboring optimal paths) for which $d^2 J' = 0$. If $t_0 < t_{cp} < t_f$, the optimal path is not a minimum. Hence, a sufficient condition for a minimum is that $t_0 > t_{cp}$, if a conjugate point exists.

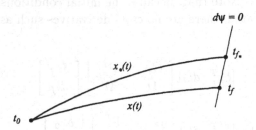

Figure 15.2: Particular Admissible Comparison Path

Next consider the class of admissible comparison paths for which $\delta x_0$ and $dt_0$ are not zero (see Fig. 15.3). If $t_{cp} < t_0$, then $x_*(t)$ can be chosen to be a neighboring optimal path so that $F = 0$ and the second differential becomes

$$d^2 J' = [\delta x_0^T \quad dt_0] \begin{bmatrix} G_{x_0 x_0} + \hat{S}_0 & \Gamma_{x_0}^T \\ \Gamma_{x_0} & \Gamma' \end{bmatrix} \begin{bmatrix} \delta x_0 \\ dt_0 \end{bmatrix}, \tag{15.18}$$

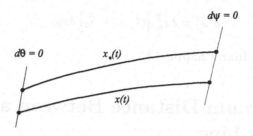

Figure 15.3: Particular Admissible Comparison Path

Since $d\theta = 0$, the changes $\delta x_0$ and $dt_0$ must satisfy the relation

$$\theta_{x_0} \delta x_0 + \dot{\theta}_0 dt_0 = 0 \tag{15.19}$$

where

$$\dot{\theta}_0 \triangleq \theta_{t_0} + \theta_{x_0} \dot{x}_0. \tag{15.20}$$

This means that there are $n+1$ parameters $\delta x_0$ and $dt_0$, $q+1$ constraints (15.19), and $n - q$ independent parameters. Hence, if the dependent variations are eliminated between Eqs. (15.18) and (15.19) and if the independent variations are denoted by $\delta y_0$, the second differential can be written as

$$d^2 J' = \delta y_0^T Y_0 \delta y_0, \tag{15.21}$$

where $Y_0$ is a square matrix of dimension $n - q$. As a consequence, another sufficient condition for a minimum is that

$$Y_0 > 0. \tag{15.22}$$

The discussion presented here is identical with that of Section 5.5 where dependent and independent variables are given separate names and a specific form of $Y_0$ is obtained [see Eq.(5.60)]. While this can be done here, the amount of extra symbology is excessive.

For the class of problems where $H_{uu} > 0$ over the interval of integration, the sufficient conditions for a minimum are the following:

$$\hat{S} \text{ finite}, \quad t_0 < t < t_f, \quad \text{and} \quad Y_0 > 0. \tag{15.23}$$

For a fixed initial time problem, $dt_0 = 0$ so that Eq.(15.18) becomes

$$d^2 J' = \delta x_0^T \left[ G_{x_0 x_0} + \bar{S}_0 \right] \delta x_0, \qquad (15.24)$$

where $\bar{S}$ is obtained from Chapter 13.

# 15.6  Minimum Distance Between a Parabola and a Line

To illustrate the use of the first and second differential conditions, the problem of minimizing the distance between a parabola and a line as shown in Fig. 15.4 is considered. Formally, the problem is to find the control history $u(t)$ that minimizes the performance index

$$J = \int_{t_0}^{t_f} \sqrt{1 + u^2} \, dt, \qquad (15.25)$$

subject to the differential constraint

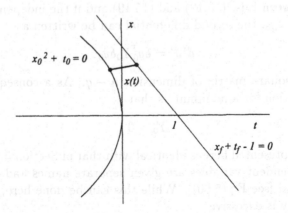

Figure 15.4: Minimum Distance Between a Parabola and a Line

$$\dot{x} = u, \tag{15.26}$$

the prescribed initial condition

$$x_0^2 + t_0 = 0, \tag{15.27}$$

and the prescribed final condition

$$x_f + t_f - 1 = 0. \tag{15.28}$$

Hence, the endpoint function and the Hamiltonian are given by

$$G = \nu(x_f + t_f - 1) + \xi(x_0^2 + t_0) \tag{15.29}$$

$$H = \sqrt{1 + u^2} + \lambda u. \tag{15.30}$$

The Euler–Lagrange equations (15.8) indicate that $\lambda$ and $u$ are constants related by

$$\lambda = -\frac{u}{\sqrt{1 + u^2}} \tag{15.31}$$

so that the Hamiltonian

$$H = \frac{1}{\sqrt{1 + u^2}} \tag{15.32}$$

is also constant. Next, the natural boundary conditions (15.9) become

$$H_f = -\nu, \quad \lambda_f = \nu, \quad H_0 = \xi, \quad \lambda_0 = -2\xi x_0. \tag{15.33}$$

The natural final conditions combined with Eqs.(15.31) and (15.32) imply that

$$u = 1, \quad \nu = -\frac{1}{\sqrt{2}}, \tag{15.34}$$

and the natural initial conditions lead to

$$x_0 = \frac{1}{2}, \quad \xi = \frac{1}{\sqrt{2}}. \tag{15.35}$$

Finally, the prescribed boundary conditions (15.27) and (15.28) yield

$$t_0 = -\frac{1}{4}, \quad t_f = \frac{1}{8}, \quad x_f = \frac{7}{8}. \tag{15.36}$$

Incidentally, the Weierstrass condition (15.12) is given by

$$\sqrt{1 + u_*^2} + \lambda u_* - \sqrt{1 + u^2} - \lambda u > 0, \tag{15.37}$$

which, when combined with Eq.(15.31), can be rewritten as

$$\sqrt{(1 + uu_*)^2 + (u_* - u)^2} - (1 + uu_*) > 0 \tag{15.38}$$

and is satisfied at every point of the optimal path for all $u_* \neq u$. Hence, the Weierstrass necessary condition for a strong relative minimum is satisfied. Also, since

$$H_{uu} = \frac{1}{(1 + u^2)^{3/2}} = \frac{1}{2^{3/2}} > 0, \tag{15.39}$$

the Legendre–Clebsch condition for a relative minimum is satisfied.

Since the strengthened Legendre–Clebsch condition ($H_{uu} > 0$) is satisfied, the next step is to apply the conjugate point condition. From the definition of the problem, it is seen that

$$f_x = 0, \quad f_u = 1, \quad H_{xx} = 0, \quad H_{xu} = 0, \quad H_{uu} = 2^{-3/2} \tag{15.40}$$

so that the matrices (13.31) are given by

$$A = 0, \quad B = 2^{3/2}, \quad C = 0. \tag{15.41}$$

Then, the solutions to the differential equations (13.41) subject to the boundary conditions (13.40) are given by

$$S = 0, \quad R = 1, \quad m = 0$$
$$Q = 2^{3/2}(t - t_f), \quad n = 2, \quad \alpha = 0. \tag{15.42}$$

Then,

$$U = [1 \ 0], \quad V = \begin{bmatrix} Q & 2 \\ 2 & 0 \end{bmatrix}, \tag{15.43}$$

so that

$$\hat{S} = S - VU^{-1}U^T = 0, \tag{15.44}$$

which is finite. Hence, a conjugate point does not exist.

From the definition (15.6) of $G$,

$$G_{x_0 x_0} = 2\xi = \sqrt{2}. \tag{15.45}$$

Next, since

$$\Gamma = \xi - \sqrt{1 + u^2} - 2\xi x_0 u, \tag{15.46}$$

it is seen that

$$\Gamma_{t_0} = 0, \quad \Gamma_{x_0} = -2\xi u = -\sqrt{2}, \quad \Gamma' = -2\xi u^2 = -\sqrt{2}. \tag{15.47}$$

Hence, Eq.(15.18) becomes

$$\delta^2 J' = [\delta x_0 \quad dt_0] \begin{bmatrix} \sqrt{2} & -\sqrt{2} \\ -\sqrt{2} & -\sqrt{2} \end{bmatrix} \begin{bmatrix} \delta x_0 \\ dt_0 \end{bmatrix}. \tag{15.48}$$

The changes $\delta x_0$ and $dt_0$ are not independent but are related by $d\theta = 0$, that is,

$$2x_0 \delta x_0 + (1 + 2x_0 u)dt_0 = 0, \tag{15.49}$$

or by

$$\delta x_0 = -2dt_0. \tag{15.50}$$

Then, choosing $dt_0$ to be the independent change, Eq. (15.48) reduces to

$$\delta^2 J' = 7\sqrt{2}dt_0^2, \tag{15.51}$$

and indicates that $Y_0 = 7\sqrt{2} > 0$.

In conclusion, since a conjugate point does not exist and $Y_0 > 0$, the optimal path is a minimum.

## 15.7 Parameters as States

In Chap. 14, the optimal control problem is extended to the case where there are unknown parameters in the performance index, the differential constraints, and the prescribed final conditions. It is also shown that free final time problems can be reformulated as fixed final time problems in which the final time appears as a parameter. In these problems, the parameters can be converted into states by introducing the constraints

$$\dot{a} = 0 \tag{15.52}$$

and adding $a$ to the state vector as $x_{n+1}, \ldots, x_{n+r}$. Since the state $a$ is constant and the final value of $a$ is being optimized, the initial value of $a$ must be free. Hence, converting parameters to states creates a free initial condition problem.

## 15.8   Navigation Problem

To demonstrate the use of $t_f$ as a state, consider the navigation problem as defined in Section 14.6. By defining a new state as $x_3 \triangleq t_f$, the optimal control problem is to find the steering angle $\theta(\tau)$ that minimizes the performance index

$$J = x_{3_f}, \tag{15.53}$$

subject to the dynamics

$$\frac{dx_1}{d\tau} = x_3 V \cos \theta$$

$$\frac{dx_2}{d\tau} = x_3 (V \sin \theta + w) \tag{15.54}$$

$$\frac{dx_3}{d\tau} = 0$$

and the prescribed boundary conditions

$$\begin{aligned} \tau_0 &= 0, \quad x_{1_0} = 0, \quad x_{2_0} = 0 \\ \tau_f &= 1, \quad x_{1_f} = 1, \quad x_{2_f} = 0. \end{aligned} \tag{15.55}$$

Note that $x_{3_0}$ and $x_{3_f}$ are free.

For this problem,

$$\theta(x_0) = \begin{bmatrix} x_{1_0} \\ x_{2_0} \end{bmatrix} = 0, \quad \psi(x_f) = \begin{bmatrix} x_{1_f} - 1 \\ x_{2_f} \end{bmatrix} = 0, \tag{15.56}$$

so that the endpoint function and the Hamiltonian are given by

$$G = x_{3_f} + \nu_1(x_{1_f} - 1) + \nu_2 x_{2_f} + \xi_1 x_{1_0} + \xi_2 x_{2_0}$$

$$H = \lambda_1 x_3 V \cos \theta + \lambda_2 x_3 (V \sin \theta + w). \tag{15.57}$$

Note that $\lambda_3$ does not appear in $H$.

The $\lambda$ differential equations $(d\lambda/d\tau = -H_x^T)$ become

$$\frac{d\lambda_1}{d\tau} = 0$$

$$\frac{d\lambda_2}{d\tau} = 0 \tag{15.58}$$

$$\frac{d\lambda_3}{d\tau} = -\lambda_1 V \cos \theta - \lambda_2 (V \sin \theta + w)$$

and show that $\lambda_1$ and $\lambda_2$ are constant. Next, the control equation ($H_u = 0$) yields

$$- \lambda_1 \sin \theta + \lambda_2 \cos \theta = 0 \qquad (15.59)$$

because $x_3$ cannot be zero. Hence, $\theta$ is constant. Since $x_3$ is constant from its differential equation, the other state equations can be integrated and the boundary conditions applied to obtain

$$\sin \theta = -\frac{w}{V}, \quad \cos \theta = \frac{v}{V}, \quad x_3 = \frac{1}{v}, \qquad (15.60)$$

where

$$v \triangleq \sqrt{V^2 - w^2}. \qquad (15.61)$$

Hence, the state histories are given by

$$x_1 = \tau, \quad x_2 = 0, \quad x_3 = 1/v. \qquad (15.62)$$

Next, the natural boundary conditions ($\lambda_0 = -G_{x_0}^T$, $\lambda_f = G_{x_f}^T$) require that

$$\lambda_{1_0} = -\xi_1, \quad \lambda_{2_0} = -\xi_2, \quad \lambda_{3_0} = 0$$
$$\lambda_{1_f} = \nu_1, \quad \lambda_{2_f} = \nu_2, \quad \lambda_{3_f} = 1. \qquad (15.63)$$

Then, Eqs.(15.58) and (15.59) lead to

$$\lambda_1 = -\frac{1}{v}, \quad \lambda_2 = \frac{w}{v^2}, \quad \lambda_3 = \tau \qquad (15.64)$$

so that Eqs.(15.63) determine $\nu_1, \nu_2, \xi_1$, and $\xi_2$.

To apply the second differential conditions, it is seen that

$$f_x = \begin{bmatrix} 0 & 0 & v \\ 0 & 0 & 0 \\ 0 & 0 & 0 \end{bmatrix}, \quad f_\theta = \begin{bmatrix} \frac{w}{v} \\ 0 \\ 0 \end{bmatrix},$$

$$H_{xx} = \begin{bmatrix} 0 & 0 & 0 \\ 0 & 0 & 0 \\ 0 & 0 & 0 \end{bmatrix}, \quad H_{x\theta} = \begin{bmatrix} 0 \\ 0 \\ 0 \end{bmatrix}, \quad H_{\theta\theta} = \frac{V^2}{v^3}. \qquad (15.65)$$

Since $H_{\theta\theta} > 0$, the second differential conditions developed in this chapter apply.

Next, the coefficients $A, B,$ and $C$ from Chapter 9 become

$$A = \begin{bmatrix} 0 & 0 & v \\ 0 & 0 & 0 \\ 0 & 0 & 0 \end{bmatrix}, \quad B = \frac{v^3}{V^2}\begin{bmatrix} \frac{w^2}{v^2} & \frac{w}{v} & 0 \\ \frac{w}{v} & 1 & 0 \\ 0 & 0 & 0 \end{bmatrix}, \quad C = \begin{bmatrix} 0 & 0 & 0 \\ 0 & 0 & 0 \\ 0 & 0 & 0 \end{bmatrix}, \quad (15.66)$$

so that the solutions for the sweep variables are given by

$$S = \begin{bmatrix} 0 & 0 & 0 \\ 0 & 0 & 0 \\ 0 & 0 & 0 \end{bmatrix}, \quad R = \begin{bmatrix} 1 & 0 \\ 0 & 1 \\ R_{31} & 0 \end{bmatrix}, \quad Q = -\frac{v^3(1-\tau)}{V^2}\begin{bmatrix} \frac{w^2}{v^2} & \frac{w}{v} \\ \frac{w}{v} & 1 \end{bmatrix}, \quad (15.67)$$

where $R_{31} = v(1 - \tau)$. Since $\det Q = 0$, its inverse does not exist, and the conjugate point condition in the form of Section 13.6 cannot be applied.

If the scalar forms of Eqs.(13.36) and (13.38) are examined, it is seen that the third multiplier equation must be used to obtain $d\nu_1$. Hence, $d\nu$ cannot be obtained from Eq.(13.43), and this is the reason $Q^{-1}$ does not exist. Note that the third multiplier equation is the one associated with the parameter. It appears that it is necessary to partition the state vector.

Before continuing, it is noted that the initial and final times are fixed and that $G_{x_0 x_0} = 0, G_{x_f x_f} = 0$. Hence, the second differential for this problem is given by

$$\delta^2 J' = \int_{\tau_0}^{\tau_f} H_{\theta\theta}\delta\theta^2 d\tau \qquad (15.68)$$

and is positive for $H_{\theta\theta} > 0$. Since this simplification does not happen for all problems, it is important to continue the analysis.

## 15.9   Partitioning the Parameter Problem

In order to expose the equations that are needed to solve for $d\nu$ as stated in the previous section, the states $x$ are partitioned into the regular states $x$ and the parameter states $y$. The optimal control problem is now to find the control $u(t)$ that minimizes the performance index

$$J = \phi(x_f, y_f) + \int_{t_0}^{t_f} L(t, x, y, u)dt \qquad (15.69)$$

subject to the differential constraints

$$\dot{x} = f(t, x, y, u)$$
$$\dot{y} = 0 \qquad (15.70)$$

and the prescribed boundary conditions

$$t_0 = t_{0_s}, \quad \theta(x_0) = 0$$
$$t_f = t_{f_s}, \quad \beta(x_f, y_f) = 0. \tag{15.71}$$

For prescribed initial states, $\theta = x_0 - x_{0_s}$ and does not contain $y_0$ which is free. If the actual final time is free, it has been made a parameter.

The endpoint function and the Hamiltonian are given by

$$G = \phi + \nu^T \psi + \xi^T \theta, \quad H = L + \lambda^T f, \tag{15.72}$$

where the multiplier $\lambda$ of Chapters 9 and 11 is partitioned into the regular state multiplier $\lambda$ and the parameter state multiplier $\mu$. Because of the differential equation for $y$, $\mu$ does not appear in $H$.

The first differential conditions for the multipliers and the control are obtained from Eqs.(9.31) through (9.36) by partitioning and are given by

$$\dot{\lambda} = -H_x^T, \quad \dot{\mu} = -H_y^T, \quad 0 = H_u^T$$
$$\lambda_f = G_{x_f}, \quad \mu_f = G_{y_f}^T, \quad \lambda_0 = -G_{x_0}^T, \quad \mu_0 = -G_{y_0}^T = 0. \tag{15.73}$$

To investigate the second differential, all first differential conditions are varied, the control perturbation is obtained, and $\delta u$ is eliminated to form the differential equations for the state and multiplier perturbations. These equations are given by the partitioned forms of Eqs.(11.25) through (11.29). Then, the partitioned forms of the final conditions (11.22) are given by

$$\delta \lambda_f = G_{x_f x_f} \delta x_f + G_{x_f y_f} \delta y_f + \psi_{x_f}^T d\nu \tag{15.74}$$

$$\delta \mu_f = G_{y_f x_f} \delta x_f + G_{y_f y_f} \delta y_f + \psi_{y_f}^T d\nu \tag{15.75}$$

$$\delta \psi = \psi_{x_f} \delta x_f + \psi_{y_f} \delta y_f = 0. \tag{15.76}$$

These equations motivate the definitions of the sweep variables as

$$\delta \lambda = S_{11} \delta x + S_{12} \delta y + R_1 d\nu \tag{15.77}$$

$$\delta \mu = S_{21} \delta x + S_{22} \delta y + R_2 d\nu \tag{15.78}$$

$$0 = R_1^T \delta x + R_2^T \delta y + Q d\nu, \tag{15.79}$$

where the $S$ and $R$ terms are the $x, y$ and $\lambda, \mu$ partitions of Eqs.(11.35) and (11.36).

At this point, Eqs.(15.78) and (15.79) are evaluated at the initial point ($\mu_0 = 0$) and solved simultaneously for $\delta y_0$ and $d\nu$ to obtain

$$\begin{bmatrix} \delta y_0 \\ d\nu \end{bmatrix} = -\tilde{V}_0^{-1}\tilde{U}_0\delta x_0, \tag{15.80}$$

where

$$\tilde{U} = \begin{bmatrix} S_{21} \\ R_1 \end{bmatrix}, \quad \tilde{V} = \begin{bmatrix} S_{22} & R_2 \\ R_2^T & Q \end{bmatrix}. \tag{15.81}$$

Then,

$$\delta\lambda_0 = \tilde{S}_0\delta x_0, \tag{15.82}$$

where

$$\tilde{S} = S_{11} - \tilde{U}\tilde{V}^{-1}\tilde{U}^T. \tag{15.83}$$

The forming of the second differential as a perfect square follows the developments of Chapter 11 and Section 15.5 and leads to

$$\delta^2 J' = \delta x_0^T \tilde{S}_0 \delta x_0 + \int_{t_0}^{t_f} F^T H_{uu} F dt, \tag{15.84}$$

where $F$ is the partitioned form of Eq.(11.43). Note that there is no initial point term due to $\theta(x_0) = 0$ because $G$ does not contain $y_0$ and $G_{x_0 x_0} = 0$. As a consequence, the sufficient condition for a minimum is stated in terms of the finiteness of $\tilde{S}$.

For the optimal control problem involving parameters and having fixed initial regular states, the parameter states affect the conjugate point analysis. However, even though it is a free initial condition problem, there is no optimal initial point condition like Eq.(15.22).

## 15.10   Navigation Problem

To complete the solution of the navigation problem of Section 15.8, it is seen from Eq.(15.67) that

$$S_{11} = \begin{bmatrix} 0 & 0 \\ 0 & 0 \end{bmatrix}, \quad S_{12} = \begin{bmatrix} 0 \\ 0 \end{bmatrix}, \quad S_{21} = [0 \ 0], \quad S_{22} = 0$$

$$\tag{15.85}$$

$$R_1 = \begin{bmatrix} 1 & 0 \\ 0 & 1 \end{bmatrix}, \quad R_2 = [v(1-\tau) \ 0], \quad Q = -\frac{v^3(1-\tau)}{V^2}\begin{bmatrix} \frac{w^2}{v^2} & \frac{w}{v} \\ \frac{w}{v} & 1 \end{bmatrix}$$

so that

$$\tilde{S} = \frac{V^2}{v^3(1-\tau)} \begin{bmatrix} 0 & 0 \\ 0 & 1 \end{bmatrix}. \tag{15.86}$$

Hence, $\tilde{S}$ is finite, no conjugate point exists, and the optimal trajectory defined in Section 15.8 is a minimum.

# Problems

15.1 Find the control history that minimizes the distance between the line $t_0 = 0$ and the parabola $x_f^2 - t_f + 1 = 0$. Show that $u = 0$, $x = 0$, $\lambda = 0$, and $\nu = 0$. Also, show that $H_{uu} = 1 > 0$, there is no conjugate point, and that $Y_0 = 2/3 > 0$, so that the optimal path is a minimum.

15.2 Find the curve that minimizes the distance between the line $x_0 = t_0$ and the parabola $x_f^2 - t_f + 1 = 0$. Show that $u = -1$, $t_0 = 7/8$, $x_0 = 7/8$, $t_f = 5/4$, and $x_f = 1/2$. Also, show that the sufficient conditions for a weak relative minimum are satisfied ($H_{uu} > 0$, no conjugate point, and $Y_0 = \hat{S}_0 > 0$).

15.3 Find the curve that minimizes the distance between the parabola $x_0^2 + t_0 = 0$ and the parabola $x_f^2 - t_f + 1 = 0$. Show that $u = 0$, $t_0 = 0$, $x_0 = 0$, $\xi = 1$, $t_f = 1$, $x_f = 0$, and $\nu = 1$. Also, show that the sufficient conditions for a weak relative minimum are satisfied, that is, show that $H_{uu} = 1 > 0$, $t'$ does not exist, and $Y_0 = 8/3 > 0$.

15.4 Form a free initial condition problem by considering an aircraft navigation problem (Problem 9.18) in which the aircraft flies from the line $x_0 = 0, t_0 = 0$ to the point $x_f = 1, y_f = 0$ in minimum time. Note that the initial $y$ coordinate is free. Show that the optimal heading angle is $\theta = 0$.

15.5 Solve Problem 15.4 again using the coordinates of Section 15.9.

15.6 A periodic optimal control problem is that of minimizing the performance index

$$J = \frac{1}{t_f} \int_{t_0}^{t_f} L(t, x, u)dt,$$

subject to the dynamics

$$\dot{x} = f(t, x, u)$$

and the boundary conditions

$$t_0 = t_{0_s}, \quad x_f = x_0.$$

Here, the time-averaged value of $L$ is being minimized, and the final state must equal the initial state. Find the first differential conditions for a minimum. Note that $t_f$ is free, it appears in the performance index, its differential must be taken, and $dt_f$ is a constant, that is, it can be moved out of an integral.

# 16

# Control Discontinuities

## 16.1   Introduction

In the previous developments, it has been assumed that the optimal control is continuous over the time interval of the problem. However, there exist problems in which the control can be discontinuous at a point or points.

Consider the problem of minimizing the distance from a point to a curve to another point. The answer is a straight line from the first point to the curve followed by a straight line from the curve to the second point. Since the control is the slope of the line, there is a discontinuity in the control at the curve. This problem is characterized by an internal point constraint.

Another example is the aircraft navigation problem, Chapter 1, in which the velocity of the air changes from one constant speed to another at a fixed point between the initial and final points. Since the steering angle, the control, is constant along each part of the solution but has different values, a discontinuity occurs in the control at the point the wind speed changes. This problem is characterized by a change in the dynamics accompanied by an internal point constraint.

In this chapter, the first and second differential conditions are derived for an optimal control problem with a change in the dynamics and/or an internal point constraint allowing for a discontinuity in the control at an internal point. The new first differential conditions are known as the Erdmann-Weierstrass corner conditions, whereas the second differential conditions are like those of previous problems.

## 16.2    Problem Statement

The optimal control problem is now formulated for the case where there is
a change in the dynamics at a point that must satisfy some constraints. A
discontinuity in the control is allowed to occur at the internal point, now called
a corner point. Only one corner point is considered because if multiple corners
exist the conditions would be the same for each corner. Each part of the optimal
path is called a subarc. While a discontinuity in the control is allowed to exist,
the state is still assumed to be continuous. A schematic showing an optimal
control $u(t)$ and path $x(t)$ as well as a comparison control $u_*(t)$ and path $x_*(t)$
is contained in Fig. 16.1.

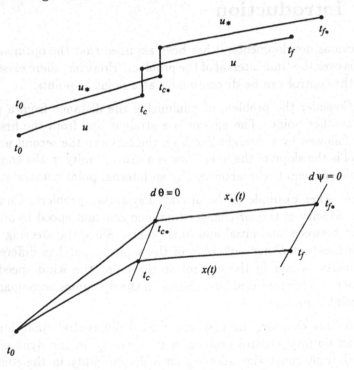

Figure 16.1: Optimal Control/Path and Comparison Control/Path

Since the function $L(t, x, u)$ is not single valued at the corner, the
integral appearing in the performance index is written over each subarc in

which the control is continuous and $L$ is single valued. The optimal control problem being solved here is the following: Find the control history $u(t)$ that minimizes the performance index

$$J = \phi(t_f, x_f) + \int_{t_0}^{t_c} L(t, x, u)dt + \int_{t_c}^{t_f} L(t, x, u)dt \qquad (16.1)$$

subject to the differential constraints

$$\dot{x} = f(t, x, u), \quad t_0 \le t \le t_c \qquad (16.2)$$
$$\dot{x} = f(t, x, u), \quad t_c \le t \le t_f \qquad (16.3)$$

over each interval and to the prescribed boundary conditions

$$t_0 = t_{0_s}, \quad x_0 = x_{0_s} \qquad (16.4)$$
$$\theta(t_c, x_c) = 0 \qquad (16.5)$$
$$\psi(t_f, x_f) = 0. \qquad (16.6)$$

The function $f$ (and $L$ for that matter) can have different forms over each subarc, but it is not necessary to give them separate names because the interval is explicit. In general, there are $q + 1$ internal point constraints where $q \le n$.

In what follows, the subscripts $c-$ and $c+$ are used to define quantities evaluated immediately before the corner and immediately after the corner, that is, the subscript $c-$ means a quantity evaluated with the control just before the discontinuity. In other words, $f_{c-}$ denotes $f(t_c, x_c, u_{c-})$.

## 16.3 First Differential Conditions

If the constraints (16.2), (16.3), (16.5), and (16.6) are adjoined by Lagrange multipliers to the performance index (16.1), the augmented performance index is given by

$$\begin{aligned} J' &= G(t_c, x_c, \xi, t_f, x_f, \nu) + \int_{t_0}^{t_c} [H(t, x, u, \lambda) - \lambda^T \dot{x}]dt \\ &+ \int_{t_c}^{t_f} [H(t, x, u, \lambda) - \lambda^T \dot{x}]dt, \end{aligned} \qquad (16.7)$$

where the endpoint function and the Hamiltonian are defined as

$$G = \phi(t_f, x_f) + \nu^T \psi(t_f, x_f) + \xi^T \theta(t_c, x_c) \qquad (16.8)$$
$$H = L(t, x, u) + \lambda^T f(t, x, u). \qquad (16.9)$$

Taking the differential of $J'$ (see Section 13.2), performing the integration by parts, and relating $\delta$ to $d$ outside the integral leads to

$$dJ' = G_{t_f}dt_f + G_{x_f}dx_f + G_\nu d\nu + G_{t_c}dt_c + G_{x_c}dx_c + G_\xi d\xi$$

$$+ [Hdt - \lambda^T dx]_{t_0}^{t_{c-}} + \int_{t_0}^{t_c}[(H_x + \dot{\lambda}^T)\delta x + (H_u)\delta u + (f^T - \dot{x}^T)\delta\lambda]dt$$

$$[Hdt - \lambda^T dx]_{t_{c+}}^{t_f} + \int_{t_c}^{t_f}[(H_x + \dot{\lambda}^T)\delta x + (H_u)\delta u + (f^T - \dot{x}^T)\delta\lambda]dt.$$

$$(16.10)$$

Since the time and state are required to be continuous at the corner, the relations

$$t_{c-} = t_{c+} = t_c \tag{16.11}$$

$$x_{c-} = x_{c+} = x_c$$

must hold and, by taking the differential, imply that

$$dt_{c-} = dt_{c+} = dt_c \tag{16.12}$$

$$dx_{c-} = dx_{c+} = dx_c.$$

Hence, the first differential can be rewritten as

$$dJ' = (G_{t_f} + H_f)dt_f + (G_{x_f} - \lambda_f^T)dx_f + G_\nu d\nu$$

$$+ (H_{c-} - H_{c+} + G_{t_c})dt_c - (\lambda_{c-}^T - \lambda_{c+}^T - G_{x_c})dx_c + G_\xi d\xi$$

$$+ \int_{t_0}^{t_c}[(H_x + \dot{\lambda}^T)\delta x + (H_u)\delta u + (f^T - \dot{x}^T)\delta\lambda]dt \tag{16.13}$$

$$+ \int_{t_c}^{t_f}[(H_x + \dot{\lambda}^T)dx + (H_u)\delta u + (f^T - \dot{x}^T)\delta\lambda]dt.$$

The first step in deriving the first differential conditions for a minimum is to note that $G_\nu = \psi = 0$, $G_\xi = \theta = 0$, $f^T - \dot{x}^T = 0$, and deleting those terms from Eq. (16.13). Next, the variations $\delta x$ are eliminated by choosing the Lagrange multipliers to satisfy the differential equations

$$\dot{\lambda} = -H_x^T, \quad t_0 \le t \le t_c \tag{16.14}$$

$$\dot{\lambda} = -H_x^T, \quad t_c \le t \le t_f$$

and boundary conditions

$$H_f = -G_{t_f}$$

$$\lambda_f = G_{x_f}^T$$

$$H_{c-} = H_{c+} - G_{t_c} \tag{16.15}$$

$$\lambda_{c-} = \lambda_{c+} + G_{x_c}^T.$$

At this point, the first differential has the form

$$dJ' = \int_{t_0}^{t_c} H_u \delta u \, dt. + \int_{t_c}^{t_f} H_u \delta u \, dt \tag{16.16}$$

Working backward from the final point, consider the class of admissible comparison paths shown in Fig. 16.2 for which $\delta u = 0$ for $t_0 \leq t \leq t_p$, and a pulse occurs at $t_p$. Here, the first differential becomes

$$dJ' = \int_{t_p}^{t_f} H_u \delta u \, dt \tag{16.17}$$

This is exactly the scenario for the free final time problem of Chapter 13 and leads to

$$H_u = 0, \quad t_c \leq t \leq t_f. \tag{16.18}$$

Next, consider the class of admissible comparison paths shown in Fig. 16.3 where a pulse occurs at $t_p < t_c$. By requiring the system to also be controllable (Section 7.4) to the internal point constraint, the reasoning of Chapter 13 applies and leads to

$$H_u = 0, \quad t_0 \leq t \leq t_c. \tag{16.19}$$

In summary, the first differential conditions to be satisfied for a problem with one internal point are given by the Euler-Lagrange equations

$$\dot{x} = f(t, x, u), \quad \dot{\lambda} = -H_x^T(t, x, u, \lambda), \quad 0 = H_u^T(t, x, u, \lambda), \quad t_0 \leq t \leq t_c \tag{16.20}$$

$$\dot{x} = f(t, x, u), \quad \dot{\lambda} = -H_x^T(t, x, u, \lambda), \quad 0 = H_u^T(t, x, u, \lambda), \quad t_c \leq t \leq t_f \tag{16.21}$$

the initial conditions

$$t_0 = t_{0_s}, \quad x_0 = x_{0_s}, \tag{16.22}$$

the corner conditions

$$\theta(t_c, x_c) = 0 \tag{16.23}$$

$$H(t_c, x_c, u_{c-}, \lambda_{c-}) = H(t_c, x_c, u_{c+}, \lambda_{c+}) - G_{t_c}(t_c, x_c, \xi) \tag{16.24}$$

$$\lambda_{c-} = \lambda_{c+} + G_{x_c}^T(t_c, x_c, \xi), \tag{16.25}$$

and the final conditions

$$\psi(t_f, x_f) = 0 \tag{16.26}$$

$$H(t_f, x_f, u_f, \lambda_f) = -G_{t_f}(t_f, x_f, \nu) \tag{16.27}$$

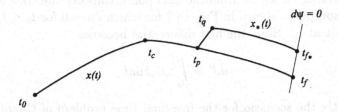

Figure 16.2: Admissible Comparison Path $t_p > t_c$

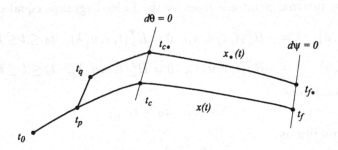

Figure 16.3: Admissible Comparison Path $t_p < t_c$

$$\lambda_f = G_{x_f}^T(t_f, x_f, \nu). \tag{16.28}$$

The new conditions (16.23) through (16.25) constitute $q + 2 + n$ equations for determining the optimal corner time $t_c$, the $n \times 1$ optimal corner state $x_c$, and the $(q+1) \times 1$ Lagrange multiplier $\xi$. They are known as the Erdmann–Weierstrass corner conditions and are used to determine $t_c$ and $x_c$. They imply that jumps in $H$ and $\lambda$ can occur; however, two subarcs can be joined optimally with no jump in the control. Should there be more than one corner point, the corner conditions are applied at each point.

A corner can exist even if there is no internal point constraint or no change in the dynamics at an internal point. In this case, the Euler–Lagrange equations or the natural boundary conditions will indicate multiple solutions. Then, the correct sequence of subarcs must be determined by applying the first differential necessary conditions as well as the Weierstrass and Legendre–Clebsch conditions.

If $L$ and $f$ do not explicitly contain the time, the first integral $H = $ Const exists over each subarc. In other words, the constant can have different values over each subarc.

## 16.4 Tests for a Minimum

The derivation of the Weierstrass condition for a strong relative minimum is similar to that of Chapter 13 and leads to

$$H(t, x, u_*, \lambda) - H(t, x, u, \lambda) > 0, \tag{16.29}$$

which must be satisfied for all $u_* \neq u$ and for all $u_* \neq u_{c_+}$. For weak variations, the Weierstrass condition reduces to the Legendre–Clebsch condition

$$H_{uu}(t, x, u, \lambda) \geq 0. \tag{16.30}$$

Both of these conditions must be satisfied at every point of a minimal path.

Consider a problem with an unrestricted corner ($G_{t_c} = G_{x_c} = 0$). Here, the corner conditions require that

$$\lambda_{c-} = \lambda_{c+} = \lambda_c \tag{16.31}$$

and that

$$H(t_c, x_c, u_{c+}, \lambda_c) = H(t_c, x_c, u_{c-}, \lambda_c). \tag{16.32}$$

The latter indicates that at an unrestricted corner the Weierstrass condition is satisfied by equality if $u = u_{c-}$ and $u_* = u_{c+}$. The Weierstrass condition can play an important role in determining the existence of an unrestricted corner in an optimal path.

## 16.5   Example

While the following example does not have a unique minimum, it provides an interesting application of the corner conditions. Consider the problem of minimizing the performance index

$$J = \int_{t_0}^{t_f} (u^2 - 1)^2 dt, \tag{16.33}$$

subject to the dynamics

$$\dot{x} = u \tag{16.34}$$

and the prescribed boundary conditions

$$t_0 = 0, \quad x_0 = 1$$
$$t_f = 1, \quad x_f = 1. \tag{16.35}$$

The endpoint function and the Hamiltonian are given by

$$G = \nu(x_f - 1), \quad H = (u^2 - 1)^2 + \lambda u \tag{16.36}$$

so that the Euler–Lagrange equations become

$$\dot{x} = u$$
$$\dot{\lambda} = -H_x = 0 \tag{16.37}$$
$$0 = H_u = 4u(u^2 - 1) + \lambda = 0$$

and imply that $\lambda = $ Const and $u = $ Const. Then, integration of the state equation and application of the prescribed boundary conditions leads to the solution

$$u = 0, \quad x = 1. \tag{16.38}$$

However, if the Legendre–Clebsch condition is applied, it is seen that

$$H_{uu} = 12u^2 - 4 = -4. \tag{16.39}$$

Hence, the Legendre–Clebsch condition is not satisfied, and there does not exist a continuous control solution for this problem.

Consider a discontinuous control with one discontinuity. Over each subarc, the optimal control and the Lagrange multiplier are constant. Then, the corner conditions (16.24) and (16.25) with $G_{t_c} = 0$ and $G_{x_c} = 0$ require that the Lagrange multiplier and the Hamiltonian be continuous at the corner. From the definition of $H$ and $H_u = 0$, these conditions become

$$4u_{c-}(u_{c-}^2 - 1) = 4u_{c+}(u_{c+}^2 - 1)$$
$$(u_{c-}^2 - 1)(3u_{c-}^2 + 1) = (u_{c+}^2 - 1)(3u_{c+}^2 + 1)$$

$$(16.40)$$

and are satisfied if $u_{c-} = \pm 1$ and $u_{c+} = \pm 1$. However, if $u_{c-} = +1$, then, $u_{c+} = -1$; otherwise, there is no corner. In conclusion, the two-subarc solution is given by either of the two paths in Fig. 16.4. Note that the Legendre–Clebsch condition is satisfied along both of these paths.

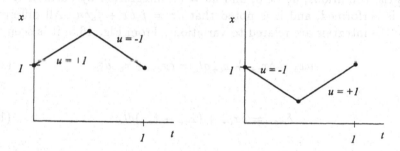

Figure 16.4: Optimal Two-Subarc Paths

A peculiarity of this problem is that there are an infinite number of minimal paths, that is, any combination of $u = \pm 1$ subarcs which satisfy the boundary conditions. This is apparent from the performance index (16.33) whose lowest possible value is zero, and it occurs along such paths.

# 16.6   Second Differential

If the first differential is developed for the case where $\delta x_0 \neq 0$ and differentials outside the integral are written in terms of variations ($dx_f = \delta x_f + \dot{x}_f dt_f$ and

$dx_c = \delta x_{c-} + \dot{x}_{c-} dt_c)$, the result is

$$
\begin{aligned}
dJ' = \ & \Omega dt_f + (G_{x_f} - \lambda_f^T)\delta x_f + \psi^T d\nu + \lambda_0^T \delta x_0 \\
& + \Gamma dt_c + (G_{x_c} - \lambda_{c-}^T + \lambda_{c+}^T)\delta x_{c-} + \theta^T d\xi \\
& + \int_{t_0}^{t_c} [(H_x + \dot{\lambda}^T)\delta x + (H_u)\delta u + (f^T - \dot{x}^T)\delta\lambda]dt \\
& + \int_{t_c}^{t_f} [(H_x + \dot{\lambda}^T)\delta x + (H_u)\delta u + (f^T - \dot{x}^T)\delta\lambda]dt,
\end{aligned}
\tag{16.41}
$$

where

$$
\Omega = G_{t_f} + L_f + G_{x_f} f_f
\tag{16.42}
$$

$$
\Gamma = G_{t_c} + L_{c-} + G_{x_c} f_{c-} - L_{c+} + \lambda_{c+}^T (f_{c-} - f_{c+}).
\tag{16.43}
$$

Next, the differential of the transpose of $dJ'$ is taken and use is made of the first differential conditions, $d\psi = 0$, and $d\theta = 0$. Integration by parts of the term $\delta x^T \delta\lambda$ is performed, and it is noted that $\delta\dot{x} = f_x \delta x + f_u \delta u$. All differentials outside the integrals are related to variations. From Fig. 16.5, it is seen that

$$
dx_c = \delta x_{c+} + \dot{x}_{c+} dt_c = \delta x_{c-} + \dot{x}_{c-} dt_c
\tag{16.44}
$$

so that

$$
\delta x_{c-} = \delta x_{c+} + (\dot{x}_{c+} - \dot{x}_{c-})dt_c.
\tag{16.45}
$$

The resulting form of the second differential is given by

$$
\begin{aligned}
d^2 J' = \ & [\delta x_f^T \ dt_f] \begin{bmatrix} G_{x_f x_f} & \Omega_{x_f}^T \\ \Omega_{x_f} & \Omega' \end{bmatrix} \begin{bmatrix} \delta x_f \\ dt_f \end{bmatrix} \\
& + [\delta x_{c-}^T \ dt_c] \begin{bmatrix} G_{x_c x_c} & \Gamma_{x_c}^T \\ \Gamma_{x_c} & \Gamma' + \lambda_{c+}^T (f_{c-} - f_{c+}) \end{bmatrix} \begin{bmatrix} \delta x_{c-} \\ dt_c \end{bmatrix} \\
& + \int_{t_0}^{t_c} [\delta x^T \ \delta u^T] \begin{bmatrix} H_{xx} & H_{xu} \\ H_{ux} & H_{uu} \end{bmatrix} \begin{bmatrix} \delta x \\ \delta u \end{bmatrix} dt \\
& + \int_{t_c}^{t_f} [\delta x^T \ \delta u^T] \begin{bmatrix} H_{xx} & H_{xu} \\ H_{ux} & H_{uu} \end{bmatrix} \begin{bmatrix} \delta x \\ \delta u \end{bmatrix} dt,
\end{aligned}
\tag{16.46}
$$

where

$$
\Omega' \triangleq \Omega_{t_c} + \Omega_{x_c}\dot{x}_{c-}, \quad \Gamma' \triangleq \Gamma_{t_c} + \Gamma_{x_c}\dot{x}_{c-}.
\tag{16.47}
$$

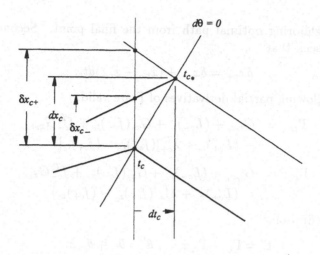

Figure 16.5: Variations and Differentials at a Corner

# 16.7  Neighboring Optimal Path

The differential equations and the final conditions for the sweep variables associated with the neighboring optimal path are the same as those for the free final time problem, that is, Eqs.(13.41) and (13.40). What needs to be established here are the boundary conditions for propagating the sweep variables backwards in time from the corner point constraint surface. These boundary conditions are obtained by taking the differential of the corner conditions (16.23) through (16.25). For convenience, these equations are rewritten here as

$$\lambda_{c-} = \lambda_{c+} + G_{x_c}^T(t_c, \, x_c, \, \xi) \tag{16.48}$$

$$\theta = \theta(t_c, x_c) = 0 \tag{16.49}$$

$$\Gamma(t_c, x_c, \xi, u_{c-}, u_{c+}, \lambda_{c+}) = G_{t_c} + L_{c-} + G_{x_c} f_{c-} - L_{c+} \tag{16.50}$$
$$+ \lambda_{c+}^T(f_{c-} - f_{c+})$$

where, in the definition of $\Gamma$, $\lambda_{c-}$ has been eliminated from $H$.

Before taking the differentials of the corner conditions, it is noted that

$$\delta\lambda_{c+} = \hat{S}_{c+}\delta x_{c+} \tag{16.51}$$

from the neighboring optimal path from the final point.  Second, from Eq. (16.45), it is seen that

$$\delta x_{c+} = \delta x_{c-} + (\dot{x}_{c-} - \dot{x}_{c+})dt_c. \tag{16.52}$$

Third, the following partial derivatives of $\Gamma$ are valid:

$$\begin{aligned}
\Gamma_{t_c} &= G_{t_c t_c} + (L_{c-})_{t_c} + G_{x_c}(f_{c-})_{t_c} + f_{c-}^T G_{x_c t_c} \\
&\quad - (L_{c+})_{t_c} + \lambda_{c+}^T[(f_{c-})_{t_c} - (f_{c+})_{t_c}]
\end{aligned} \tag{16.53}$$

$$\begin{aligned}
\Gamma_{x_c} &= G_{t_c x_c} + (L_{c-})_{x_c} + G_{x_c}(f_{c-})_{x_c} + f_{c-}^T G_{x_c x_c} \\
&\quad - (L_{c+})_{x_c} + \lambda_{c+}^T[(f_{c-})_{x_c} - (f_{c+})_{x_c}]
\end{aligned} \tag{16.54}$$

Finally, by definition,

$$\Gamma' = \Gamma_{t_c} + \Gamma_{cc}\dot{x}_{c-}, \quad \theta' = \theta_{t_c} + \theta_{x_c}\dot{x}_{c-}. \tag{16.55}$$

At this point, taking the differential of Eqs. (16.48) through (16.50) and converting to variations leads to

$$\begin{bmatrix} \delta\lambda_{c-} \\ d\theta \\ d\Gamma \end{bmatrix} = \begin{bmatrix} S_{c-} & R_{c-} & m_{c-} \\ R_{c-}^T & Q_{c-} & n_{c-} \\ m_{c-}^T & n_{c-}^T & \alpha_{c-} \end{bmatrix} \begin{bmatrix} \delta x_{c-} \\ d\xi \\ dt_c \end{bmatrix}, \tag{16.56}$$

where

$$S_{c-} = G_{x_c x_c} + \hat{S}_{c+}, \quad R_{c-} = \theta_{x_c}^T$$

$$m_{c-} = \Gamma_{x_c}^T + \hat{S}_{c+}(f_{c-} - f_{c+}), \quad Q_{c-} = 0, \quad n_{c-} = \theta' \tag{16.57}$$

$$\alpha_{c-} = \Gamma' - (f_{c-} - f_{c+})^T(H_x^T)_{c+} + (f_{c-} - f_{c+})^T \hat{S}_{c+}(f_{c-} - f_{c+}).$$

Equations (16.57) are the boundary conditions for propagating the neighboring optimal path backward in time from $t_c$.

Given $\delta x_0$, $d\xi$ and $dt_c$ can be determined as

$$\begin{bmatrix} d\xi \\ dt_c \end{bmatrix} = -V_0^{-1}U_0^T \delta x_0 \tag{16.58}$$

so that

$$\delta\lambda_0 = \hat{S}_0 \delta x_0. \tag{16.59}$$

where $U, V$, and $\hat{S}$ are defined in Section 13.5.  Then, Eqs.(13.36) can be integrated forward in time to obtain $\delta x$ and $\delta\lambda$, and $\delta u$ comes from Eq.(13.35). With $\delta x_{c-}$ known, Eqs.(16.52) and (16.51) yield $\delta x_{c+}$ and $\delta\lambda_{c+}$ so the calculation of $\delta x$, $\delta\lambda$, and $\delta u$ can continue to $t_f$.

## 16.8   Second Differential Conditions

Consider the forms

$$
\begin{aligned}
E &= H_{uu}^{-1}[(H_{ux} + f_u^T S)\delta x + f_u^T U dq] + \delta u, \quad t_0 \leq t \leq t_c \\
F &= H_{uu}^{-1}[(H_{ux} + f_u^T S)\delta x + f_u^T U dp] + \delta u, \quad t_c \leq t \leq t_f,
\end{aligned}
\tag{16.60}
$$

where

$$
dq = \begin{bmatrix} d\xi \\ dt_c \end{bmatrix}, \quad dp = \begin{bmatrix} d\nu \\ dt_f \end{bmatrix}.
\tag{16.61}
$$

By substitution, it can be shown that

$$
\begin{aligned}
E^T H_{uu} E &= [\delta x^T \; \delta u^T]\begin{bmatrix} H_{xx} & H_{xu} \\ H_{ux} & H_{uu} \end{bmatrix}\begin{bmatrix} \delta x \\ \delta u \end{bmatrix} \\
&+ \frac{d}{dt}\left\{ [\delta x^T \; dq^T]\begin{bmatrix} S & U \\ U^T & V \end{bmatrix}\begin{bmatrix} \delta x \\ dq \end{bmatrix}\right\}
\end{aligned}
\tag{16.62}
$$

and that

$$
\begin{aligned}
F^T H_{uu} F &= [\delta x^T \; \delta u^T]\begin{bmatrix} H_{xx} & H_{xu} \\ H_{ux} & H_{uu} \end{bmatrix}\begin{bmatrix} \delta x \\ \delta u \end{bmatrix} \\
&+ \frac{d}{dt}\left\{ [\delta x^T \; dp^T]\begin{bmatrix} S & R \\ R^T & Q \end{bmatrix}\begin{bmatrix} \delta x \\ dp \end{bmatrix}\right\}.
\end{aligned}
\tag{16.63}
$$

If these expressions are substituted into the second differential (16.51) along with

$$
d\xi = -V_0^{-1}U_0^T \delta x_0, \quad d\nu = -V_{c+}^{-1}U_{c+}^T \delta x_{c+},
\tag{16.64}
$$

the corner point and final point terms cancel, leaving

$$
\begin{aligned}
d^2 J' &= \delta x_0^T \hat{S}_0 \delta x_0 + \int_{t_0}^{t_c} E^T H_{uu} E \, dt \\
&+ \int_{t_c}^{t_f} F^T H_{uu} F dt.
\end{aligned}
\tag{16.65}
$$

As in the free final time problem, the optimal path is a minimum if $\hat{S}$ is finite over the interval of integration.

# 16.9  Supersonic Airfoil of Minimum Pressure Drag

It is desired to find the shape of a supersonic airfoil symmetric top and bottom that minimizes the pressure drag for the case where the chord and the maximum thickness are prescribed (see Fig. 16.6). In nondimensional coordinates, the pressure drag of a slender supersonic airfoil is given by

$$D_p \sim \int_{t_0}^{t_f} \dot{x}^2 \, dt \tag{16.66}$$

and the conditions of given chord and maximum thickness are expressed as $t_f = 1$ and $x_c = \tau$ where $\tau$ denotes the thickness ratio divided by two.

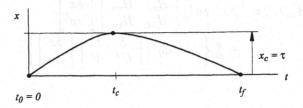

Figure 16.6: Supersonic Airfoil Shape.

Formally, the optimal control problem is to find the control history $u(t)$ which minimizes the performance index

$$J = \int_{t_0}^{t_f} u^2 \, dt, \tag{16.67}$$

subject to the differential constraint

$$\dot{x} = u, \tag{16.68}$$

the prescribed initial conditions

$$t_0 = 0, \qquad x_0 = 0, \tag{16.69}$$

the prescribed internal point condition

$$x_c = \tau, \tag{16.70}$$

and the prescribed final conditions

$$t_f = 1, \quad x_f = 0. \tag{16.71}$$

For this problem, the endpoint function and the Hamiltonian are given by

$$G = vx_f + \xi(x_c - \tau) \tag{16.72}$$

$$H = u^2 + \lambda u. \tag{16.73}$$

The application of the Euler–Lagrange equations (16.20) and (16.21) lead to

$$\dot{\lambda} = -H_x = 0 \tag{16.74}$$

$$0 = H_u = 2u + \lambda \tag{16.75}$$

and imply that $\lambda$ and $u$ are constants. As a consequence, the one-subarc solution is a straight line, but it cannot satisfy the boundary conditions. A shape defined by two straight-line segments can satisfy the boundary conditions, but a jump in the control must exist at $t_c$.

With $\theta = x_c - \tau$, it is seen that $\theta_{t_c} = 0$ and $\theta_{x_c} = 1$ so that $G_{t_c} = 0$ and $G_{x_c} = \xi$. Then, the corner conditions (16.24) and (16.25) become

$$u_{c+}^2 + \lambda_{c+} u_{c+} = u_{c-}^2 + \lambda_{c-} u_{c-} \tag{16.76}$$

$$\lambda_{c+} = \lambda_{c-} - \xi. \tag{16.77}$$

Equation (16.77) yields the value of the Lagrange multiplier $\xi$, and since $\lambda = -2u$ along each subarc, Eq.(16.76) leads to $u_{c+}^2 = u_{c-}^2$. This, in turn, says that $u_{c+} = \pm u_{c-}$ where

$$u_{c+} = -u_{c-} \tag{16.78}$$

is the only solution that can satisfy the boundary conditions.

Next, Eq. (16.68) is integrated over each subarc ($u = \text{Const}$ over each subarc), and the boundary conditions (16.69) through (16.71) and Eq. (16.78) are applied. The results are the following:

$$t_0 \le t \le t_c \quad u = 2\tau, \quad x = 2\tau t, \quad \lambda = -4\tau$$

$$t_c \le t \le t_f \quad u = -2\tau, \quad x = 2\tau(1 - t), \quad \lambda = 4\tau \tag{16.79}$$

$$t_c = \tfrac{1}{2}, \quad \xi = -4\tau, \quad v = 4\tau.$$

However, this is just an optimal solution. To verify that it is a minimum, the second differential conditions must be applied.

The Legendre–Clebsch condition (16.30) leads to

$$H_{uu} = 2 \qquad (16.80)$$

and is satisfied. What remains is to show that there is no conjugate point or that, if there is a conjugate point, it is outside the interval of integration.

From the definitions (13.33) of $A$, $B$, and $C$, it is seen that

$$A = 0, \quad B = \frac{1}{2}, \quad C = 0. \qquad (16.81)$$

Hence, the solutions of Eq.(13.41) subject to the boundary conditions (13.40) are given by

$$S = 0, \quad R = 1, \quad m = 0$$
$$Q = \tfrac{1-t}{2}, \quad n = -2\tau, \quad \alpha = 0 \qquad (16.82)$$

so that

$$V = \begin{bmatrix} Q & -2\tau \\ -2\tau & 0 \end{bmatrix}, \quad U = [1 \ \ 0] \qquad (16.83)$$

and

$$\hat{S} = S - UV^{-1}U^T = 0. \qquad (16.84)$$

This means that there is no conjugate point in $[t_c, t_f]$.

To continue the neighboring optimal path for $t \le t_c$, Eqs. (16.57) require that

$$S_{c-} = 0, \quad R_{c-} = 1, \quad m_{c-} = 0$$
$$Q_{c-} = 0, \quad n_{c-} = 2\tau, \quad \alpha_{c-} = 0. \qquad (16.85)$$

Then, the solution of Eqs. (13.41) subject to these boundary conditions leads to

$$S = 0, \quad R = 1, \quad m = 0$$
$$Q = \tfrac{1-2t}{4}, \quad n = -2\tau, \quad \alpha = 0 \qquad (16.86)$$

so that

$$U = [1 \ \ 0], \quad V = \begin{bmatrix} Q & 2\tau \\ 2\tau & 0 \end{bmatrix} \qquad (16.87)$$

and

$$\hat{S} = S - UV^{-1}U^T = 0. \qquad (16.88)$$

Hence, there is no conjugate point in $[t_0, t_c]$.

In conclusion, the optimal shape (16.79) is a mimimum pressure drag shape.

# Problems

16.1 Derive the Weierstrass condition for the case of a single jump in the control.

16.2 Find the control history $u(t)$ that minimizes the distance from the point $t_0 = 0$, $x_0 = 0$ to the line $x_c = 1 + t_c$ to the point $t_f = 1$, $x_f = 0$. Apply both first and second differential conditions.

16.3 Find the control history $u(t)$ that minimizes the distance from the point $t_0 = 0$, $x_0 = 0$ to the line $x_c = 1 - t_c$ to the point $t_f = 1$, $x_f = 1$. Apply both first and second differential conditions.

16.4 Examine the Weierstrass condition for the example problem in Soution 16.5. Show that at a point on a $u = +1$ subarc the Weierstrass condition is satisfied as $\Delta H = 0$ with $u_* = -1$. This means that a switch to the $u = -1$ subarc is optimal.

16.5 Consider the problem of minimizing the performance index

$$J = t_f + k \int_{t_0}^{t_f} u^2 dt,$$

subject to the differential constraint

$$\dot{x} = u$$

and the prescribed boundary conditions

$$t_0 = 0, \quad x_0 = 0, \quad x_f = 1.$$

Show that the natural boundary condition $H_f = -G_{t_f}$ admits the solution $u = \pm 1/\sqrt{k}$ but that the corner conditions require that $u_{c+} = u_{c-}$ so that no corner can exist.

16.6 Consider the following optimal control problem:

$$J = \phi + \int_{t_0}^{t_f} L \, dt$$

$$\dot{x} = f$$

$$t_0 = t_{0_s}, \quad x_0 = x_{0_s}$$

$$\psi = 0.$$

where the control is a pulse of given magnitude and time $t_p$, that is,

$$u = 0, \qquad t_0 \leq t \leq t_1$$
$$u = 1, \quad t_1 \leq t \leq t_2, \quad t_2 = t_1 + t_p$$
$$u = 0, \qquad t_2 \leq t \leq t_f.$$

The time of pulse initiation $t_1$ is unknown. One can think of a missile powered by a solid propellant motor that is dropped from an aircraft and whose optimal ignition time is to be determined. Derive the first differential and show that the corner conditions require the multiplier $\lambda$ to be continuous at $t_1$ and $t_2$ and

$$H_{1-} - H_{1+} + H_{2-} - H_{2+} = 0.$$

16.7 To illustrate the handling of a state discontinuity, consider the following optimal control problem:

$$J = \phi + \int_{t_0}^{t_f} L \, dt$$

$$\dot{x} = f$$

$$t_0 = t_{0_s}, \quad x_0 = x_{0_s}$$

$$t_c = t_{c_s}, \quad x_{c+} = x_{c-} - \Delta x$$

$$\psi = 0,$$

where $\Delta x$ is prescribed. One can think of a two-stage rocket where the burnout time of the first stage ($t_c$) is known, and the mass of the rocket undergoes a jump as the first stage motor is dropped. The way the problem is stated, all states change at $t_c$, but the actual state changes are governed by the elements of $\Delta x$. Derive the first differential, and show that the corner condition gives $\lambda_{c-} = \lambda_{c+}$.

# 17

# Path Constraints

## 17.1  Introduction

In this chapter, several path constraints are considered. These are the integral constraint, the control equality constraint, the state equality constraint, the control inequality constraint, and the state inequality constraint. With the exception of the Legendre–Clebsch condition, which is obtained from the Weierstrass condition, only first differential conditions are derived. For bounded controls, the optimality conditions plus the Weierstrass condition are equivalent to the Pontryagin Minimum Principle (Refs. PB and He).

Because there are so many types of path constraints and because some path constraints lead to optimal controls with a variety of configurations, no attempt is made to create a general sufficient condition for a minimum. One must rely on the Weierstrass and the Legendre–Clebsch conditions to get some assurance that the optimal control is a minimum. Nevertheless, for a particular problem that is not too complicated, it should be possible using theory developed in previous chapters to derive a sufficient condition for a minimum.

## 17.2 Integral Constraint

A scalar integral constraint has the form

$$\int_{t_0}^{t_f} F(t, x, u) dt = K, \qquad (17.1)$$

where $K$ is a constant. There are two equivalent ways to apply this constraint.

First, it can be converted into a differential constraint and two pre-scribed boundary conditions. If a new state, say $x_{n+1}$, is defined as

$$\dot{x}_{n+1} = F(t, x, u), \quad x_{n+1,0} = 0, \qquad (17.2)$$

the integral constraint becomes the final condition

$$x_{n+1,f} = K. \qquad (17.3)$$

In this way, the integral constraint fits directly into the standard optimal control problem. The Lagrange multiplier associated with the differential equation (17.2) is constant because the state $x_{n+1}$ does not occur on the right-hand side.

Second, the original method for handling such a constraint has been to adjoin it to the performance index by a constant Lagrange multiplier. After the first differential conditions are applied, the value of the Lagrange multiplier is determined by applying the integral constraint.

An integral constraint was originally called an isoperimetric constraint because it was used to fix the perimeter of a curve.

## 17.3 Control Equality Constraint

A control equality constraint explicitly contains the control and has the form

$$C(t, x, u) = 0, \qquad (17.4)$$

where $C$ is an $r \times 1$ vector. The effect of such a constraint is to reduce the number of independent controls. If there are $m$ controls, then there are $m - r$ independent controls.

For the optimal control problem of Chapter 13 with the added constraint (17.4), the augmented performance index is written as

$$J' = G(t_f, x_f, \nu) + \int_{t_0}^{t_f} [\hat{H}(t, x, u, \lambda, \mu) - \lambda^T \dot{x}] dt, \qquad (17.5)$$

where $\mu(t)$ is the Lagrange multiplier used to adjoin the new constraint. The end-point function $G$ is defined as before, but the extended Hamiltonian is defined as

$$\hat{H} = L + \lambda^T f + \mu^T C = H + \mu^T C, \qquad (17.6)$$

where $H(t, x, u, \lambda)$ is the regular Hamiltonian, that is, $H = L + \lambda^T f$.

Taking the differential of Eq.(17.5) and performing the standard manipulations leads to

$$\begin{aligned} dJ' = {} & (G_{t_f} + \hat{H}_f) dt_f + (G_{x_f} - \lambda_f^T) dx_f + \psi d\nu \\ & + \int_{t_0}^{t_f} [(\hat{H}_x + \dot{\lambda}^T) \delta x + \hat{H}_u \delta u + (f - \dot{x})^T \delta\lambda + C^T \delta\mu] dt. \end{aligned} \qquad (17.7)$$

To obtain the first differential conditions, it is seen that the terms involving the multiplier variations are zero. Next, the multipliers $\lambda$ and $\nu$ are chosen so that the coefficients of $dt_f, dx_f$, and $\delta x$ are zero. Then, since the controls are not independent, the multipliers $\mu$ are chosen so that the coefficients of the dependent control variations vanish. Finally, the reasoning of Chapter 13 indicates that the coefficients of the independent control variations must vanish. Hence, the first differential conditions are given by

$$\dot{x} = f, \quad \dot{\lambda} = -\hat{H}_x^T, \quad 0 = \hat{H}_u^T, \quad C = 0$$
$$\psi = 0, \quad \hat{H}_f = -G_{t_f}, \quad \lambda_f = G_{x_f}^T. \qquad (17.8)$$

Note that any time $\hat{H}$ appears without differentiation it becomes $H$ since $C = 0$ at every point of the optimal path. If there is a control discontinuity, the corner conditions of Chapter 16 written in terms of $\hat{H}$ are valid here. Finally, if $L, f$ and $C$ do not explicitly contain the time, the first integral $\hat{H} = $ Const holds and becomes $H = $ Const since $C = 0$.

The development of the Weierstrass condition follows the same reasoning as in Chapter 10 with the exception that only the $m - r$ independent controls are given a pulse over $[t_p, t_q]$. The dependent controls are obtained from the constraint (17.4). The final result for a minimum is that

$$\hat{H}(t, x, u_*, \lambda, \mu) - \hat{H}(t, x, u, \lambda, \mu) > 0 \qquad (17.9)$$

for all $u_* \neq u$ at every point of the optimal path for all $-\infty < u_* < \infty$ that satisfy the constraint

$$C(t, x, u_*) = 0. \tag{17.10}$$

Because of the constraints (17.4) and (17.10), Eq.(17.9) can be rewritten as

$$H(t, x, u_*, \lambda) - H(t, x, u, \lambda) > 0. \tag{17.11}$$

Hence, the Hamiltonian must be an absolute minimum with respect to the set of $m - r$ independent controls.

For weak variations in the control, that is, $u_* = u + \delta u$, the Weierstrass condition (17.10) leads to

$$\delta u^T \hat{H}_{uu}(t, x, u) \delta u \geq 0 \tag{17.12}$$

for all $\delta u$ satisfying

$$C_u(t, x, u) \delta u = 0. \tag{17.13}$$

Hence, once the dependent control variations are eliminated from the quadratic form (17.12), the reduced quadratic form must be positive semidefinite for a minimum.

Note that Eq.(17.9) must be used as the starting point for deriving the Legendre–Clebsch condition because $\hat{H}_u = 0$. Equation (17.10) cannot be used because $H_u \neq 0$.

If the control equality constraint (17.4) is imposed only over a subinterval $[t_1, t_2]$ of $[t_0, t_f]$, then three cases must be considered: (a) $t_1 = t_0$ and $t_2 < t_f$, (b) $t_1 > t_0$ and $t_2 = t_f$, and (c) $t_1 > t_0$ and $t_2 < t_f$. In each case, the integral in Eq.(17.4) must be divided into the sum of integrals over each subinterval. For case (a), the corner conditions of Chapter 16 must be applied at $t_2$. For case (b), the corner conditions must be applied at $t_1$. For case (c), the corner conditions must be applied at $t_1$ and $t_2$. If there exist multiple intervals with $C = 0$, then each interval is handled as indicated above.

## 17.4   State Equality Constraint

A state equality constraint has the form

$$S(t, x) = 0 \tag{17.14}$$

where $S$ is an $s \times 1$ vector and $S$ does not explicitly contain the control. The effect of such a constraint is to reduce the number of independent controls and to affect the prescribed boundary conditions.

To demonstrate this, consider the dynamical system $\dot{x} = f$ and a scalar constraint of the form (17.14). First, Eq.(17.14) is differentiated with respect to time, that is,

$$\dot{S} = S_t + S_x \dot{x} = 0. \tag{17.15}$$

Then, the state equation $\dot{x} = f$ is substituted to obtain

$$\dot{S} = S_t + S_x f = 0. \tag{17.16}$$

While this result appears to contain the control, $S_t, S_x$, and $f$ may be such that $\dot{S}$ does not contain the control. Then, $\dot{S}$ has the functional form $\dot{S}(t, x)$. With $S^{(1)} \triangleq \dot{S}$, the procedure of differentiation and substitution of $\dot{x} = f$ is continued until

$$S^{(q)} = S^{(q)}(t, x, u) = 0, \tag{17.17}$$

that is, until the $q$th derivative explicitly contains the control. Such a constraint is called a $q$th-order state equality constraint. For dynamics problems originating from Newton's second law ($F - ma$), these constraints are no more than second order.

Even though Eq.(17.17) contains the control, its satisfaction does not guarantee satisfaction of Eq.(17.14). Note that by integration with respect to time

$$
\begin{aligned}
S^{(q-1)} &= C_1 \\
S^{(q-2)} &= C_1 t + C_2 \\
&\vdots \\
S^{(0)} &= C_1 \frac{t^{q-1}}{(q-1)!} + C_2 \frac{t^{q-2}}{(q-2)!} + \cdots + C_{q-1} t + C_q.
\end{aligned} \tag{17.18}
$$

Hence, the only way to satisfy $S \triangleq S^{(0)} = 0$ is to require that $C_1, C_2, \ldots, C_q$ be zero. This can be achieved by requiring that $S^{(q-1)}, S^{(q-2)}, \ldots, S^{(0)}$ be zero at a point. However, within the context of the optimization problem, these conditions must be satisfied at every point where additional point constraints are being applied, that is, the initial point, the final point, and at any intermediate point.

In summary, the state equality constraint (17.14) can be replaced by the control equality constraint

$$S^{(q)}(t, x, u) = 0 \tag{17.19}$$

and the point constraints

$$S^{(q-1)}(t_p,\ x_p) = 0,\quad S^{(q-2)}(t_p,\ x_p) = 0,\ \ldots,\ S^{(0)}(t_p,\ x_p) = 0. \qquad (17.20)$$

If the state equality constraint (17.14) is imposed only over a subinterval $[t_1,\ t_2]$ of $[t_0,\ t_f]$, three cases must be considered: (a) $t_1 = t_0$ and $t_2 < t_f$; (b) $t_1 > t_0$ and $t_2 = t_f$, and (c) $t_1 > t_0$ and $t_2 < t_f$. For case (a), the prescribed initial conditions must satisfy the point constraints (17.20). For case (b), the final conditions must satisfy these constraints. For case (c), the point constraints need only be imposed at $t_1$.

## 17.5   Control Inequality Constraint

In almost all physical problems, the controls are subjected to inequality constraints. For example, for aircraft motion in a plane perpendicular to the surface of the earth, the control variables are the thrust $T$ and the lift coefficient $C_L$. Because of physical limitations of the engines, $T_{\min} \leq T \leq T_{\max}$. Also, because of the lifting characteristics of airplanes, the lift coefficient is bounded as $C_{L\min} \leq C_L \leq C_{L\max}$. For modern airplanes where $T_{\max}$ is greater than the weight of the airplane, the latter constraint can be omitted.

The general form of the control inequality constraint is

$$\bar{C}(t,x,u) \leq 0, \qquad (17.21)$$

where $\bar{C}$ is an $r \times 1$ vector and $r < m$. The inequality constraint is formed as $\leq 0$ so that the condition for a minimal boundary subarc has the sign $\geq 0$ as do all conditions for a minimum.

To illustrate the effect of a control inequality constraint, only one control and one control inequality constraint are considered. For multiple controls and multiple constraints, the procedures would be similar, but the solution process would be more complicated.

Formally, the optimal control problem being considered here is that of finding the scalar control $u(t)$ which minimizes the performance index

$$J = \phi(t_f, x_f) + \int_{t_0}^{t_f} L(t, x, u)dt, \qquad (17.22)$$

subject to the state dynamics

$$\dot{x} = f(t, x, u), \tag{17.23}$$

the prescribed initial conditions

$$t_0 = t_{0_s}, \quad x_0 = x_{0_s}, \tag{17.24}$$

the prescribed final conditions

$$\psi(t_f, \ x_f) = 0 \tag{17.25}$$

and the scalar control inequality constraint

$$\bar{C}(t, x, u) \leq 0. \tag{17.26}$$

Again, $\phi$ and $L$ are scalars, $x$ and $f$ are $n \times 1$ vectors, and $\psi$ is a $(p + 1) \times 1$ vector. This is the optimal control problem of Chapter 13 with a scalar control and one control inequality constraint.

There are some simple forms of $\bar{C}$ where the bounded control can be replaced by an unbounded control. For example, for the case where $u \geq u_1$ is required, a slack variable $\alpha$ (what Valentine called an auxiliary variable in Ref. Va) can be introduced as

$$u = u_1 + \alpha^2. \tag{17.27}$$

Then, $u$ can be eliminated from the problem in favor of the unbounded control $\alpha(t)$. Another example is the constraint $u_1 \leq u \leq u_2$. Here, the form

$$u = u_1 + (u_2 - u_1) \sin^2 \alpha \tag{17.28}$$

can be used to eliminate the bounded control.

In general, the approach used here is to introduce a slack variable $\alpha$ and convert the inequality constraint (17.26) into an equality constraint, that is,

$$C(t, x, u, \alpha) = \bar{C}(t, x, u) + \alpha^2 = 0. \tag{17.29}$$

Note that there are now two control variables, $u$ and $\alpha$. The augmented performance index is written as

$$J' = G(t_f, x_f, \nu) + \int_{t_0}^{t_f} [\hat{H}(t, x, u, \lambda, \alpha, \mu) - \lambda^T \dot{x}] dt. \tag{17.30}$$

While the functions $G$ and $H$ have the standard definitions, the extended Hamiltonian is defined as

$$\hat{H} = H + \mu(\bar{C} + \alpha^2). \tag{17.31}$$

Since, this problem just involves a control equality constraint, the results of Section 17.3 are valid and lead to

$$\dot{x} = f, \quad \dot{\lambda} = -\hat{H}_x^T, \quad 0 = \hat{H}_u, \quad 0 = \hat{H}_\alpha, \quad \bar{C} = -\alpha^2$$
$$\psi = 0, \quad \hat{H}_f = -G_{t_f}, \quad \lambda_f = G_{x_f}^T. \tag{17.32}$$

The equation $\hat{H}_\alpha = 0$ implies that $2\mu\alpha = 0$. Hence, either $\mu = 0$ which is an off-boundary subarc or $\alpha = 0$ which is an on-boundary subarc. For the off-boundary subarc ($\mu = 0$), $u(t)$ is determined from $\hat{H}_u = 0$, and $\alpha(t)$, from $\alpha^2 = -\bar{C}$. For the on-boundary subarc ($\alpha = 0$), $u(t)$ is obtained from $\bar{C} = 0$, and $\mu(t)$, from $\hat{H}_u = 0$.

The point where two subarcs join is a corner point, and the corner conditions of Chapter 16 hold. Since there are no conditions imposed on the location of the corner point, the corner conditions become

$$\hat{H}_{c+} = \hat{H}_{c-}, \quad \lambda_{c+} = \lambda_{c-} \tag{17.33}$$

and are used to find the optimal corner time and state.

To test for a minimal path, the Weierstrass condition for a minimum requires that

$$\hat{H}(t, x, u_*, \lambda, \alpha_*, \mu) - \hat{H}(t, x, u, \lambda, \alpha, \mu) > 0 \tag{17.34}$$

at every point of the optimal path and for all $u_*$ and $\alpha_*$ satisfying the constraint

$$\bar{C}(t, x, u_*) = -\alpha_*^2. \tag{17.35}$$

Since $\bar{C} + \alpha^2 = 0$, the definition (17.31) of $\hat{H}$ leads to

$$H(t, x, u_*, \lambda) - H(t, x, u, \lambda) > 0 \tag{17.36}$$

for all $u_* \neq u$ satisfying

$$\bar{C}(t, x, u_*) \leq 0. \tag{17.37}$$

This means that the minimal control makes the Hamiltonian an absolute minimum with respect to the control at each point of the minimal path.

At a corner, $H(t, x, u_*, \lambda) - H(t, x, u, \lambda)$ vanishes if $u_* = u_{c-}$ and $u = u_{c+}$. This is due to the corner conditions (17.33).

It is noted that the first differential conditions (17.32) plus the Weierstrass condition (17.36) are equivalent to the Pontryagin Minimum Principle (Refs. PB and He).

The Legendre–Clebsch condition for a weak minimum is obtained by setting $u_* = u + \delta u$ and $\alpha_* = \alpha + \delta\alpha$ in Eqs.(17.34) and (17.35) and expanding in a Taylor series. Since $\hat{H}_{u\alpha} = 0$ and $\hat{H}_{\alpha\alpha} = 2\mu$, the results are

$$\hat{H}_{uu}\delta u^2 + 2\mu\delta\alpha^2 \geq 0,  \tag{17.38}$$

and

$$\bar{C}_u\delta u + 2\alpha\delta\alpha = 0.  \tag{17.39}$$

Off the boundary ($\mu = 0$), Eq.(17.38) requires that

$$\hat{H}_{uu}(t, x, u, \lambda, \alpha, 0) \geq 0  \tag{17.40}$$

or

$$H_{uu}(t, x, u, \lambda) \geq 0.  \tag{17.41}$$

Equation (17.39) just gives the relation between $\delta\alpha$ and $\delta u$ which is not needed. On the boundary ($\alpha = 0$), Eq.(17.39) leads to $\delta u = 0$ (which requires that $\bar{C}_u \neq 0$) so that Eq. (17.38) gives

$$\mu \geq 0.  \tag{17.42}$$

Hence, the Legendre–Clebsch condition has the standard form $H_{uu} \geq 0$ off the boundary and requires that the Lagrange multiplier associated with the equality constraint be nonnegative on the boundary. Often, the Legendre–Clebsch condition is helpful in determining which subarcs are contained in the minimal path and their sequence.

For a problem in which $H$ is linear in $u$ ($H_u \neq 0$), and the control inequality constraint has the form $u_1 \leq u \leq u_2$, the optimal control is on the boundary. Since $H_{uu} = 0$, the Weierstrass condition (17.36) must be used to determine the control sequence. Since the control must be on one boundary or the other with the switch occurring when $H_{uu} = 0$, this is called a *bang-bang control problem*. On the other hand, if $H_u$ vanishes over a finite portion of the path, the control is called *singular* ($H_{uu} = 0$); these problems are discussed in Refs. BJ and CA.

## 17.6  Example

A simple example has been created in order to demonstrate the solution process for the case where a control inequality constraint exists. It is to find the control $u(t)$ that minimizes the performance index

$$J = \int_{t_0}^{t_f} (u - t)^2 dt \qquad (17.43)$$

subject to the differential constraint

$$\dot{x} = u, \qquad (17.44)$$

the prescribed initial conditions

$$t_0 = 0, \quad x_0 = \frac{1}{8}, \qquad (17.45)$$

the prescribed final conditions

$$t_f = 2, \quad x_f = 1, \qquad (17.46)$$

and the inequality constraint

$$u \leq 1. \qquad (17.47)$$

In the format of Section 17.5, the following functions are defined:

$$\bar{C} = u - 1 + \alpha^2 \qquad (17.48)$$

$$\hat{H} = (u - t)^2 + \lambda u + \mu(u - 1 + \alpha^2), \qquad (17.49)$$

where $\bar{C}_u = 1 \neq 0$. Then, the first differential conditions (17.32) for an off-boundary segment give

$$
\begin{aligned}
\mu &= 0 \\
\lambda &= \text{Const} \\
x &= \frac{t^2}{2} - \frac{\lambda}{2}t + \text{Const} \\
u &= t - \frac{\lambda}{2} \\
\alpha^2 &= 1 - u
\end{aligned}
\qquad (17.50)
$$

and for an on-boundary segment

$$
\begin{aligned}
\alpha &= 0 \\
\lambda &= \text{Const} \\
x &= t + \text{Const} \\
u &= 1 \\
\mu &= 2(t - 1) - \lambda.
\end{aligned}
\qquad (17.51)
$$

Note that the Lagrange multiplier $\lambda$ is constant on each subarc, but it may not be the same value. However, the corner conditions (17.33) state that, since there are no constraints on the location of the corner, the Hamiltonian (17.46) and the Lagrange multiplier $\lambda$ must be continuous across a corner. Hence, $\lambda$ has the same value along the entire optimal path.

Before continuing with the solution of the problem as stated, neglect the inequality constraint, and find the solution to the subproblem. In this connection, the off-boundary solution (17.50) for $x$ combined with the prescribed boundary conditions (17.45) and (17.46) leads to

$$x = \frac{t^2}{2} - \frac{9}{16}t + \frac{1}{8}$$

(17.52)

and, since $u = \dot{x}$,

$$u = t - \frac{9}{16}.$$

(17.53)

Hence, the control inequality constraint is violated at $t = 25/16$, and there is no one-subarc solution off the boundary. Similarly, there is no one-subarc solution on the boundary $u = 1$ because the prescribed boundary conditions cannot be satisfied. From this discussion, it seems reasonable to assume that the solution is composed of an off-boundary subarc followed by an on boundary subarc.

To gain additional insight to the solution of the problem, consider the corner conditions for the junction of an off-boundary subarc to an on-boundary subarc. From the corner conditions $H_{c+} = H_{c-}$, the definition (17.49) of $H$, and the result that $\lambda$ is the same constant over the entire optimal path, it is seen that

$$(u_{c+} - t)^2 + \lambda u_{c+} = (u_{c-} - t)^2 + \lambda u_{c-}.$$

(17.54)

Then, if

$$u_{c+} = 1, \quad u_{c-} = t_c - \frac{\lambda}{2},$$

(17.55)

Eq.(17.54) can be written as a perfect square and gives the following result:

$$\lambda = -2(1 - t_c).$$

(17.56)

Since there is only one value of $\lambda$, this equation shows that there is only one value of $t_c$ or that there can exist only one corner. Next, if Eq.(17.56) is combined with Eq.(17.51) for $\mu$, it is seen that

$$\mu = 2(t - t_c).$$

(17.57)

Since $\mu$ must be nonnegative along the on-boundary subarc, this equation requires that the $u = 1$ subarc follow the corner $(t \geq t_c)$. In summary, there can be only one corner, and the on-boundary subarc must follow the corner.

The next step is to apply the boundary conditions. Combining Eq.(17.50) for $x$ with the prescribed initial conditions (17.45) leads to

$$x = \frac{1}{2}t^2 - \frac{\lambda}{2}t + \frac{1}{8}. \tag{17.58}$$

Also, the on-boundary equation (17.51) for $x$ combined with the prescribed final conditions (17.46) gives

$$x = t - 1. \tag{17.59}$$

Then, both equations for $x$ evaluated at $t_c$ combined with Eq. (17.56) for $\lambda$ yields

$$t_c = \frac{3}{2}, \quad x_c = \frac{1}{2}, \quad \lambda = 1 \tag{17.60}$$

so that the optimal path becomes

$$0 \leq t \leq \tfrac{3}{2}, \quad x = \tfrac{1}{2}t^2 - \tfrac{1}{2}t + \tfrac{1}{8}, \quad u = t - \tfrac{1}{2}$$
$$\tfrac{3}{2} \leq t \leq 2, \quad x = t - 1, \quad u = 1 \tag{17.61}$$

The last step is to check the Legendre–Clebsch condition and the Weierstrass condition. Since

$$0 \leq t \leq \tfrac{3}{2}, \quad H_{uu} = 2 > 0$$
$$\tfrac{3}{2} \leq t \leq 2, \quad \mu = 2t - 3 \geq 0, \tag{17.62}$$

the Legendre–Clebsch conditions (17.41) and (17.42) are satisfied. Next, the Weierstrass condition (17.36) becomes

$$(u_* - t)^2 + \lambda u_* - (u - t)^2 - \lambda u > 0 \tag{17.63}$$

for all $u_* \leq 1$ and can be rewritten as

$$(u_*^2 - u^2) + (\lambda - 2t)(u_* - u) > 0. \tag{17.64}$$

Off the boundary, $\lambda = 2t - 2u$ so that this equation can be put in the form

$$(u_* - u)^2 \geq 0. \tag{17.65}$$

Hence, the Weierstrass condition is satisfied along the off-boundary subarc. On the boundary, $u = 1$ and $\lambda = 1$ so that Eq. (17.64) can be rewritten as

$$(u_* - 1)^2 + (2t - 3)(1 - u_*) + 2 > 0 \qquad (17.66)$$

and is satisfied for $u_* \leq 1$ and $3/2 \leq t \leq 2$.

Even though the Legendre–Clebsch test for a weak relative minimum and the Weierstrass test for a strong relative minimum are satisfied, a relative minimum cannot be assured until a conjugate point condition is established.

## 17.7 Acceleration Problem

Consider the problem of moving a car from zero position and velocity to another position and zero velocity in minimum time assuming that the acceleration is controllable and bounded. In other words, minimize the performance index

$$J = t_f, \qquad (17.67)$$

subject to the differential constraints

$$\dot{x}_1 = x_2 \qquad (17.68)$$
$$\dot{x}_2 = u, \qquad (17.69)$$

the prescribed boundary conditions

$$t_0 = 0, \quad x_{1_0} = 0, \quad x_{2_0} = 0 \qquad (17.70)$$

$$x_{1_f} = 1, \quad x_{2_f} = 0, \qquad (17.71)$$

and the control inequality constraint

$$-1 \leq u \leq 1. \qquad (17.72)$$

The double inequality constraint can be adjoined as two separate constraints

$$-1 - u \leq 0, \quad u - 1 \leq 0 \qquad (17.73)$$

using two slack variables. However, by multiplying them together, the single inequality constraint

$$\bar{C} \triangleq u^2 - 1 \leq 0 \qquad (17.74)$$

can be formed.  It is converted to an equality constraint by using a slack variable, that is.

$$C \triangleq u^2 - 1 - \alpha^2 = 0. \tag{17.75}$$

To begin the solution of the problem, the following functions are defined:

$$H = \lambda_1 x_2 + \lambda_2 u \tag{17.76}$$
$$\hat{H} = H + \mu(u^2 - 1 + \alpha^2) \tag{17.77}$$
$$G = t_f + \nu_1(x_{1_f} - 1) + \nu_2 x_{2_f}. \tag{17.78}$$

The next step is to list all of the first differential conditions.

The Euler–Lagrange equations defining the optimal path are given by

$$\dot{x} = u \;\Rightarrow\; \dot{x}_1 = x_2 \tag{17.79}$$
$$\dot{x}_2 = u \tag{17.80}$$
$$\dot{\lambda} = -\hat{H}_x^T \;\Rightarrow\; \dot{\lambda}_1 = 0 \quad \Rightarrow \lambda_1 = \text{Const} \tag{17.81}$$
$$\dot{\lambda}_2 = -\lambda_1 \Rightarrow \lambda_2 = -\lambda_1 t + \text{Const} \tag{17.82}$$
$$0 = \hat{H}_u^T \;\Rightarrow\; \lambda_2 + 2\mu u = 0 \tag{17.83}$$
$$2\mu\alpha = 0 \Rightarrow \mu = 0 \quad \text{or} \quad \alpha = 0. \tag{17.84}$$

Equation (17.84) indicates that there are three possible subarcs:

$$u = +1, \;\; +1 \geq u \geq -1, \;\; u = -1. \tag{17.85}$$

The only natural boundary condition that contributes to the solution is that associated with free final time:

$$\hat{H}_f + G_{t_f} = 0 \;\; \Rightarrow \;\; H_f = -1. \tag{17.86}$$

Since $t$ is not explicitly present in the problem, the first integral

$$H = \text{Const} \tag{17.87}$$

holds. However, at this point the constant has a different value along each subarc.

Since there are no prescribed constraints on the location of a junction or a jump in the control between subarcs, the natural corner conditions are given by

$$\hat{H}_{c+} = \hat{H}_{c-} \quad \Rightarrow \quad H_{c+} = H_{c-} \tag{17.88}$$

$$\lambda_{c+} = \lambda_{c-} \quad \Rightarrow \quad \lambda_{1c+} = \lambda_{1c-}, \quad \lambda_{2c+} = \lambda_{2c-}. \tag{17.89}$$

Hence, the Hamiltonian and each Lagrange multiplier must be continuous across a corner. This means that $H$ and $\lambda_1$ have the same constant values along the optimal path.

Next, the Weierstrass condition is given by

$$(\lambda_1 x_2 + \lambda_2 u_*) - (\lambda_1 x_2 + \lambda_2 u) > 0 \tag{17.90}$$

or

$$\lambda_2(u_* - u) > 0, \tag{17.91}$$

where the comparison control and the optimal control must satisfy the inequality constraints

$$u_*^2 - 1 \le 0, \quad u^2 - 1 \le 0. \tag{17.92}$$

Consider a point on the upper boundary where $u = +1$ and $u_* < +1$. Since $u_* - u < 0$, the relation $\lambda_2 > 0$ must hold on a $u = +1$ subarc. Similarly, on a $u = -1$ subarc, it must be that $\lambda_2 < 0$. Finally, on an off-boundary subarc, $-1 < u < 1$, the Weierstrass condition cannot be satisfied because $u_* - u$ is plus and minus. Such a subarc is not minimizing. Hence, the only possible subarcs now are $u = +1$ and $u = -1$.

At a corner where there are no constraints prescribed on the location of the corner, the Weierstrass condition vanishes if $u_* = u_{c-}$ and $u = u_{c+}$ or vice versa (Section 16.4). Hence, it becomes

$$\lambda_{2_c}(u_{c+} - u_{c-}) = 0, \tag{17.93}$$

which is the same result obtained from the corner conditions (17.88) and (17.89). Since $u_{c+} - u_{c-} = \pm 2$, it is seen that

$$\lambda_{2_c} = 0 \tag{17.94}$$

and that

$$\lambda_2 = \lambda_1(t_c - t) \tag{17.95}$$

from Eq.(17.82). This result shows that there is only one corner because $\lambda_2$ can vanish only once. The order of the subarcs is still unknown.

The multiplier $\lambda_2$ takes on the character of a *switching function*. When $\lambda_2$ changes sign, the subarc changes.

The geometry of each subarc can be obtained from the system equations as follows:

$$
\begin{aligned}
u &= 1 & u &= -1 \\
\dot{x}_1 &= x_2 & \dot{x}_1 &= x_2 \\
\dot{x}_2 &= 1 & \dot{x}_2 &= -1 \\
\frac{dx_1}{dx_2} &= x_2 & \frac{dx_1}{dx_2} &= -x_2 \\
x_1 &= \frac{x_2^2}{2} + \text{Const} & x_1 &= -\frac{x_2^2}{2} + \text{Const} \\
x_2 &= \pm\sqrt{2(x_1 - \text{Const})} & x_2 &= \pm\sqrt{2(\text{Const} - x_1)}.
\end{aligned}
\tag{17.96}
$$

These equations represent two families of parabolas as shown in Fig. 17.1. Furthermore, $\dot{x}_2 > 0$ along a $u = 1$ subarc, and $\dot{x}_2 < 0$ along a $u = -1$ subarc.

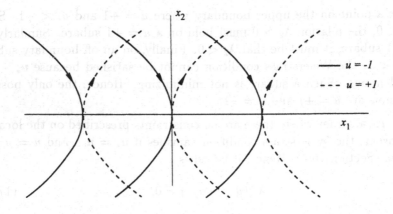

Figure 17.1: Families of Optimal Paths

Hence, the parabolas can only be traveled in the directions indicated by the arrows. Given the initial state $x_{1_0} = 0$, $x_{2_0} = 0$ and the final state $x_{1_f} = 1$, $x_{2_f} = 0$, the solution of the minimum time problem is shown in Fig. 17.2.

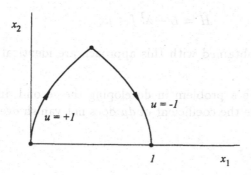

Figure 17.2: Particular Optimal Path

To complete the solution of the problem, it can be shown that Eq.(17.86) with $x_{2_f}$ gives $\lambda_{2_f} = 1$. Then, from Eq.(17.95),

$$\lambda_1 = \frac{-1}{t_f - t_c} < 0. \qquad (17.97)$$

This result shows that $\lambda_2 < 0$ before the corner and $\lambda_2 > 0$ after the corner. Since $u = +1$ before the corner and $u = -1$ after the corner, Eq.(17.83) shows that $\mu \geq 0$ along both boundary subarcs. Hence, the Legendre–Clebsch condition (17.42) is satisfied along the optimal path.

The results of this section also apply to the problem of changing the orientation of a rigid body when the equation of motion about a single axis is given by $I\ddot{\theta} = M$. The symbol $\theta$ denotes the angular orientation about the axis, $I$ is the moment of inertia, and $M$ is the controllable moment.

## 17.8   Alternate Approach for $\bar{C}(t, x, u) \leq 0$

Another way to handle the control inequality constraint is to write

$$\mu\bar{C} = 0 \qquad (17.98)$$

with the understanding that either $\mu = 0$ or $\bar{C} = 0$ at each point of the optimal path. Then, Eq.(17.98) is integrated over $[t_0,\ t_f]$ and added to the performance index so that the extended Hamiltonian becomes

$$\hat{H} = L + \lambda^T f + \mu \bar{C}. \tag{17.99}$$

All of the conditions obtained with this approach are identical with those of the slack variable.

There may be a problem in developing the second differential with this symbology because the coefficient of $d\mu$ does not vanish everywhere along the optimal path.

## 17.9  State Inequality Constraint

The formal statement of the optimal control problem with a state inequality constraint is to find the control history that minimizes the performance index (17.23), subject to the dynamics (17.24), the prescribed initial conditions (17.25), the prescribed final conditions (17.26), and the inequality constraint

$$\bar{S}(t, x) \leq 0. \tag{17.100}$$

It is desired to obtain the first differential conditions to be satisfied by an optimal path that contains a subarc along the boundary $\bar{S} = 0$. To simplify the discussion, consider again the case of a scalar control and one inequality constraint. As shown in Section 17.4, the state equality constraint $\bar{S}(t, x) = 0$ can be replaced by a control equality constraint $\bar{S}^{(q)}(t, x, u) = 0$ and a set of point conditions $\bar{S}_p^{(q-1)} = S_p^{(q-2)} = \cdots = \bar{S}_p^{(0)} = 0$. If the optimal path begins with a boundary subarc, the point conditions must be applied at the initial point, and the prescribed initial conditions must satisfy these constraints. If the optimal path ends with a boundary subarc, the point conditions must be applied at the final point, and the prescribed final conditions must be consistent with these constraints. Finally, if the boundary subarc occurs between the end points, the point conditions need be applied only at the beginning of the subarc.

To illustrate the approach, consider the case where the boundary subarc occurs between the endpoints, that is, over the interval $[t_{c_1}, t_{c_2}]$. Then, the optimal control problem is to find the control history $u(t)$ that minimizes the performance index (17.23), subject to the dynamics (17.24), the prescribed initial conditions (17.25), the prescribed final conditions (17.26), the control equality constraint (17.19) along the boundary, and the point constraints

$$\begin{aligned}
\theta_1 &= \bar{S}^{(q-1)}(t_{c_1}, x_{c_1}) = 0 \\
\theta_2 &= \bar{S}^{(q-2)}(t_{c_1}, x_{c_1}) = 0 \\
&\vdots \\
\theta_q &= \bar{S}^{(0)}(t_{c_1}, x_{c_1}) = 0
\end{aligned} \qquad (17.101)$$

at the beginning of the boundary.

For this problem, the augmented performance index becomes

$$\begin{aligned}
J' &= G(t_f, x_f, \nu, t_c, x_c, \xi) + \int_{t_0}^{t_{c_1}} [H(t, x, u, \lambda) - \lambda^T \dot{x}] dt \\
&+ \int_{t_{c_1}}^{t_{c_2}} [\hat{H}(t, x, u, \lambda, \mu) - \lambda^T \dot{x}] dt + \int_{t_{c_2}}^{t_f} [H(t, x, u, \lambda) - \lambda^T \dot{x}] dt,
\end{aligned} \qquad (17.102)$$

where

$$\begin{aligned}
G &= \phi + \nu^T \psi + \xi^T \theta \\
t_0 \leq t \leq t_{c_1} &: \quad H = L + \lambda^T f \\
t_{c_1} \leq t \leq t_{c_2} &: \quad \hat{H} = L + \lambda^T f + \mu^T S^{(q)} \\
t_{c_2} \leq t \leq t_f &: \quad H = L + \lambda^T f.
\end{aligned} \qquad (17.103)$$

Then, standard differential techniques lead to the following first differential conditions:

$$\begin{aligned}
t_0 \leq t \leq t_{c_1} &: \quad \dot{x} = f, \quad \dot{\lambda} = -H_x^T, \quad 0 = H_u^T \\
t_{c_1} \leq t \leq t_{c_2} &: \quad \dot{x} = f, \quad \dot{\lambda} = \hat{H}_x^T, \quad 0 = \hat{H}_u^T \\
t_{c_2} \leq t \leq t_f &: \quad \dot{x} = f, \quad \dot{\lambda} = -H_x^T, \quad 0 = H_u^T \\
t_{c_1} &: \quad H_{c_1+} = H_{c_1-} + G_{t_{c_1}}, \quad \lambda_{c_1+} = \lambda_{c_1-} - G_{x_{c_1}}^T \\
t_{c_2} &: \quad H_{c_2+} = H_{c_2-}, \quad \lambda_{c_2+} = \lambda_{c_2-} \\
t_f &: \quad \psi = 0, \quad H_f = -G_{t_f}, \quad \lambda_f = G_{x_f}^T.
\end{aligned} \qquad (17.104)$$

Note that while a jump can occur in $H$ and $\lambda$ going on the boundary, $H$ and $\lambda$ are continuous coming off the boundary.

Off the boundary, the Hamiltonian must be an absolute minimum with respect to the control. On the boundary, $H$ must be a minimum with respect to the controls that satisfy the constraint $\bar{S}^{(q)}(t, x, u_*) = 0$. With regard to the Legendre–Clebsch condition, the inequality $H_{uu} \geq 0$ must be satisfied off the boundary and $\mu \geq 0$ on the boundary.

## 17.10    Example

As an example of a state inequality constraint, consider the problem of minimizing the performance index (17.44), subject to the dynamics (17.45), the prescribed initial conditions (17.46), the prescribed final conditions (17.47), and the inequality constraint

$$x \geq 0. \tag{17.105}$$

For this problem, $\bar{S} = -x \leq 0$.

If the constraint (17.105) is ignored temporarily, the off-boundary solution (17.51) subject to the boundary conditions (17.46) and (17.47) leads to

$$x = \frac{1}{2}t^2 - \frac{9}{16}t + \frac{1}{8} \tag{17.106}$$

and can be shown to satisfy the inequality constraint at the initial and final points but violate it between $t_0$ and $t_f$. Hence, the optimal path subject to the inequality constraint must contain a subarc along the boundary.

To convert the state inequality constraint into a control equality constraint and internal point constraints, the process of Section 17.9 is applied. Differentiating $\bar{S} = -x$ once leads to

$$\bar{S}^{(1)} = -\dot{x} = -u. \tag{17.107}$$

Hence, the state inequality constraint (17.105) is a first-order constraint and along the boundary is equivalent to the control equality constraint

$$u = 0 \tag{17.108}$$

and the intermediate point constraint

$$x_{c_1} = 0. \tag{17.109}$$

In the terminology of Section 17.9, this means that

$$\begin{aligned} \hat{H} &= (u - t)^2 + \lambda u + \mu(-u) \\ G &= \xi(-x_{c_1}), \end{aligned} \tag{17.110}$$

where only the corner terms are listed in $G$. Since the final point is specified, the natural final conditions just give the values of the multipliers used to adjoin the prescribed boundary conditions.

The Euler–Lagrange equations (17.104) yield the following results off the boundary:

$$\mu = 0$$
$$x = \tfrac{1}{2}t^2 - \tfrac{\lambda}{2}t + \text{Const}$$
$$\lambda = \text{Const} \tag{17.111}$$
$$u = t - \tfrac{\lambda}{2}$$

and the following results on the boundary:

$$u = 0$$
$$x = \text{Const}$$
$$\lambda = \text{Const} \tag{17.112}$$
$$\mu = -2t + \lambda.$$

Application of the prescribed boundary conditions to an off-on-off optimal path leads to

$$t_0 \leq t \leq t_{c_1}: \quad x = \tfrac{1}{2}t^2 - \tfrac{\lambda}{2}t + \tfrac{1}{8}, \quad u = t - \tfrac{\lambda}{2}$$
$$t_{c_1} \leq t \leq t_{c_2}: \quad x = 0, \quad u = 0, \quad \mu = \lambda - 2t \tag{17.113}$$
$$t_{c_2} \leq t \leq t_f: \quad x = \tfrac{1}{2}t^2 - \tfrac{\lambda}{2}t + \lambda - 1, \quad u = t - \tfrac{\lambda}{2}.$$

While $\lambda$ is constant along each subarc, it may have different values from subarc to subarc.

The conditions needed to join the subarcs are obtained from Eq. (17.103) and at $t_{c_1}$ are given by

$$x_{c_1} = 0$$
$$(u_{c_1+} - t_{c_1})^2 + \lambda_{c_1+}u_{c_1+} = (u_{c_1-} - t_{c_1})^2 + \lambda_{c_1-}u_{c_1-} \tag{17.114}$$
$$\lambda_{c_1+} = \lambda_{c_1-} + \xi$$

and at $t_{c_2}$, by

$$(u_{c_2+} - t_{c_2})^2 + \lambda_{c_2+}u_{c_1+} = (u_{c_2-} - t_{c_2})^2 + \lambda_{c_2-}u_{c_2-}$$
$$\lambda_{c_2+} = \lambda_{c_2-}. \tag{17.115}$$

At $t_{c_1}$, Eqs.(17.113) and (17.114) yield

$$\tfrac{1}{2}t_{c_1}^2 - \tfrac{\lambda_{c_1-}}{2}t_{c_1} + \tfrac{1}{8} = 0$$
$$2t_{c_1} - \lambda_{c_1-} = 0 \tag{17.116}$$

or

$$t_{c_1} = \frac{1}{2}, \quad \lambda_{c_1-} = 1.$$  (17.117)

At $t_{c_2}$, Eqs.(17.113) and (17.115) can be combined to obtain

$$\tfrac{1}{2}t_{c_2}^2 - \tfrac{\lambda_{c_2}}{2}t_{c_2} + \lambda_{c_2} - 1 = 0$$  (17.118)
$$2t_{c_2} - \lambda_{c_2} = 0$$

or

$$t_{c_2} = 2 - \sqrt{2}, \quad \lambda_{c_2} = 2(2 - \sqrt{2}).$$  (17.119)

Note that there is a jump in $\lambda$ at $t_{c_1}$.

Off the boundary, $H_{uu} = 2$, and on the boundary, $\mu = -2t + \lambda_{c_2} = 2(t_{c_2} - t) \geq 0$ since $t \leq t_{c_2}$. Hence, the Legendre–Clebsch condition for a minimum is satisfied, but a conjugate point test must be made before a minimum is assured. It is not difficult to verify that the Weierstrass condition is satisfied.

# Problems

17.1 An interesting problem that introduced the integral constraint can be simply stated as follows: Find the curve of given length passing between fixed endpoints that maximizes the enclosed area. Formally, find the function $y(x)$ that minimizes the performance index

$$J = -\int_{x_0}^{x_f} y \, dx,$$

subject to the constant-length constraint

$$\int_{x_0}^{x_f} \sqrt{1 + y'^2} \, dx = l,$$

and the prescribed final conditions

$$x_0 = 0, \quad y_0 = 0, \quad x_f = 1, \quad y_f = 0.$$

Formulate this problem as a standard optimal control problem, and show that the solution is an arc of a circle. This problem is the origin of the integral constraint.

17.2 Find the supersonic airfoil shape of given chord $l$ and enclosed area $A$
that minimizes the pressure drag. In other words, find the curve $y(x)$
that minimizes the performance index

$$J = \int_{x_0}^{x_f} y'^2 dx,$$

subject to the prescribed boundary conditions

$$x_0 = 0, \quad y_0 = 0, \quad x_f = l, \quad y_f = 0$$

and the integral constraint

$$\int_{x_0}^{x_f} y\,dx = A.$$

Show that the optimal shape is a parabola and that the maximum thickness $t = 3A/2l$ occurs at midchord.

17.3 Solve the example problem of Section 17.6 for the case where $0 \le u \le 1$.
Show that the correct sequence of subarcs is $u = 0$ followed by $u = t - 5/8$
followed by $u = 1$ and that $t_{c_1} = 5/8$ and $t_{c_2} = 13/8$.

17.4 Find the control history that maximizes the performance index

$$J = x_f,$$

for the dynamical system

$$\dot{x} = u^2,$$

subject to the prescribed boundary conditions

$$t_0 = 0, \quad x_0 = 0, \quad t_f = 1$$

and the control inequality constraint

$$0 \le u \le 1.$$

First, show that there can be no off-boundary subarc. Second, show that there can be no jump between the boundary controls. Finally, using the Weierstrass condition, show that the optimal control is $u = 1$ all the way; also, show that the Legendre–Clebsch condition is satisfied.

17.5 Find the control history that minimizes the time to transfer the system

$$\dot{x} = u$$

from the initial conditions

$$t_0 = 0, \quad x_0 = 0$$

to the final condition

$$x_f = 1.$$

The control is required to satisfy the inequality constraint

$$u \le u_{\max}.$$

17.6 Find the control history that minimizes the performance index

$$J = x_f^2$$

subject to the differential constraint

$$\dot{x} = u,$$

the prescribed boundary conditions

$$t_0 = 0, \quad x_0 = 1, \quad t_f = 1,$$

and to the control inequality constraint

$$u \ge 0.$$

Show that the optimal control is $u = 0$ everywhere and that the Weier-strass and Legendre–Clebsch conditions for a minimum are satisfied.

17.7 Problem 17.6 can be reformulated by eliminating the bounded control $u$ in favor of the unbounded control $\alpha$. The new problem is to find the control history $\alpha(t)$ that minimizes the performance index

$$J = x_f^2,$$

subject to the differential constraint

$$\dot{x} = \alpha^2,$$

and the prescribed boundary conditions

$$t_0 = 0, \quad x_0 = 1, \quad t_f = 1.$$

Show that the optimal control is $\alpha = 0$ everywhere and that the Weier-strass and Legendre–Clebsch conditions for a minimum are satisfied.

# Part III Approximate Solutions

Part III has several purposes. One is to discuss the approximate analytical solution of perturbation problems, that is, problems involving a small parameter. Another is to discuss the conversion of an optimal control problem into a parameter optimization problem. An underlying purpose is to use differentials to carry out whatever Taylor series expansions are needed.

For perturbation problems, the perturbed point or path and the nominal point or path are related by a system of small changes. Differentials are used to derive the equations for the first-order part, the second-order part, and so on, of the total change. For many problems, the zeroth-order equations have an analytical solution and only a first-order correction is being sought.

For the algebraic perturbation problem (Chapter 18), both the Taylor series approach and the differential approach are applied, and the efforts required by each are compared. For the perturbation problem involving differential equations (Chapter 19), differentials are used to derive the various-order differential equations and boundary conditions (fixed final time and free but constrained final time) needed to get an approximate analytical solution. For both algebraic and differential systems, the interest is in the derivation of the various-order equations, not in their solution. In Chapter 20, the approximate analytical solution of an optimal control problem involving a small parameter is discussed. Here, a particular problem is analyzed, and the various-order solutions are obtained. Also, if the zeroth-order solution is almost good enough and only the first-order correction is wanted, it is shown that the first-order correction is the neighboring optimal path, and its solution is obtained by using the sweep method.

In Chapter 21, the conversion of optimal control problems into parameter optimization problems is discussed. The conversion techniques depend on what quantities are guessed (control values and/or state values), the kind of numerical integration method to be used (explicit or implicit), and the order of the integrator. While this chapter is more concerned with numerical optimization, it is here because the differential is used to derive the equations of condition for the numerical integrators.

# 18

# Approximate Solutions of Algebraic Equations

## 18.1 Introduction

The overall theme of this and the next two chapters is the approximate analytical solution of algebraic, differential, and optimal control equations. There are several ways to get an approximate analytical solution of an equation. First, approximations can be made in the equation such that the remaining equation has an analytical solution. If the approximate solution is accurate enough, work stops. If not, something more must be done. Second, instead of discarding the small terms, they can be replaced by a small parameter, thereby creating a perturbation problem. The effect of the size of the small parameter can be investigated by simulation. Third, if the original equation contains a small parameter and if the equation with the small parameter set equal to zero has an analytical solution, a perturbation problem results.

The usual procedure for solving a perturbation problem is to recognize that the solution can be expressed as a power series in the small parameter, substitute the series into the equation, expand in a Taylor series, and equate the coefficients of the powers of the small parameter to zero. This process leads to the equations for the zeroth-order part of the solution, the first-order part of the solution, and so on.

An algebraic perturbation problem has a nominal point (solution with the small parameter equal to zero) and a perturbed point defined by a given value of the small parameter. In Section 2.2 in association with parameter optimization, the nominal point is the minimal point and the perturbed point is the comparison point. It has been shown that the Taylor series process is equivalent to taking differentials with the differential of an independent differential equal to zero. The specific theme of Chapters 18 through 20 is the application of differentials to obtain the equations of the various-order solutions of a perturbation problem. It is shown that the differential approach may have advantages over the expansion approach in the derivation of these equations.

First, the algebraic perturbation problem is defined. Then, the expansion process and the differential process are both used to derive the equations for the zeroth-order part, the first-order part, and so on of the algebraic perturbation problem in general and in particular for Kepler's equation with small eccentricity. The efforts required by both processes are compared.

While only a scalar equation is considered here, the differential process can be applied to vector equations if indicial notation is employed. If only the first-order correction to the zeroth-order solution is needed, matrix notation can be used.

## 18.2  Algebraic Perturbation Problem

The algebraic perturbation problem is to find an approximate analytical solution of the scalar equation

$$f(z, \varepsilon) = 0, \tag{18.1}$$

where the scalar $z$ is the unknown and $\varepsilon$ is a given, small, scalar parameter. Hence, $\varepsilon$ is the independent variable, and $z$ is the dependent variable. It is assumed that an analytical solution exists for the equation $f(z,0) = 0$. The solution of Eq. (18.1) for a particular value of $\varepsilon$ is denoted by $z_*$, that is, $f(z_*, \varepsilon) = 0$. In general, the solution of this equation has the functional form

$$z_* = g(\varepsilon). \tag{18.2}$$

For small $\varepsilon$, a Taylor series expansion leads to

$$z_* = g(0) + g_\varepsilon(0)\varepsilon + \frac{1}{2!}g_{\varepsilon\varepsilon}(0)\varepsilon^2 + \cdots, \tag{18.3}$$

meaning that the solution can be expressed as a power series in $\varepsilon$.

In what follows, the equations for finding an approximate analytical solution of Eq.(18.1) are obtained four ways: applying the expansion and differential processes to the general problem (18.1) and then to a particular problem (Kepler's equation with small eccentricity). The objective is to demonstrate the use of the differential process.

## 18.3    Expansion Process for the General Problem

For the expansion process, Eq.(18.3) is rewritten in the form

$$z_* = z_0 + z_1 \varepsilon + z_2 \varepsilon^2 + \cdots , \tag{18.4}$$

where $z_0$ is called the zeroth-order solution, $z_1 \varepsilon$ is called the first-order solution, and so on. Equation (18.4) can be rewritten as

$$z_* = z_0 + \Delta z \tag{18.5}$$

where

$$\Delta z = z_1 \varepsilon + z_2 \varepsilon^2 + \cdots . \tag{18.6}$$

Then, $z_*$ is substituted into Eq.(18.1) which in turn is expanded in terms of the small quantities $\Delta z$ and $\varepsilon$ to give

$$f(z_0 + \Delta z, \varepsilon) = f(z_0, 0) + f_z(z_0, 0)\Delta z + f_\varepsilon(z_0, 0)\varepsilon$$
$$\tfrac{1}{2!}[f_{zz}(z_0, 0)\Delta z^2 + 2f_{\varepsilon z}(z_0, 0)\varepsilon\Delta z + f_{\varepsilon\varepsilon}(z_0, 0)\varepsilon^2] + \cdots = 0. \tag{18.7}$$

Then, with Eq.(18.6), Eq.(18.7) is rewritten in terms of powers of $\varepsilon$, and coefficients of the various powers of $\varepsilon$ are equated to zero. This process gives the equations

$$f(z_0, 0) = 0$$

$$f_z(z_0, 0)z_1 + f_\varepsilon(z_0, 0) = 0$$

$$f_z(z_0, 0)z_2 + \tfrac{1}{2!}[f_{zz}(z_0, 0)z_1^2 + 2f_{\varepsilon z}(z_0, 0)z_1 + f_{\varepsilon\varepsilon}(z_0, 0)] = 0 \tag{18.8}$$

$$\vdots$$

which can be solved sequentially for $z_0$, $z_1$, $z_2$,.... Then, the approximate analytical solution of Eq.(18.1) is given by Eq.(18.4).

To obtain the next-order equation, another order is added to Eqs.(18.6) and (18.7); Eq.(18.6) is substituted into Eq.(18.7); and some rearrangment is required to get the desired equation. Note that the entire process is repeated.

## 18.4   Differential Process for the General Problem

For the differential process (Section 2.2), Eq.(18.3) is rewritten as

$$z_* = z + dz + \frac{1}{2!}d^2z + \cdots, \tag{18.9}$$

where $z$ is the zeroth-order solution, $dz$ is the first-order solution, and so on. Then, starting from Eq.(18.1) and taking differentials leads to the following equations:

$$f(z,\varepsilon) = 0$$
$$f_z(z,\varepsilon)dz + f_\varepsilon(z,\varepsilon)d\varepsilon = 0$$
$$f_z(z,\varepsilon)d^2z + f_{zz}(z,\varepsilon)dz^2 + 2f_{\varepsilon z}(z,\varepsilon)dzd\varepsilon + f_{\varepsilon\varepsilon}(z,\varepsilon)d\varepsilon^2 = 0 \tag{18.10}$$
$$\vdots$$

where $d(d\varepsilon) = 0$ because $d\varepsilon$ is an independent differential. Finally, these equations are evaluated at $\varepsilon = 0$ ($z$ becomes the zeroth-order solution) to obtain

$$f(z,0) = 0$$
$$f_z(z,0)dz + f_\varepsilon(z,0)d\varepsilon = 0$$
$$f_z(z,0)d^2z + f_{zz}(z,0)dz^2 + 2f_{\varepsilon z}(z,0)dzd\varepsilon + f_{\varepsilon\varepsilon}(z,0)d\varepsilon^2 = 0 \tag{18.11}$$
$$\vdots$$

These equations can be solved sequentially for $z$, $dz$, $d^2z$, ..., and the solution of Eq.(18.1) is given by Eq.(18.9).

To get the next-order equation, it is only necessary to take the variation of the last-order equation in (18.10) and evaluate the result on $\varepsilon = 0$.

Note that the total change in $\varepsilon$ equals the first-order change, that is, $\Delta\varepsilon = d\varepsilon$, because $d\varepsilon$ is an independent differential and differentials of $d\varepsilon$ are zero. Then, with $\Delta\varepsilon = \varepsilon - 0$, it is seen that $\varepsilon = d\varepsilon$.

To show that Eqs.(18.11) are equivalent to Eqs.(18.8), set

$$d\varepsilon = \varepsilon, \quad z = z_0, \quad dz = z_1\varepsilon, \quad \frac{1}{2!}d^2z = z_2\varepsilon^2, \quad \ldots \qquad (18.12)$$

## 18.5   Kepler's Equation

The particular problem used to demonstrate these techniques is the solution of Kepler's equation (Ref. Ro, p. 85) with small eccentricity, that is,

$$E - e\sin E - M = 0, \qquad (18.13)$$

where $E$ is the eccentric anomaly of a point in an orbit, $M$ is the known mean anomaly of the point, and $e$ is the known eccentricity of the orbit. The orbit is assumed to be nearly circular so that the eccentricity $e$ is small.

The equations for finding an approximate analytical solution of Kepler's equation (18.13) can be obtained from the general expansion approach or the general differential approach with about the same effort. These equations are stated in the next two sections. The only difference between the two approaches is the amount of effort required to get the equation for the next order of the solution, if it is needed. In both cases the equations become quite complex as the desired order of the solution increases.

In practice, the equations for the approximate analytical solution of an equation would probably be obtained by working directly with the particular equation because many of the derivatives in Eq.(18.8) or (18.11) would be zero. In this connection, the equations for solving Kepler's equation approximately are now derived directly.

## 18.6   Expansion Process for a Particular Problem

To derive the equations for an approximate analytical solution of Kepler's equation (18.13), $E$ is expressed as

$$E = E_0 + \Delta E \qquad (18.14)$$

so that
$$\sin E = \sin E_0 \cos \Delta E + \cos E_0 \sin \Delta E. \qquad (18.15)$$

Then, with the expansions for sine and cosine, Kepler's equation becomes

$$E_0 + \Delta E - e\left[\sin E_0 \left(1 - \frac{\Delta E^2}{2!} + \cdots\right) + \cos E_0 \left(\Delta E - \frac{\Delta E^3}{3!} + \cdots\right)\right] - M = 0.$$
$$(18.16)$$

Finally, substitute
$$\Delta E = E_1 e + E_2 e^2 + \cdots \qquad (18.17)$$

into Eq.(18.16), rearrange terms in descending powers of $e$, and set the co-efficients of the various-order terms to zero.  This process gives the following results:

$$E_0 = M$$

$$E_1 = \sin E_0$$
$$(18.18)$$
$$E_2 = -E_1 \cos E_0$$
$$\vdots$$

The expansion process involves a number of operations. (a) All of the mathematical manipulations indicated by the particular problem must be performed with series. (b) Care must be exercised to ensure that all terms of a given order are collected. (c) The effort required to get the equations for another-order solution is significant.

# 18.7  Differential Process for a Particular Problem

Taking differentials of Eq.(18.13) leads to

$$E - e \sin E - M = 0$$

$$dE - de \sin E + e \cos E \, dE = 0$$

$$d^2 E + de \cos E \, dE + de \cos E \, dE - e \sin E \, dE^2 + e \cos E \, d^2 E = 0$$

$$\vdots$$

$$(18.19)$$

where $d(de) = 0$. Then, evaluating the coefficients at $e = 0$ ($E$ becomes the zeroth-order solution) gives the following:

$$E = M$$
$$dE = \sin E \, de$$
$$d^2E = -2 \cos E \, dE \, de \qquad (18.20)$$

$$\vdots$$

That these equations are the same as Eqs.(18.18) can be shown by making the connections

$$de = e, \quad E = E_0, \quad dE = E_1 e, \quad \frac{1}{2!}d^2E = E_2 e^2, \quad \ldots \qquad (18.21)$$

The differential process entails a number of benefits over the expansion process. (a) Only differentiation is required. (b) All terms of a given order are automatically included. (c) The equation for the next order can be obtained from the last-order equation in a straightforward manner.

# 18.8  Another Example

It is desired to obtain an equation for the radial distance $r$ to an orbiting object at a given time. The equations that must be solved are (Ref. Ro, p. 85)

$$r = a(1 - e \cos E)$$
$$E - e \sin E = M, \qquad (18.22)$$

where $a$ is the semimajor axis of the orbit and $M$ is linear in the given time. Because Kepler's equation cannot be solved for $E(M)$, the radius cannot be obtained analytically as $r(M)$.

For small eccentricity $e$, the solution can be obtained approximately as a power series in $e$. Here, the zeroth-order solution ($e = 0$) is given by

$$r = a$$
$$E = M. \qquad (18.23)$$

Then, the equations for the first-order solution are obtained by taking the differential of Eqs.(18.22) as

$$dr = -ade \sin E + ae \sin E \, dE$$
$$dE - de \sin E - e \cos E \, dE = 0. \tag{18.24}$$

Applied on the nominal path ($e = 0$), these equations become

$$dr = -a \sin E \, de$$
$$dE - \sin E \, de = 0, \tag{18.25}$$

where $E$ is now the zeroth-order solution.

The equations for the second-order part of the solution come from the differential of Eqs.(18.24), that is,

$$d^2 r = -ade \cos E \, dE + ade \sin E \, dE + ae \cos E \, dE^2 + ae \sin E \, d^2 E$$
$$d^2 E - de \cos E \, dE - de \cos E \, dE + e \sin E \, dE^2 - e \cos E \, d^2 E = 0, \tag{18.26}$$

where $d(de) = 0$. On $e = 0$,

$$d^2 r = -ade \cos E \, dE + ade \sin E \, dE$$
$$d^2 E - 2de \cos E \, dE = 0. \tag{18.27}$$

From Eqs.(18.23), (18.25), and (18.27), it is seen that

$$r = a$$
$$dr = -a \sin M \, de \tag{18.28}$$
$$d^2 r = a \sin M (\cos M + \sin M) de^2.$$

Finally, the approximate solution to second order is given by

$$r_* = r + dr + \frac{1}{2!} d^2 r + \dots, \tag{18.29}$$

and it is recalled that $de = e$.

# 18.9  Remarks

To establish their relative merits, the expansion process and the differential process have been applied to the algebraic perturbation problem in the form of Kepler's equation with small eccentricity. If the equations for solving the general perturbation problem are obtained first and then applied to Kepler's equation, there is not much difference between the expansion process and the differential process. On the other hand, when applied directly to Kepler's equation, the differential process is superior to the expansion process. In the expansion process, all of the mathematical operations must be carried out in terms of series; care must be taken to ensure that all terms of a given order are included; and it is laborious to derive the equations for another-order solution. With the differential process, only differentiation is required; all terms of a given order are included; and the equations for another order can be obtained easily from the general form of the previous-order equations.

Usually, it is difficult to ensure that the various-order equations are correct regardless of the method being used. Having two ways to derive the equations helps alleviate this problem or, perhaps, make it worse.

While the algebraic perturbation problem has been used to demonstrate the benefits of calculus of differentials, any problem involving a Taylor series expansion could have been used. Unfortunately, differential calculus has been used in Ref. (H1) to expand the potential of an oblate spheroid planet in a Taylor series of the small eccentricity of the cross-sectional shape of the planet. In this case, the expansion process and the differential process have been equally difficult.

# 19

# Approximate Solutions of Differential Equations

## 19.1 Introduction

The subject of this chapter is the derivation of the equations that are to be solved for an approximate analytical solution of a perturbation problem involving a differential equation. There are several ways to obtain an approximate analytical solution. First, approximations can be made in the equation such that the remaining equation has an analytical solution. If the approximate solution is accurate enough, work stops. If not, something more must be done. Second, instead of discarding the small terms, they can be replaced by a small parameter, thereby creating a perturbation problem. The effect of the size of the parameter can be investigated by simulation. Third, if the original equation contains a small parameter and if the equation with the parameter equal to zero has an analytical solution, a perturbation problem results. Fourth, a perturbation problem also arises when a small change in the initial conditions is introduced.

A perturbation problem has a nominal path and a perturbed path. In Chapter 6 in association with optimal control theory, the nominal path is the minimal path, and the perturbed path is an admissible comparison path. The perturbed path is caused by a small change in the control history relative to that of the minimal path. In this chapter, the nominal path is the solution of the differential equation with no perturbations, and the perturbed path comes from a parameter perturbation or an initial condition perturbation.

The usual procedure for solving a perturbation problem is to recognize that the solution can be expressed as a power series in the small parameter, substitute the series into the differential equation, expand in a Taylor series, and equate the coefficients of the powers of the small parameter to zero. This process leads to the equations for the zeroth-order part of the solution, the first-order part, and so on. Another way to get the equations for the perturbed path in terms of the properties of the nominal path is to use differential calculus.

The calculus of differentials is demonstrated here by deriving the equations for the various-order solutions of the perturbation problem for two problems: the initial value problem with fixed final time and the initial value problem with free but constrained final time. Particular applications are made to the problem of satellite motion in the equatorial plane of an oblate spheroid earth with small eccentricity and to the derivation of the Clohessy-Wiltshire equations that arise because of a small change in initial conditions.

While only a scalar differential equation is considered here, the differential process can be applied to vector equations if indicial notation is employed. Matrix notation can be employed if only the first-order correction to the zeroth-order solution is being sought.

## 19.2   Regular Perturbation Problem

It is desired to find an approximate analytical solution of the scalar ordinary differential equation

$$\dot{z} = f(t, z, \varepsilon), \tag{19.1}$$

where $\varepsilon$ is a small parameter whose value is known. A regular perturbation problem occurs when $f(t, z, 0)$ is finite. An example of such a function is given by

$$f(t, z, \varepsilon) = \alpha(t, z) + \varepsilon \beta(t, z). \tag{19.2}$$

It is assumed that $\dot{z} = f(t, z, 0)$ has an analytical solution; otherwise, there is no reason to proceed. The objective is to derive the equations that define an approximate analytical solution of Eq.(19.1) for small $\varepsilon$.

The continued discussion of this problem must include the initial conditions and the final conditions. Two classes of problems are considered here:

the initial value problem with fixed final time and the initial value problem with free but constrained final time.

## 19.3 Initial Value Problem with Fixed Final Time

For an initial value problem, the initial conditions are prescribed, that is,

$$t_i = t_{i_s}, \quad z_i = z_{i_s}, \tag{19.3}$$

where the subscript $s$ denotes a specific value. The final condition is taken to be

$$t_f = t_{f_s}. \tag{19.4}$$

The general solution of Eq.(19.1) subject to the initial conditions (19.3) has the functional form

$$z = g(t, \varepsilon, t_i, z_i). \tag{19.5}$$

Given a nominal path generated by $t_i$, $z_i$, and $\varepsilon = 0$, a perturbed path can be generated by perturbing $\varepsilon$ or by perturbing the initial conditions . From here on, the only concern is an $\varepsilon$ perturbation.

Consider two neighboring paths (Fig. 19.1): the nominal path $z(t) = g(t, 0.t_i, z_i)$ and the perturbed path $z_*(t) = g(t, \varepsilon, t_i, z_i)$. In variational terminology (Section 6.3), the two paths are related as

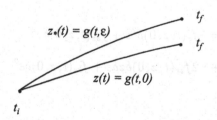

Figure 19.1: Neighboring Paths for Fixed Final Time

$$z_* = z + \delta z + \frac{1}{2!}\delta^2 z + \cdots, \tag{19.6}$$

where $\delta$ is a variation (differential taken while holding the time fixed). The quantity $z$ denotes the zeroth-order solution; $\delta z$ denotes the first-order solution; and so on.

The differential equations for the various-order parts of the solution are obtained by taking variations of Eq.(19.1), interchanging $\delta$ and $d$, and evaluating the results on the nominal path. Taking variations and interchanging $\delta$ and $d$ leads to the following:

$$\dot{z} = f(t, z, \varepsilon) \tag{19.7}$$

$$\frac{d}{dt}\delta z = f_z(t, z, \varepsilon)\delta z + f_\varepsilon(t, z, \varepsilon)\delta\varepsilon \tag{19.8}$$

$$\frac{d}{dt}\delta^2 z = f_z(t, z, \varepsilon)\delta^2 z + f_{zz}(t, z, \varepsilon)\delta z^2 \tag{19.9}$$

$$2 \ f_{z\varepsilon}(t, z, \varepsilon)\delta z \delta\varepsilon + f_{\varepsilon\varepsilon}(t, z, \varepsilon)\delta\varepsilon^2$$

$$\vdots$$

Finally, if these equations are evaluated on the nominal ($\varepsilon = 0$), the differential equations for $z$, $\delta z$, $\delta^2 z$, ... become

$$\dot{z} = f(t, z, 0) \tag{19.10}$$

$$\frac{d}{dt}\delta z = f_z(t, z, 0)\delta z + f_\varepsilon(t, z, 0)\delta\varepsilon \tag{19.11}$$

$$\frac{d}{dt}\delta^2 z = f_z(t, z, 0)\delta^2 z + f_{zz}(t, z, 0)\delta z^2 \tag{19.12}$$

$$+ \ 2f_{z\varepsilon}(t, z, 0)\delta z \delta\varepsilon + f_{\varepsilon\varepsilon}(t, z, 0)\delta\varepsilon^2$$

$$\vdots$$

The argument $z$ is now the nominal $z$, that is, the zeroth-order solution. Note that the equations are solved from first to last and that the equations for the various perturbations are linear but with time-dependent coefficients. Also, they have the same transition matrix.

In Eqs.(19.11), (19.12), and on, $\delta\varepsilon = d\varepsilon$ from $\delta\varepsilon = d\varepsilon - \dot{\varepsilon}\delta t$ and $\dot{\varepsilon} = 0$. Next, $\Delta\varepsilon = d\varepsilon$ because $d\varepsilon$ is an independent differential $(d^2\varepsilon = 0, d^3\varepsilon = 0, \ldots)$. Hence, if $\Delta\varepsilon = \varepsilon - 0 = \varepsilon$, then $d\varepsilon = \varepsilon$ and $\delta\varepsilon = \varepsilon$.

To obtain the initial conditions for $\delta z, \delta^2 z, \ldots$, the differential is taken of Eq.(19.3). Because $t_i$ is given $(dt_i = 0)$, $d(\ ) = \delta(\ )$. Hence, taking the differential and taking the variation are the same operation. Taking the variation of Eq.(19.3) leads to

$$\delta(t_i) = 0 \ (t_i = t_{i_s}), \quad \delta(z_i) = 0, \quad \delta^2(z_i) = 0, \quad \cdots. \tag{19.13}$$

Note that

$$\delta(z_i) = \text{fop}(z_{*_i} - z_i) \tag{19.14}$$

where fop stands for first-order part. Then,

$$\text{fop}(z_{*_i} - z_i) = [\text{fop}(z_* - z)]_i = (\delta z)_i \tag{19.15}$$

so that

$$\delta(z_i) = (\delta z)_i. \tag{19.16}$$

Hence, the initial conditions are given by

$$t_i = t_{i_s}, \quad (\delta z)_i = 0, \quad (\delta^2 z)_i = 0, \quad \ldots. \tag{19.17}$$

At this point several comments are in order. First, the usual way of getting the equations for an approximate solution is to work with the particular equation, not to apply the general equations to the particular problem. The reasons for this are that the general equations become very complicated with order and there may be many zero derivatives. Second, the variational approach is superior to the expansion approach because the next-order equations can be obtained from the previous-order equations and only differentiation is required. The expansion process requires that an additional term be added to the assumed solution and that all of the mathematical operations in the particular equation be carried out again with the new series. However, having both approaches makes it possible to check the equations by deriving them two different ways. Third, the differential process can be applied to a system of differential equations, but indicial notation must be used. On the other hand, if the zeroth-order solution is almost satisfactory and if only the first-order correction is needed, then matrix notation can be employed.

## 19.4   Initial Value Problem with Free Final Time

The simplest example of a regular perturbation problem that has a free final time is the solution of the scalar differential equation

$$\dot{z} = f(t, z, \varepsilon),$$   (19.18)

where $\varepsilon$ is a small parameter, subject to the initial conditions

$$t_i = t_{i_s}, \quad z_i = z_{i_s}$$   (19.19)

and the final condition

$$h(t_f, z_f) = 0.$$   (19.20)

The problem is shown schematically in Fig. 19.2. Note that the final time changes from the nominal path to the perturbed path.

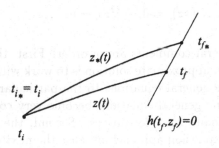

Figure 19.2: Neighboring Paths for Free Final Time

It is assumed that the zeroth-order solution $z(t)$ can be obtained analytically. It must satisfy the differential equation

$$\dot{z} = f(t, z, 0),$$   (19.21)

and the boundary conditions (19.19) and (19.20). Having the solution means that $z(t), t_f$, and $z_f$ are known.

The first-order solution must satisfy the differential equation (19.11) subject to the initial conditions

$$t_i = t_{i_*}, \quad (\delta z)_i = 0. \tag{19.22}$$

Since the final time is free, the differential of Eq.(19.20) leads to

$$dh = h_{t_f}(t_f, z_f)dt_f + h_{z_f}(t_f, z_f)dz_f = 0 \tag{19.23}$$

where the coefficients are to be evaluated on the zeroth-order path. Next, the differential is converted into a variation (Section 6.7). Substitution gives

$$(h_{t_f} + h_{z_f}\dot{z}_f)dt_f + h_{z_f}\delta z_f = 0 \tag{19.24}$$

or

$$dt_f = -\frac{h_{z_f}}{h_{t_f} + h_{z_f}\dot{z}_f}\delta z_f. \tag{19.25}$$

Having $\delta z_f$ from Eq.(19.11), $dt_f$ from Eq.(19.25), and $dz_f = \delta z_f + \dot{z}_f dt_f$ gives the solution to first order as

$$\begin{aligned} t_{f_*} &= t_f + dt_f \\ z_{*f} &= z_f + dz_f. \end{aligned} \tag{19.26}$$

The second-order correction is easily obtained by continuing this process. The differential equation for $\delta^2 z$ is given by Eq.(19.12) which is to be solved subject to the initial conditions

$$t_i = t_{i_*}, \quad (\delta^2 z)_i = 0. \tag{19.27}$$

The differential of Eq.(19.23) leads to

$$\begin{aligned} d^2 h &= h_{t_f t_f}dt_f^2 + 2h_{t_f z_f}dt_f dz_f + h_{z_f z_f}dz_f^2 \\ &\quad + h_{t_f}d^2 t_f + h_{z_f}d^2 z_f = 0, \end{aligned} \tag{19.28}$$

where from Eq.(6.29)

$$d^2 z_f = \delta^2 z_f + \dot{z}_f d^2 t_f + 2\delta \dot{z}_f dt_f + \ddot{z}_f dt_f^2. \tag{19.29}$$

Equations (19.28) and (19.29) can be solved for $d^2 t_f$ and $d^2 z_f$. Finally, Eqs.(19.26) can be extended to second order as

$$
\begin{aligned}
t_{f_*} &= t_f + dt_f + \tfrac{1}{2!}d^2 t_f \\
z_{*_f} &= z_f + dz_f + \tfrac{1}{2!}d^2 z_f.
\end{aligned}
\tag{19.30}
$$

At this point, what exists is $z_*(t)$ from $t_i$ to $t_f$ and $z_*(t_{f_*})$. To get the values of $z_*$ from $t_f$ to $t_{f_*}$, it is seen that

$$
z_*(t) = z_*(t_f + \Delta t),
\tag{19.31}
$$

where

$$
\Delta t = t - t_f.
\tag{19.32}
$$

Then, Eq.(19.31) can be rewritten as

$$
z_*(t) = z_*(t_f) + \dot{z}_*(t_f)\Delta t + \frac{1}{2!}\ddot{z}_*(t_f)\Delta t^2 + \cdots.
\tag{19.33}
$$

Each of the quantities on the right hand side can be obtained from previous formulas.

# 19.5   Motion Relative to an Oblate Spheroid Earth

In the previous sections, the use of differentials has been demonstrated by applying it to general problems. In practice, it would be applied directly to the problem being solved. To illustrate the approach for a fixed final time, consider a satellite moving in the equatorial plane of an oblate spheroid earth. The motion of the satellite is governed by the equation (Ref. Ro, pp. 75 and 209)

$$
\ddot{r} - \frac{h^2}{r^3} = -\frac{\mu}{r^2}\left[1 + \left(\frac{R^2}{r^2}\right)\varepsilon\right],
\tag{19.34}
$$

where $r$ is the radial distance to the satellite, $h$ is the constant angular momentum of the satellite, $R$ is the equatorial radius of the earth, $\mu$ is the gravitational constant, and $\varepsilon = (3/2)J_2 = .001624$. The boundary conditions are given by

$$
t_i = t_{i_s}, \quad r_i = r_{i_s}, \quad \dot{r}_i = \dot{r}_{i_s}, \quad t_f = t_{f_s}.
\tag{19.35}
$$

The objective is to derive the equations that govern the approximate solution of Eq.(19.34) using $\varepsilon$ as a small parameter. This problem can be discussed using two first-order equations, but it is easier to leave it as a second-order equation.

Taking the variation of Eq.(19.34) leads to

$$\delta\ddot{r} + \tfrac{3h^2}{r^4}\delta r = \tfrac{2\mu}{r^3}\delta r + \tfrac{4\mu R^2}{r^5}\varepsilon\delta r - \tfrac{\mu R^2}{r^4}\delta\varepsilon. \qquad (19.36)$$

With $\varepsilon = 0$, Eq.(19.34) gives the equation for the zeroth-order solution to be

$$\ddot{r} - \frac{h^2}{r^3} = -\frac{\mu}{r^2}, \qquad (19.37)$$

and Eq.(19.36) gives the equation for the first-order solution, that is,

$$\frac{d^2\delta r}{dt^2} + \frac{3h^2 - 2\mu r}{r^4}\delta r = -\frac{\mu}{r^4}R^2\delta\varepsilon, \qquad (19.38)$$

after $\delta$ and $d$ are interchanged. The equation for $\delta^2 r$ can be derived by taking the variation of Eq.(19.36) with $\delta(\delta\varepsilon) = 0$, and setting $\varepsilon = 0$. The last step is to recognize that $\delta\varepsilon = d\varepsilon = \varepsilon$ in these equations.

The boundary conditions for these equations are obtained by taking the variation of Eq.(19.35). They are given by

$$\delta t_i = 0, \quad \delta r_i = 0, \quad \delta\dot{r}_i = 0, \quad \delta t_f = 0 \qquad (19.39)$$

and become

$$t_i = t_{i_s}, \quad (\delta r)_i = 0, \quad (\tfrac{d}{dt}\delta r)_i = 0, \quad t_f = t_{f_s}. \qquad (19.40)$$

When the solutions of all orders have been obtained, the overall solution $r_*$ becomes

$$r_* = r + \delta r + \frac{1}{2!}\delta^2 r + \cdots. \qquad (19.41)$$

## 19.6  Clohessy–Wiltshire Equations

The Clohessy–Wiltshire equations (Ref. CW) are the small perturbation equations for motion relative to an object moving on a prescribed orbit. The perturbed path is caused by a perturbation in the initial conditions. The perturbation equations are derived here for a circular orbit using the differential approach. For planar motion, the nonlinear equations are given by

$$\ddot{x} - 2\omega\dot{z} - \omega^2 x + \mu\frac{x}{r^3} = 0$$

$$\ddot{z} + 2\omega\dot{x} - \omega^2(z - r_0) + \mu\frac{z-r_0}{r^3} = 0,$$

(19.42)

where

$$r = \sqrt{x^2 + (z - r_0)^2}.$$

(19.43)

In these formulas, $\mu$ is the gravitational constant, $x$ and $z$ are the coordinates of a point mass in the local-horizontal local-vertical reference frame whose origin is moving on a circular orbit, and $r$ is the distance from the point mass to the center of the earth. The radius of the orbit is $r_0$, and

$$\omega^2 = \frac{\mu}{r_0^3}.$$

(19.44)

The initial conditions of the point mass are given by

$$x_i = 0, \quad z_i = \eta, \quad \dot{x}_i = 0, \quad \dot{z}_i = 0,$$

(19.45)

where $\eta$ is a small perturbation in $z_i$. The final time is fixed. The objective is to derive the small perturbation equations of motion for the point mass using variations.

Taking the variation of Eqs.(19.42) leads to

$$\delta\ddot{x} - 2\omega\delta\dot{z} - \omega^2\delta x + \mu\frac{r^3\delta x - x3r^2\delta r}{r^6} = 0$$

$$\delta\ddot{z} + 2\omega\delta\dot{x} - \omega^2\delta z + \mu\frac{r^3\delta z - (z-r_0)3r^2\delta r}{r^6} = 0,$$

(19.46)

where

$$\delta r = \frac{x\delta x + (z - r_0)\delta z}{r}.$$

(19.47)

Next, the coefficients of the variations are evaluated on the reference orbit $(x = z = 0, r = r_0)$ so that $\delta r = -\delta z$, and the CW equations become

$$\frac{d^2}{dt^2}\delta x - 2\omega\frac{d}{dt}\delta z = 0$$

$$\frac{d^2}{dt^2}\delta z + 2\omega\frac{d}{dt}\delta x - 3\omega^2\delta z = 0.$$

(19.48)

Finally, the initial conditions are obtained by taking the variation of Eq.(19.45) and are given by

$$(\delta x)_i = 0, \quad (\delta z)_i = \delta\eta, \quad (\delta\dot{x})_i = 0, \quad (\delta\dot{z})_i = 0.$$

(19.49)

If second-order terms are needed, they can be obtained by taking the variation of Eqs.(19.46), (19.47) and (19.49) with $\delta(\delta\eta) = 0$. Once the various-order solutions have been obtained, the overall solution is given by

$$
\begin{aligned}
x_* &= x + \delta x + \tfrac{1}{2!}\delta^2 x + \cdots \\
z_* &= z + \delta z + \tfrac{1}{2!}\delta^2 z + \cdots,
\end{aligned}
\tag{19.50}
$$

where $x$ and $z$ are zero.

## 19.7 Remarks

It has been shown that Taylor series expansions can be made on a term by term basis by applying differentials and variations (differentials taken while holding the time constant). Variations interchange with derivatives, and variations of independent variations are zero.

To demonstrate the application of differentials, the equations to be solved for the various-order solutions of the regular perturbation problem have been derived for two classes of problems: initial value problems with fixed final time and initial value problems with free but constrained final time. Also, differentials have been applied to the particular problems of satellite motion in the equatorial plane of an oblate spheroid earth with small eccentricity and motion near a circular orbit.

The advantages of the differential approach relative to the Taylor series expansion approach are that only differentiation is used, all terms of a given order are automatically included, and the development of the equations for another order solution is straightforward.

Usually, it is difficult to ensure that the various-order equations are correct regardless of the method being used. Having two ways to derive these equations may be helpful.

While only scalar equations are considered here, differentials can be applied to vector equations by employing indicial notation. If only the first-order correction to the zeroth-order solution is being sought, matrix notation can be employed.

The regular perturbation problem has been used to demonstrate the use of differentials. However, any problem involving a Taylor series expansion could have been used.

# 20

# Approximate Solutions of Optimal Control Problems

## 20.1 Introduction

In the previous two chapters, differentials are used to derive the equations for the various-order terms of the perturbation problem first involving algebraic equations then involving differential equations. In this chapter, differentials are used to derive the corresponding equations for the optimal control perturbation problem that involves both algebraic and differential equations.

As an example of applying differentials and solving a problem, the navigation problem is considered, and the wind speed is considered to be a small parameter. In many cases, only the first-order correction is desired. Here, the formulas for the first-order correction are derived in matrix notation and applied to the navigation problem for verification.

## 20.2 Optimal Control Problem with a Small Parameter

Consider the following optimal control problem: Find the control history $u(t)$ that minimizes the performance index

$$J = \phi(t_f, x_f) + \int_{t_0}^{t_f} L(t, x, u)dt, \qquad (20.1)$$

subject to the state differential constraints

$$\dot{x} = f(t, x, u, \varepsilon), \tag{20.2}$$

to the prescribed initial conditions

$$t_0 = t_{0_s}, \quad x_0 = x_{0_s}, \tag{20.3}$$

and to the prescribed final conditions

$$\psi(t_f, x_f) = 0. \tag{20.4}$$

Here, $\phi$ and $L$ are scalars, and the dimensions of $u$, $x$, and $\psi$ are $m \times 1, n \times 1$, and $(p + 1) \times 1$, respectively. The state equations are assumed to contain a small scalar parameter $\varepsilon$, and the function $f$ is assumed to be finite as $\varepsilon$ goes to zero. An example of such a function is

$$f = \alpha(x, u) + \varepsilon\beta(x). \tag{20.5}$$

After the Hamiltonian and the endpoint function are formed, that is,

$$H = L(t, x, u) + \lambda^T f(t, x, u, \varepsilon)$$
$$G = \phi(t_f, x_f) + \nu^T \psi(t_f, x_f), \tag{20.6}$$

the equations that define the solution of the optimal control problem are given by

$$\dot{x} = f(t, x, u, \varepsilon) \tag{20.7}$$

$$\dot{\lambda} = -H_x^T(t, x, u, \lambda, \varepsilon) \tag{20.8}$$

$$0 = H_u^T(t, x, u, \lambda, \varepsilon) \tag{20.9}$$

and the boundary conditions

$$t_0 = t_{0_s}, \quad x_0 = x_{0_s} \tag{20.10}$$

$$\lambda_f = G_{x_f}^T(t_f, x_f, \nu) \tag{20.11}$$

$$\psi(t_f, x_f) = 0 \tag{20.12}$$

$$L(t_f, x_f, u_f) + G_{x_f}(t_f, x_f, \nu) f(t_f, x_f, u_f, \varepsilon)$$
$$+ G_{t_f}(t_f, x_f, \nu) = 0. \tag{20.13}$$

If Eqs.(20.11) and (20.13) are combined, the more familiar $H_f + G_{t_f} = 0$ results.

## 20.3    Application to a Particular Problem

When trying to solve a particular problem, the usual approach is to work on the particular problem directly rather than deriving the conditions for a general problem and applying them to the particular problem. In this section, the intent is to derive the equations to be solved by the zeroth-order , first-order, and so on parts of the approximate analytical solution of the navigation problem where the small parameter is the wind speed. The solutions of these equations are stated for later use. The navigation problem has been selected because it has an analytical solution.

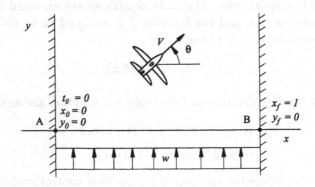

Figure 20.1: Navigation Problem

The navigation problem (see Fig. 20.1) is to fly an airplane from point A to point B in a crosswind in minimum time. Points A and B are on the $x$-axis, and the wind is moving up the $y$-axis. The velocity of the airplane $V$ relative to the wind and the velocity $w$ of the wind relative to the ground are assumed constant. The states are $x(t)$ and $y(t)$, and the control $\theta(t)$ is the angle of the airplane velocity vector relative to the $x$-axis. The optimal control problem is to find the control history $\theta(t)$ that minimizes the final time

$$J = t_f, \qquad (20.14)$$

subject to the differential constraints

$$\dot{x} = V \cos\theta$$
$$\dot{y} = V \sin\theta + w, \qquad (20.15)$$

the prescribed initial conditions

$$t_0 = 0, \quad x_0 = 0, \quad y_0 = 0, \tag{20.16}$$

and the prescribed final conditions

$$x_f = 1, \quad y_f = 0. \tag{20.17}$$

To create a problem that has a small parameter, it is assumed that $w$ is small so that $\varepsilon \triangleq w$.

The Hamiltonian $H$ and the endpoint function $G$ are defined as

$$H = \lambda_1 V \cos \theta + \lambda_2 (V \sin \theta + w)$$
$$G = t_f + \nu_1(x_f - 1) + \nu_2 y_f \tag{20.18}$$

so that Eqs.(20.7) through (20.13) become

$$\dot{x} = V \cos \theta, \quad \dot{y} = V \sin \theta + w, \tag{20.19}$$

$$\dot{\lambda}_1 = 0, \quad \dot{\lambda}_2 = 0, \tag{20.20}$$

$$0 = -\lambda_1 V \cos \theta + \lambda_2 V \sin \theta, \tag{20.21}$$

$$t_0 = 0, \quad x_0 = 0, \quad y_0 = 0, \tag{20.22}$$

$$\lambda_{1_f} = \nu_1, \quad \lambda_{2_f} = \nu_2, \tag{20.23}$$

$$x_f = 1, \quad y_f = 0, \tag{20.24}$$

$$\nu_1 V \cos \theta_f + \nu_2(V \sin \theta_f + w) + 1 = 0. \tag{20.25}$$

The zeroth-order solution is the solution of these equations for $w = 0$. It is easily shown to be

$$\theta = 0, \ x = Vt, \ y = 0, \ \lambda_1 = \nu_1 = -\frac{1}{V}, \lambda_2 = \nu_2 = 0, \ t_f = \frac{1}{V}. \tag{20.26}$$

The equations for the first-order solution are obtained by taking the variation of Eqs.(20.19) through (20.21) and the differential of Eqs.(20.22) through (20.25). The latter are then converted into variations by

$$dv = \delta v + \dot{v}dt. \tag{20.27}$$

The final step is to evaluate the coefficients on the zeroth-order solution ($w = 0$).

The variation of Eqs.(20.19) through (20.21) along with the interchanging of $\delta$ and $d/dt$ gives

$$\frac{d}{dt}\delta x = -V \sin\theta \, \delta\theta$$
$$\frac{d}{dt}\delta y = -V \cos\theta \, \delta\theta + dw$$
$$\frac{d}{dt}\delta\lambda_1 = 0 \qquad\qquad\qquad\qquad (20.28)$$
$$\frac{d}{dt}\delta\lambda_2 = 0$$
$$0 = -\delta\lambda_1 V \sin\theta - \lambda_1 V \cos\theta \, \delta\theta + \delta\lambda_2 V \cos\theta - \lambda_2 V \sin\theta \, \delta\theta,$$

where $\delta w = dw$ from Eq.(20.27). Evaluated on the zeroth-order path ($w = 0$), these equations reduce to

$$\frac{d}{dt}\delta x = 0$$
$$\frac{d}{dt}\delta y = V\delta\theta + dw$$
$$\frac{d}{dt}\delta\lambda_1 = 0 \qquad\qquad\qquad\qquad (20.29)$$
$$\frac{d}{dt}\delta\lambda_2 = 0$$
$$0 = \delta\theta + V\delta\lambda_2.$$

The boundary conditions for these equations are obtained by taking the differential of Eqs.(20.22) through (20.25), that is,

$$dt_0 = 0, \quad dx_0 = 0, \quad dy_0 = 0,$$
$$d\lambda_1 = d\nu_1, \quad d\lambda_2 = d\nu_2,$$
$$dx_f = 0, \quad dy_f = 0, \qquad\qquad\qquad (20.30)$$
$$d\nu_1 V \cos\theta_f - \nu_1 V \sin\theta_f d\theta_f$$
$$+ d\nu_2(V \sin\theta_f + w) + \nu_2(V \cos\theta_f d\theta_f + dw) = 0.$$

Then, the differentials are converted into variations using Eq.(20.27), and the coefficients are evaluated on the zeroth-order path (20.25). The results are

$$t_0 = 0, \quad \delta x_0 = 0, \quad \delta y_0 = 0,$$
$$\delta\lambda_1 = d\nu_1, \quad \delta\lambda_2 = d\nu_2,$$
$$\delta x_f + V dt_f = 0, \quad \delta y_f = 0, \qquad\qquad (20.31)$$
$$d\nu_1 = 0.$$

Finally, the solution of Eqs.(20.29) subject to the conditions (20.31) gives the first-order solution, that is,

$$\delta x = 0, \quad \delta y = 0,$$
$$\delta \lambda_1 = d\nu_1 = 0, \quad \delta \lambda_2 = d\nu_2 = \tfrac{dw}{V^2}, \qquad (20.32)$$
$$\delta \theta = -\tfrac{dw}{V}, \quad dt_f = 0.$$

It is noted that the second-order solution is obtained from Eqs.(20.28) and (20.30) in the same way that the first-order solution is obtained from Eqs.(20.19) through (20.25). In doing so, $\delta(dw) = 0$, and the relationships between $d^2$ and $\delta^2$ for functions of time $v(t)$ and constants $c$ are given by

$$d^2 v = \delta^2 v + \dot{v} d^2 t + 2\delta \dot{v} dt + \ddot{v}$$
$$d^2 c = \delta^2 c. \qquad (20.33)$$

The results for the second-order solution are

$$\delta^2 x = -\tfrac{dw^2}{V} t, \quad \delta^2 y = 0,$$
$$\delta^2 \lambda_1 = d^2 \nu_1 = -\tfrac{dw^2}{V^3}, \quad \delta \lambda_2 = d\nu_2 = 0, \qquad (20.34)$$
$$\delta^2 \theta = 0, \quad d^2 t_f = \tfrac{dw^2}{V^3}.$$

If the zeroth- through second-order parts are combined, the overall solution (to second order) becomes

$$x = Vt - \tfrac{1}{2!} \tfrac{dw^2}{V} t, \quad y = 0,$$
$$\lambda_1 = \nu_1 = -\tfrac{1}{V} - \tfrac{1}{2!} \tfrac{dw^2}{V^3}, \quad \lambda_2 = \nu_2 = \tfrac{dw}{V}, \qquad (20.35)$$
$$\theta = -\tfrac{dw}{V}, \quad t_f = \tfrac{1}{V} + \tfrac{1}{2!} \tfrac{dw^2}{V^3}.$$

It is recalled that $dw = \Delta w = w$ .

The correctness of this result can be verified from the exact solution of the navigation problem, that is,

$$x = V\sqrt{1 - (w/V)^2}\, t, \quad y = 0$$
$$\lambda_1 = \nu_1 = -\frac{1}{V\sqrt{1-(w/V)^2}}, \quad \lambda_2 = \nu_2 = \frac{(w/V)}{1-(w/V)^2} \qquad (20.36)$$
$$\theta = -\arcsin(w/V), \quad t_f = \frac{1}{V\sqrt{1-(w/V)^2}}.$$

Then, by assuming that $w$ is small, a Taylor series expansion to second order gives Eqs.(20.35). It is also possible to derive Eqs.(20.35) by taking variations of Eqs.(20.36) and evaluating them on $w = 0$.

## 20.4   Application to a General Problem

If the zeroth-order solution is almost accurate enough and only the first-order correction is needed, it is worthwhile deriving and solving the general equations of the first-order solution. It is interesting to note that the first-order solution is the neighboring optimal path (Chap. 13) for the given value of $\varepsilon$.

If $\varepsilon$ is set equal to zero in Eqs.(20.7) through (20.13), the resulting equations define the zeroth-order solution.

The equations for the first-order solution are obtained by taking the variation of Eqs. (20.7) through (20.9), that is,

$$
\begin{aligned}
\delta \dot{x} &= f_x \delta x + f_u \delta u + f_\varepsilon d\varepsilon \\
\delta \dot{\lambda} &= -H_{xx}\delta x - H_{xu}\delta u - f_x^T \delta\lambda - H_{x\varepsilon}d\varepsilon \\
0 &= H_{ux}\delta x + H_{uu}\delta u + f_u^T \delta\lambda + H_{u\varepsilon}d\varepsilon,
\end{aligned} \tag{20.37}
$$

where $\delta\varepsilon = d\varepsilon$ because $\varepsilon$ is a constant. If it is assumed that the optimal control problem is nonsingular $(H_{uu} > 0)$, the control perturbation can be written as

$$
\delta u = -H_{uu}^{-1}(H_{ux}\delta x + f_u^T \delta\lambda + H_{u\varepsilon}d\varepsilon). \tag{20.38}
$$

Then, since $\delta$ and $d$ are interchangable, the differential equations for $\delta x$ and $\delta\lambda$ become

$$
\begin{aligned}
\tfrac{d}{dt}\delta x &= A\delta x - B\delta\lambda + Xd\varepsilon \\
\tfrac{d}{dt}\delta\lambda &= -C\delta x - A^T\delta\lambda - Yd\varepsilon,
\end{aligned} \tag{20.39}
$$

where the coefficients $A$ through $Y$ are given by

$$
\begin{aligned}
A &= f_x - f_u H_{uu}^{-1} H_{ux} \\
B &= f_u H_{uu}^{-1} f_u^T \\
C &= H_{xx} - H_{xu}H_{uu}^{-1}H_{ux} \\
X &= f_\varepsilon - f_u H_{uu}^{-1}H_{u\varepsilon} \\
Y &= H_{x\varepsilon} - H_{xu}H_{uu}^{-1}H_{u\varepsilon}.
\end{aligned} \tag{20.40}
$$

Since the final time is free, the boundary conditions for the differential equations are obtained by taking the differential of the boundary conditions (20.10) through (20.13) and converting to variations by using Eq.(20.27). The corresponding initial conditions are

$$
t_0 = t_{0_s}, \quad \delta x_0 = 0, \tag{20.41}
$$

and the final conditions become

$$\delta\lambda_f = G_{x_f x_f}\delta x_f + \psi_{x_f}^T dv + \Omega_{x_f}^T dt_f \tag{20.42}$$

$$d\psi = \psi_{x_f}\delta x_f + \psi' dt_f \tag{20.43}$$

$$d\Omega = \Omega_{x_f}\delta x_f + \psi'^T dv + \Omega' dt_f + \Omega_\varepsilon d\varepsilon, \tag{20.44}$$

where

$$\Omega(t_f, x_f, \nu, u_f) = L_f + G_{x_f}f_f + G_{t_f} \tag{20.45}$$

$$(\cdot)' = (\cdot)_{t_f} + (\cdot)_{x_f}f_f \tag{20.46}$$

and where $d\psi$ and $d\Omega$ are zero.

In Eqs.(20.39) and (20.42) through (20.44), the coefficients of all variations are evaluated on the zeroth-order solution.

## 20.5  Solution by the Sweep Method

Equations (20.39) through (20.44) define a linear two-point boundary-value problem that can be solved by the sweep method or the transition matrix method. Since an example of the transition matrix method is given in Ref. IS, the sweep method is used here.

As an extension of Chapter 13, it is assumed that

$$\begin{bmatrix} \delta\lambda \\ d\psi \\ d\Omega \end{bmatrix} = \begin{bmatrix} S(t) & R(t) & m(t) & p(t) \\ R^T(t) & Q(t) & n(t) & q(t) \\ m^T(t) & n^T(t) & \alpha(t) & r(t) \end{bmatrix} \begin{bmatrix} \delta x \\ dv \\ dt_f \\ d\varepsilon \end{bmatrix}, \tag{20.47}$$

where from Eqs.(20.42) through (20.44)

$$\begin{aligned} S_f &= G_{x_f x_f}, & R_f &= \psi_{x_f}^T, & m_f &= \Omega_{x_f}^T \\ Q_f &= 0, & n_f &= \psi', & \alpha_f &= \Omega' \\ p_f &= 0, & q_f &= 0, & r_f &= \Omega_\varepsilon. \end{aligned} \tag{20.48}$$

To obtain the differential equations for $S$ through $r$, Eqs.(20.47) are differentiated with respect to time and combined with Eqs.(20.39). The results are

$$
\begin{aligned}
\dot{S} &= -C - A^T S - SA + SBS \\
\dot{R} &= (SB - A^T)R \\
\dot{m} &= (SB - A^T)m \\
\dot{Q} &= R^T BR \\
\dot{n} &= R^T Bm \\
\dot{\alpha} &= m^T Bm \\
\dot{p} &= (SB - A^T)p - SX - Y \\
\dot{q} &= R^T(Bp - X) \\
\dot{r} &= m^T(Bp - X).
\end{aligned}
\tag{20.49}
$$

Hence, the solutions of Eqs.(20.49) subject to the boundary conditions (20.48) lead to the functions $S(t)$ though $r(t)$ so that Eqs.(20.47) can be used along the nominal path.

Since $\delta x_0 = 0$, the equations for $d\psi$ and $d\Omega$ (both $= 0$) are evaluated at $t_0$ to obtain

$$
\begin{aligned}
Q_0 d\nu + n_0 dt_f + q_0 d\varepsilon &= 0 \\
n_0^T d\nu + \alpha_0 dt_f + r_0 d\varepsilon &= 0.
\end{aligned}
\tag{20.50}
$$

These equations can be solved simultaneously for $d\nu$ and $dt_f$ as

$$
\begin{bmatrix} d\nu \\ dt_f \end{bmatrix} = -V_0^{-1} \begin{bmatrix} q_0 \\ r_0 \end{bmatrix} d\varepsilon
\tag{20.51}
$$

with

$$
V \triangleq \begin{bmatrix} Q & n \\ n^T & \alpha \end{bmatrix}.
\tag{20.52}
$$

Then, the expression for $\delta\lambda_0$ can be obtained from Eq.(20.47) as

$$
\delta\lambda_0 = \left\{ -U_0 V_0^{-1} \begin{bmatrix} q_0 \\ r_0 \end{bmatrix} + p_0 \right\} d\varepsilon,
\tag{20.53}
$$

where

$$
U \triangleq [R \; m].
\tag{20.54}
$$

Given $d\varepsilon$, Eq.(20.53) yields $\delta\lambda_0$ so that Eqs.(20.39) can be solved for $\delta x(t)$ and $\delta\lambda(t)$. Then, Eq.(20.38) leads to the control perturbation $\delta u(t)$ for the first-order solution, that is,

$$\delta u = -H_{uu}^{-1}(H_{ux}\delta x + f_u^T \delta\lambda + H_{u\varepsilon}d\varepsilon). \tag{20.55}$$

## 20.6 Navigation Problem

The derivation of the first-order part of the solution requires the following derivatives which are evaluated along the zeroth-order solution:

$$f_x = \begin{bmatrix} 0 & 0 \\ 0 & 0 \end{bmatrix}, \quad f_u = \begin{bmatrix} 0 \\ V \end{bmatrix}, \quad f_\varepsilon = \begin{bmatrix} 0 \\ 1 \end{bmatrix}, \quad H_{uu} = 1$$

$$H_{xx} = \begin{bmatrix} 0 & 0 \\ 0 & 0 \end{bmatrix}, \quad H_{xu} = \begin{bmatrix} 0 \\ 0 \end{bmatrix}, \quad H_{x\varepsilon} = \begin{bmatrix} 0 \\ 0 \end{bmatrix}, \quad H_{u\varepsilon} = 0. \tag{20.56}$$

Next, the coefficient matrices $A$ through $Y$ are given by

$$A = \begin{bmatrix} 0 & 0 \\ 0 & 0 \end{bmatrix}, \quad B = \begin{bmatrix} 0 & 0 \\ 0 & V^2 \end{bmatrix}, \quad C = \begin{bmatrix} 0 & 0 \\ 0 & 0 \end{bmatrix}$$

$$X = \begin{bmatrix} 0 \\ 1 \end{bmatrix}, \quad Y = \begin{bmatrix} 0 \\ 0 \end{bmatrix} \tag{20.57}$$

and the boundary conditions (20.48) become

$$S_f = \begin{bmatrix} 0 & 0 \\ 0 & 0 \end{bmatrix}, \quad R_f = \begin{bmatrix} 1 & 0 \\ 0 & 1 \end{bmatrix}, \quad m_f = \begin{bmatrix} 0 \\ 0 \end{bmatrix}$$

$$Q_f = 0, \quad n_f = \begin{bmatrix} V \\ 0 \end{bmatrix}, \quad \alpha_f = 0 \tag{20.58}$$

$$p_f = \begin{bmatrix} 0 \\ 0 \end{bmatrix}, \quad q_f = \begin{bmatrix} 0 \\ 0 \end{bmatrix}, \quad r_f = 0.$$

Then, Eqs.(20.49) lead to the following solutions for $S(t)$ through $r(t)$:

$$S = \begin{bmatrix} 0 & 0 \\ 0 & 0 \end{bmatrix}, \quad R = \begin{bmatrix} 1 & 0 \\ 0 & 1 \end{bmatrix}, \quad m = \begin{bmatrix} 0 \\ 0 \end{bmatrix}$$

$$Q = (t - t_f) \begin{bmatrix} 0 & 0 \\ 0 & V^2 \end{bmatrix}, \quad n = \begin{bmatrix} V \\ 0 \end{bmatrix}, \quad \alpha = 0 \tag{20.59}$$

$$p = \begin{bmatrix} 0 \\ 0 \end{bmatrix}, \quad q = -\begin{bmatrix} 0 \\ t - t_f \end{bmatrix}, \quad r = 0.$$

The matrices $U_0$ and $V_0$ are given by

$$U_0 = \begin{bmatrix} 1 & 0 & 0 \\ 0 & 1 & 0 \end{bmatrix}, \quad V_0 = V \begin{bmatrix} 0 & 0 & 1 \\ 0 & -1 & 0 \\ 1 & 0 & 0 \end{bmatrix}, \tag{20.60}$$

so that

$$V_0^{-1} = \frac{1}{V} \begin{bmatrix} 0 & 0 & 1 \\ 0 & -1 & 0 \\ 1 & 0 & 0 \end{bmatrix}. \tag{20.61}$$

Then, from Eq.(20.51), it is seen that

$$d\nu_1 = 0, \quad d\nu_2 = \frac{dw}{V^2}, \quad dt_f = 0. \tag{20.62}$$

Next, the change in the multipliers is given by Eq.(20.47) to be $\delta\lambda = d\nu$ or

$$\delta\lambda_1 = 0, \quad \delta\lambda_2 = \frac{dw}{V^2}. \tag{20.63}$$

The changes in the states are obtained by integrating Eqs.(20.39) subject to the boundary conditions (20.41) and are given by

$$\delta x = 0, \quad \delta y = 0. \tag{20.64}$$

Finally, the first-order change in the control is obtained from Eq.(20.55) as

$$\delta\theta = -\frac{dw}{V}. \tag{20.65}$$

The sum of the zeroth-order and first-order solutions leads to the following approximate solution:

$$\theta = -\frac{dw}{V}, \quad x = Vt, \quad y = 0$$
$$\lambda_1 = \nu_1 = -\frac{1}{V}, \quad \lambda_2 = \nu_2 = \frac{dw}{V^2}, \quad t_f = \frac{1}{V}, \tag{20.66}$$

which checks with Eq.(20.35). Recall that now $dw = w$.

## 20.7 Remarks

The thrust of this chapter has been the derivation of the equations to be solved to obtain an approximate analytical solution of an optimal control problem involving a small parameter. The usual procedure is to express the solution as a power series in the small parameter, substitute it into the optimal control equations, expand in a Taylor series, and set the coefficients of the various powers of the small parameter to zero. Here, the Taylor series is developed by taking differentials. This process is applied to the navigation problem where the small parameter is the wind speed. First, differentials have been applied directly to the navigation problem because that is probably what would be done in practice. The equations governing the various-order solutions are derived, and the solutions are stated. Second, it is assumed that the zeroth-order solution is nearly accurate enough and that the first-order correction is all that is needed. Then, the first-order correction is derived analytically for a general optimal control problem and then applied to the navigation problem. It is noted that the first-order solution is the neighboring optimal path for the given value of the small parameter.

# 21

# Conversion into a Parameter Optimization Problem

## 21.1 Introduction

A standard method for solving optimal control problems is to convert them into parameter optimization problems and solve them using one of several available nonlinear programming codes (Ref. H2). This process is called suboptimal control because the optimal control is restricted to a subclass of functions, for example, piecewise linear controls.

Several methods for performing the conversion have been proposed: direct shooting, collocation, direct transcription, and differential inclusion (see, for example, Refs. HS, HP, EC, and Se). In direct shooting, the controls are guessed, and the state differential equations are integrated explicitly. In collocation and direct transcription, the controls and states are guessed, and the differential equations are integrated implicitly. Finally, in differential inclusion, the controls are eliminated between the differential equations, the states are guessed, and integration is performed implicitly. The methods differ by what is guessed (controls and/or states), how the integration is performed (explicitly or implicitly), and the order of the integration.

The purpose of this chapter is to review these methods with the intention of unifying the conversion process. First, the optimal control problem and the parameter optimization or nonlinear programming problem are defined. Then, explicit numerical integration is reviewed, and the conversion process for the case where the controls are the unknowns is presented. To improve the integration accuracy, a method in which the states are guessed at several points in the interval of integration is also discussed. Next, implicit numerical integration is reviewed and applied to the conversion for which the controls and the states are the unknowns. Finally, implicit integration is applied to the conversion where only the states are the unknowns. In discussing numerical integration, both structure and order are emphasized.

This chapter can also be viewed as an application of differential calculus to the creation of explicit and implicit Runge–Kutta integrators. In other words, differentials are used to get the equations of condition instead of the usual Taylor series.

## 21.2   Optimization Problems

The baseline optimal control problem is to find the design parameter $a$ and the control history $u(t)$ that minimize the scalar performance index

$$J = \phi(x_f, a), \tag{21.1}$$

subject to the differential constraint

$$\dot{x} = g(t, x, u, a), \tag{21.2}$$

the prescribed initial conditions

$$t_0 = 0, \quad x_0 = x_{0_s}, \tag{21.3}$$

and the prescribed final conditions

$$t_f = t_{f_s}, \quad \beta(x_f, a) = 0, \quad \theta(x_f, a) \geq 0. \tag{21.4}$$

The dimensions of $a, u, x, \beta$, and $\theta$ are $l \times 1, m \times 1, n \times 1, p \times 1$, and $q \times 1$, respectively, and the subscript $s$ denotes a specific value. In this problem statement, the final time is fixed. If the actual final time is free, a "fixed final time" problem can be created by introducing the transformation $\tau = t/t_f$. The above format contains the resulting problem if $t$ is replaced by $\tau$ and $\tau_{f_s} = 1$. The actual final time $t_f$ is one of the elements of the design parameter $a$.

The conversion of this optimal control problem into a parameter optimization problem begins with the definition of $N$ fixed times

$$t_0 = t_1 < t_2 < \cdots < t_k < \cdots < t_{N-1} < t_N = t_f \qquad (21.5)$$

called nodes, which are usually equally spaced. Then, functions of time such as $u(t)$ and $x(t)$ are replaced by their values at the nodes ($u_k$ and $x_k$) and some form of interpolation. In general, the unknowns of the parameter optimization problem are the design parameter $a$ and some combination of control parameters $u_k$ and/or state parameters $x_k$. If $X$ denotes the vector of unknown parameters, the corresponding parameter optimization problem is to find the value of $X$ which minimizes the scalar performance index

$$J = F(X), \qquad (21.6)$$

subject to the equality constraints

$$C(X) = 0 \cdot \qquad (21.7)$$

and the inequality constraints

$$D(X) \geq 0. \qquad (21.8)$$

The functions $F, C$, and $D$ are different names for $\phi, \beta$, and $\theta$. They depend on what the unknowns are and how the differential equations (21.2) are integrated, explicitly or implicitly. The specific forms of these functions are presented later in the chapter. In general, the solution process is to guess values for the unknown parameters and use a nonlinear programming code to find the values of the parameters that minimize the performance index and satisfy the constraints.

In what follows, explicit numerical integration is reviewed, and two conversion methods based on explicit integration are discussed. Then, implicit numerical integration is reviewed, and two conversion methods of this class are discussed. The integration methods are developed through taking differentials rather than making Taylor series.

Before continuing, it is remarked that the optimal control problem can be made more complicated by including a free initial point with equality and inequality constraints, internal points with equality and inequality constraints, integral constraints, path equality constraints, and path inequality constraints. In all cases, conversion into a parameter optimization problem is possible. Path constraints can be converted to point constraints and included in the baseline problem by using penalty functions (see Ref. H2). Alternately, path constraints can be imposed at each node. However, if a path constraint is in effect at several consecutive nodes, it is not in general satisfied between the nodes. Satisfaction can be improved by adding more nodes in this area.

## 21.3  Explicit Numerical Integration

In terms of the differential equation

$$\dot{x} = f(t, x), \tag{21.9}$$

the term explicit means that each time the function $f$ is evaluated, the value of $x$ is known. In the following developments, Eq.(21.9) is assumed to be scalar, but the results apply to vector equations. Only fixed-step integration is considered.

The most commonly used technique for explicit numerical integration is the Runge-Kutta (RK) method (Refs. Ho and AM). Given the time $t_i$, the state $x_i = x(t_i)$, and the stepsize $h = t_{i+1} - t_i$, the state $x_{i+1} - x(t_i + h)$ can be obtained from the formula

$$x_{i+1} = x_i + h \sum_{j=1}^{p} c_j f_j, \tag{21.10}$$

where

$$f_1 = f(t_i, x_i) \tag{21.11}$$

and for $j \geq 2$

$$f_j = f(t_i + h\alpha_j, x_i + h \sum_{\lambda=1}^{j-1} \beta_{j\lambda} f_\lambda). \tag{21.12}$$

In these equations, $p$ is the number of function evaluations to be made, and $c, \alpha$, and $\beta$ are constants whose values are to be determined. To obtain a particular integrator, $p$ is specified; $h$ is assumed to be a small quantity; and Eq.(21.10) is expanded in a Taylor series and compared with the general Taylor series expansion of $x_{i+1} = x(t_i + h)$, that is,

$$x_{i+1} = x_i + \dot{x}_i h + \frac{1}{2!} \ddot{x}_i h^2 + \frac{1}{3!} \dddot{x}_i h^3 + \cdots, \tag{21.13}$$

where

$$\dot{x} = f$$
$$\ddot{x} = f_t + f_x \dot{x}$$
$$\dddot{x} = f_{tt} + 2f_{tx}\dot{x} + f_{xx}\dot{x}^2 + f_x\ddot{x} \tag{21.14}$$
$$\vdots$$

Equating corresponding terms of the two series leads to a set of equations of condition that can be solved for $c$, $\alpha$, and $\beta$. An $n$th-order integrator is obtained if the two series match through terms of order $h^n$.

The Taylor series can also be obtained by taking differentials. If

$$dx_i \triangleq x_{i+1} - x_i = x(t_i + h) - x(t_i) \tag{21.15}$$

and if $\Delta x_i$ is expressed as

$$\Delta x_i = dx_i + \frac{1}{2!}d^2x_i + \frac{1}{3!}d^3x_i + \cdots, \tag{21.16}$$

the different-order terms can be obtained by taking differentials of the right-hand side of Eq.(20.15) with respect to $h$ which is a small parameter. Doing so leads to

$$dx_i = \dot{x}(t_i + h)dh$$
$$d^2x_i = \ddot{x}(t_i + h)dh^2$$
$$d^3x_i = \dddot{x}(t_i + h)dh^3 \tag{21.17}$$
$$\vdots$$

where $d(dh) = 0$ because $dh$ is an independent differential (see Section 18.2). Evaluated on $h = 0$ these differentials become

$$dx_i = \dot{x}(t_i)dh = f_i dh$$
$$d^2x_i = \ddot{x}(t_i)dh^2 = [f_t + f_x\dot{x}]_i dh^2$$
$$d^3x_i = \dddot{x}(t_i)dh^3 = [f_{tt} + 2f_{tx}\dot{x} + f_{xx}\dot{x}^2 + f_x\ddot{x}]_i dh^3 \tag{21.18}$$
$$\vdots$$

After the differentials have been taken and evaluated on $h = 0$, the term $dh$ becomes $dh = \Delta h = h$. Equations (20.16) and (20.18) then give Eq.(21.13).

To illustrate the process, consider the case of one function evaluation. Equation (21.10) becomes

$$\Delta x_i = x_{i+1} - x_i = hc_1 f_1. \tag{21.19}$$

Taking the differential of the right-hand side leads to

$$dx_i = c_1 f_1 dh$$
$$d^2 x_i = 0 \tag{21.20}$$

$$\vdots$$

and, since $f_1 = f_i$, has the same results when evaluated on $h = 0$. These equations match Eq.(21.18) through first order if the equation of condition

$$c_1 = 1 \tag{21.21}$$

is satisfied. The first-order RK integrator, called the Euler integrator, is given by

$$x_{i+1} = x_i + h f_i. \tag{21.22}$$

By starting at the initial point, it is possible to propagate the state over the entire interval of integration $[t_0, t_f]$.

For two function evaluations, Eq.(21.10) becomes

$$\Delta x_i = x_{i+1} - x_i = h(c_1 f_1 + c_2 f_2), \tag{21.23}$$

where

$$f_1 = f(t_i, x_i), \quad f_2 = f(t_i + \alpha_2 h, x_i + \beta_{12} f_1 h). \tag{21.24}$$

Taking differentials of these equations leads to

$$
\begin{aligned}
dx_i &= (c_1 f_1 + c_2 f_2) dh \\
&\quad + h(c_2 f_{2_t} \alpha_2 + c_2 f_{2_x} \beta_{12} f_1) dh \\
d^2 x_i &= 2(c_2 f_{2_t} \alpha_2 + c_2 f_{2_x} \beta_{12} f_1) dh^2 \\
&\quad + h(c_2 \alpha_2 f_{2_{tt}} \alpha_2 + c_2 \alpha_2 f_{2_x} \beta_{12} f_1 \\
&\quad + c_2 \beta_{12} f_{2_{xt}} \alpha_2 + c_2 \beta_{12} f_{2_{xx}} \beta_{12} f_1^2) dh^2
\end{aligned}
\tag{21.25}
$$

$$\vdots$$

Evaluated on $h = 0$ these equations become

$$
\begin{aligned}
dx_i &= (c_1 + c_2) f_i dh \\
d^2 x_i &= 2[c_2 \alpha_2 (f_t)_i + c_2 \beta_{12} (f_x)_i f_i] dh^2
\end{aligned}
\tag{21.26}
$$

$$\vdots$$

and match Eqs.(20.18) through second order if the following equations of condition are satisfied:

$$c_1 + c_2 = 1, \quad 2c_2\alpha_2 = 1, \quad 2c_2\beta_{21} = 1. \tag{21.27}$$

Note that there is one more parameter than there are equations of condition. Hence, it is possible to choose one parameter, for example, $\alpha_2 = 1$. This choice leads to

$$\alpha_2 = 1, \quad c_1 = \frac{1}{2}, \quad c_2 = \frac{1}{2}, \quad \beta_{21} = 1, \tag{21.28}$$

and the second-order RK integrator, called the Euler–Cauchy integrator, is given by

$$x_{i+1} = x_i + \tfrac{h}{2}(f_1 + f_2)$$
$$f_1 = f(t_i, x_i) \tag{21.29}$$
$$f_2 = f(t_i + h, x_i + f_1 h).$$

A popular RK integrator is the one of fourth order defined by

$$x_{i+1} = x_i + \frac{h}{6}(f_1 + 2f_2 + 2f_3 + f_4), \tag{21.30}$$

where

$$f_1 = f(t_i, x_i)$$
$$f_2 = f(t_i + \tfrac{h}{2}, x_i + \tfrac{h}{2}f_1)$$
$$f_3 = f(t_i + \tfrac{h}{2}, x_i + \tfrac{h}{2}f_2) \tag{21.31}$$
$$f_4 = f(t_i + h, x_i + f_3).$$

Runge–Kutta integrators of all orders exist. Through order four, each additional function evaluation increases the order by one. A fifth-order integrator, however, requires six function evaluations.

## 21.4   Conversion with $a, u_k$ as Unknowns

In the first conversion method that uses explicit numerical integration (see, for example, Ref. HS), the design parameter $a$ and the control parameters $u_k, k = 1, \ldots, N$ are the unknowns, and the control history $u(t)$ is formed by linear interpolation, that is,

$$u = u_k + \frac{u_{k+1} - u_k}{t_{k+1} - t_k}(t - t_k), \quad t_k < t < t_{k+1}. \tag{21.32}$$

This method is called *direct shooting* because the state differential equations are integrated from $t_0$ to $t_f$ in one pass. As many integration steps can be used between nodes as are needed to achieve a desired accuracy.

Assume that values for the design parameter and the control parameters are given and placed in the unknown parameter vector $X$, that is,

$$X = [\, a^T \; u_1^T \; \cdots \; u_N^T \,]^T, \qquad (21.33)$$

which contains $l+mN$ elements. Since the design parameter $a$ and the piecewise linear control history $u(t)$ are known, the state equation (21.2) has the form of Eq. (21.9) and can be integrated numerically from $t_0, x_0$ to $t_f$ to obtain the final state

$$x_f = x_f(X), \qquad (21.34)$$

meaning that the final state is a function only of the unknown parameters. Substitution of Eq.(21.34) into the performance index (21.1) and constraints (21.4) reduces the optimal control problem to the parameter optimization problem defined by Eqs.(21.6) through (21.8). In other words, the performance index function is defined as

$$F(X) = \phi(x_f(X), a) \; ; \qquad (21.35)$$

the equality constraint function is

$$C(X) = \beta(x_f(X), a) \; ; \qquad (21.36)$$

and the inequality constraint function is

$$D(X) = \theta(x_f(X), a) \qquad (21.37)$$

since $a$ is contained in $X$.

Note that any order integrator can be used but that considerable accuracy is needed for computing the partial derivatives needed by nonlinear programming. Accuracy is achieved by a trade-off between order and stepsize. Derivatives are computed by finite differences, and fixed step integration is employed on the premise that the difference used in computing a derivative cancels error.

# 21.5 Conversion with $a, u_k, x_j$ as Unknowns

The method defined in the previous section is characterized by a single integration over the time interval $[t_0, t_f]$. If the time interval is very long, the accuracy of the integration is affected, and the accuracy of numerical derivatives is even worse. One way to increase accuracy is to guess the value of the state $x$ at the end of every $\nu$ intervals ($N-1$ must be an integer multiple of $\nu$) and to start a new integration. Then, as a part of the optimization process the difference of the computed and guessed values of $x$ at each of these nodes is driven to zero. This method is called *direct multiple shooting*.

Specifically, let $\bar{x}$ denote the computed value and

$$r = x - \bar{x} \tag{21.38}$$

denote the difference between the guessed and computed values of $x$. Then, Eq. (21.33) becomes

$$X = [\, a^T \; u_1^T \; \cdots \; u_N^T \; x_{\nu+1}^T \; x_{2\nu+1}^T \; \cdots \; x_{N-\nu}^T \,]^T \tag{21.39}$$

so that there are now $l + mN + n((N-1)/\nu - 1)$ unknown parameters. Finally, the equality constraint (21.36) becomes the following:

$$C(X) = \begin{bmatrix} \beta(x_f(X), a) \\ r_1(x_{\nu+1}) \\ r_2(x_{2\nu+1}) \\ \vdots \\ r_{\frac{N-1}{\nu}-1}(x_{N-\nu}) \end{bmatrix} = 0 \tag{21.40}$$

and contains $p + n((N-1)/\nu - 1)$ elements. There is no change in $F(X)$ or $D(X)$.

In the limit, the value of $x$ can be guessed at every node, with explicit integration taking place between the nodes. If the nodes are sufficiently close together, a single integration step can be taken. In this case, it is possible to use implicit integration which has more accuracy per function evaluation.

## 21.6   Implicit Numerical Integration

Relative to the differential equation (21.9), the term implicit means that the value of $x$ needed for evaluating $f$ is not known and that, to perform one integration step, a predictor-corrector approach must be used (Ref. Ho, p. 248). This local iteration approach can be replaced by a global iteration approach

(Ref.  AM, p.194) where the value of $x$ at the end of every integration step is guessed and then iterated upon until the correct values are obtained. This means that $t_1, \ldots, t_N$ are known, and values for $x_1, \ldots, x_N$ are available, $t_1$ and $x_1$ being the prescribed initial conditions. In what follows, the differential equation (21.9) is assumed to be scalar, but the results can be applied to the vector equation. In the integration literature, the integration formulas developed in this section have their individual names (implicit Euler, midpoint rule, trapezoid rule, and so on), but collectively they are also called implicit Runge–Kutta, finite–difference, Newton–Coates, and collocation methods. The implicit Runge–Kutta approach is used here (Refs. Bu and Hem).

Given the time $t_i$, the state $x_i$, and the stepsize $h$, the state $x_{i+1}$ is written in the form

$$x_{i+1} = x_i + h \sum_{j=1}^{p} c_j f_j, \qquad (21.41)$$

where

$$f_j = f(t_i + h\alpha_j, x_i + h \sum_{\lambda=1}^{p} \beta_{j\lambda} f_\lambda). \qquad (21.42)$$

Note that the value of $f$ is needed to compute the value of $f$. Also, the explicit RK integrator is a particular case of the implicit RK integrator in which $\beta_{j\lambda} = 0$ for $j \leq \lambda$.

For the case of one function evaluation, the implicit RK integrator is defined by

$$\Delta x_i = x_{i+1} - x_i = hc_1 f_1, \qquad (21.43)$$

where

$$f_1 = f(t_i + h\alpha_1, x_i + h\beta_{11} f_1). \qquad (21.44)$$

Taking variations for small $h$ leads to

$$\begin{aligned}
dx_i &= c_1 f dh + hc_1(f_t\alpha_1 + f_x\beta_{11}f)dh \\
d^2x &= 2c_1(f_t\alpha_1 + f_x\beta_{11}f)dh^2 \\
&\quad + hc_1(f_{tt}\alpha_1^2 + f_{tx}\alpha_1\beta_{12}f + f_{xt}\alpha_1\beta_{11}f + f_{xx}\beta_{11}f^2)dh^2
\end{aligned} \qquad (21.45)$$

$$\vdots$$

Evaluated on $h = 0$ these equations become

$$
\begin{aligned}
dx_i &= c_1 f_i dh \\
d^2 x_i &= (2c_1\alpha_1(f_t)_i + 2c_1\beta_{11}(f_x)_i f_i)dh^2
\end{aligned}
\tag{21.46}
$$

$$\vdots$$

and match the general expansion (21.18) through second order if

$$
c_1 = 1, \quad c_1\alpha_1 = \frac{1}{2}, \quad c_1\beta_{11} = \frac{1}{2}.
\tag{21.47}
$$

Hence, a second-order integrator results if

$$
c_1 = 1, \quad \alpha_1 = \frac{1}{2}, \quad \beta_{11} = \frac{1}{2}
\tag{21.48}
$$

and is written as

$$
x_{i+1} = x_i + h f_1,
\tag{21.49}
$$

where

$$
f_1 = f(t_i + \frac{h}{2}, \; x_i + \frac{1}{2} f_1 h).
\tag{21.50}
$$

Note that $f_1$ is needed to compute $f_1$ so that an iteration (predictor-corrector) procedure is needed to perform a single integration step. The iteration can be performed on $x_{i+1}$ by solving Eq.(21.49) for $f_1$ and rewriting the integrator in the form

$$
x_{i+1} = x_i + h f(t_m, x_m),
\tag{21.51}
$$

where

$$
t_m = t_i + \frac{h}{2}, \quad x_m = \frac{x_i + x_{k+1}}{2}.
\tag{21.52}
$$

This integrator is called the midpoint rule.

　　　　To convert the local iteration process for a single integration step into an iteration process over the entire interval of integration, the values of $x_i$ are guessed at the end of each integration step. Because the $x_i$s are guessed, an integration formula such as the midpoint rule (21.52) is not satisfied. It is rewritten in the residual form

$$
R_i = x_{i+1} - x_i - h f(t_m, x_m)
\tag{21.53}
$$

for every integration step. Then, all the $x_i$s are iterated upon until all the $R_i$s are zero.

If two function evaluations $f_1, f_2$ are used, it is possible to develop a fourth-order integrator, but it is not possible to put the integrator in the form of Eq.(21.51) where the iteration is on $x_{i+1}$. On the other hand, if $f_1 = f_i$ and $f_2$ are used, a third-order integrator of this form can be created. However, for the global iteration process, if $f_i$ is available, then so is $f_{i+1}$ from the next integration step. Hence, three function evaluations are available for developing the integrator, even though only two are made. It is shown later in this section that a fourth-order integrator can be developed with three function evaluations.

An integrator involving the two function evaluations $f_i$ and $f_{i+1}$ can be developed by first writing Eqs.(21.41) and (21.42) as

$$x_{i+1} = x_i + h(c_1 f_1 + c_2 f_2), \tag{21.54}$$

where

$$f_1 = f(t_i, x_i)$$
$$f_2 = f[t_i + h, x_i + h(\beta_{21} f_1 + \beta_{22} f_2)]. \tag{21.55}$$

In order for $f_2$ to be evaluated at $x_{i+1}$, it is necessary to have $\beta_{21} = c_1$ and $\beta_{22} = c_2$. Then, differentials lead to the following equations of condition through second order:

$$c_1 + c_2 = 1, \quad 2c_2 = 1, \quad 2c_2(c_1 + c_2) = 1, \tag{21.56}$$

which are satisfied by $c_1 = c_2 = \frac{1}{2}$. The integrator becomes

$$x_{i+1} = x_i + \frac{1}{2}h[f(t_i, x_i) + f(t_{i+1}, x_{i+1})] \tag{21.57}$$

and is known as the trapezoid rule. While two function evaluations are used to derive the integrator, $f_{i+1}$ for this integration step becomes $f_i$ for the next step. Hence, only one function evaluation is actually made per step. This explains why the trapezoid rule has the same order as the midpoint rule.

A fourth-order integrator can be made by using the function evaluations $f_i$, $f_m$, and $f_{i+1}$. Here, only two function evaluations are actually made per integration step because $f_{i+1}$ is used in the next step. The integrator is defined as

$$x_{i+1} = x_i + h(c_1 f_k + c_2 f_m + c_3 f_{i+1}), \tag{21.58}$$

where

$$f_1 = f(t_i, x_i)$$
$$f_m = f[t_i + h\alpha_2, x_i + h(\beta_{21} f_i + \beta_{22} f_m + \beta_{23} f_{i+1})]$$
$$f_{i+1} = f[t_i + h, x_i + h(c_1 f_i + c_2 f_m + c_3 f_{i+1})]. \tag{21.59}$$

The equations of condition through fourth order yield the following results:

$$c_1 = \tfrac{1}{6}, \quad c_2 = \tfrac{2}{3}, \quad c_3 = \tfrac{1}{6}, \quad \alpha_2 = \tfrac{1}{2}$$
$$\beta_{21} = \tfrac{5}{24}, \quad \beta_{22} = \tfrac{1}{3}, \quad \beta_{23} = -\tfrac{1}{24}, \tag{21.60}$$

which show that

$$
\begin{aligned}
x_m &= x_i + h(\beta_{21} f_i + \beta_{22} f_m + \beta_{23} f_{i+1}) \\
&= \tfrac{x_i + x_{i+1}}{2} - \tfrac{h}{8}(f_{i+1} - f_i).
\end{aligned}
\tag{21.61}
$$

The final form of the integrator, called Simpson's one-third rule, is given by

$$x_{i+1} = x_i + \frac{h}{6}[f(t_i, x_i) + 4f(t_m, x_m) + f(t_{i+1}, x_{i+1})], \tag{21.62}$$

where

$$t_m = t_i + \frac{h}{2}, \quad x_m = \frac{x_i + x_{i+1}}{2} - \frac{h}{8}(f_{i+1} - f_i). \tag{21.63}$$

## 21.7   Conversion with $a, u_k, x_k$ as Unknowns

For this conversion method (Ref. HP), the design parameter $a$, the control parameters $u_1, \ldots, u_N$, and the state parameters $x_2, \ldots, x_N$, are the unknowns. Note that $x_1 = x_0$ from the prescribed initial conditions (21.3). Again, $z_k$ denotes the vector $z$ at the node $t_k$. The vector of unknown parameters is given by

$$X = [\, a^T \; u_1^T \; \cdots \; u_N^T \; x_2^T \; \cdots \; x_N^T \,]^T \tag{21.64}$$

which has $l + mN + n(N - 1)$ elements. While $a$ and $u$ are normal optimization variables, $n(N - 1)$ additional constraints must be imposed to compute $x_2, \ldots, x_N$. Note that each interval between nodes represents one integration step.

Integration of Eq.(21.2) is performed by calculating the residuals (one for each interval) and driving them to zero as a part of the optimization process. For the second-order midpoint rule that has one function evaluation at $t_m$, the integration formula is applied in the form

$$R_k = x_{k+1} - x_k - g_m(t_{k+1} - t_k), \tag{21.65}$$

where

$$g_m = g(t_m, x_m, u_m, a) \tag{21.66}$$

and

$$t_m = \frac{t_k + t_{k+1}}{2}, \quad x_m = \frac{x_k + x_{k+1}}{2}, \quad u_m = \frac{u_k + u_{k+1}}{2}. \tag{21.67}$$

For the second-order trapezoid rule that uses values of $g$ at $t_k$ and $t_{k+1}$, the residual is given by

$$R_k = x_{k+1} - x_k - \frac{1}{2}(g_k + g_{k+1})(t_{k+1} - t_k). \tag{21.68}$$

For the fourth-order Simpson one-third rule that uses the function evaluations at $t_k, t_m,$ and $t_{k+1}$, the residual is

$$R_k = x_{k+1} - x_k - \frac{1}{6}(g_k + 4g_m + g_{k+1})(t_{k+1} - t_k), \tag{21.69}$$

where $g_m$ is evaluated at

$$\begin{aligned} t_m &= \frac{t_k + t_{k+1}}{2}, \quad u_m = \frac{u_k + u_{k+1}}{2}, \\ x_m &= \frac{x_k + x_{k+1}}{2} - \frac{1}{8}(g_{k+1} - g_k)(t_{k+1} - t_k). \end{aligned} \tag{21.70}$$

Regardless of the integration method used, the parameter optimization problem is the same. The performance index and inequality constraint function are given by

$$F(X) = \phi(x_N, a), \quad D(X) = \theta(x_N, a), \tag{21.71}$$

while the equality constraint function becomes

$$C(X) = \begin{bmatrix} \beta(x_N, a) \\ R_1(a, u_1, u_2, x_2) \\ R_2(a, u_2, u_3, x_2, x_3) \\ \vdots \\ R_{N-1}(a, u_{N-1}, u_N, x_{N-1}, x_N) \end{bmatrix} \tag{21.72}$$

and now contains $p + n(N - 1)$ elements.

In the optimization literature, the Simpson one-third rule is used in Ref. HP and is called *collocation*. The use of any rule is called *direct transcription* (Refs. EC and Be).

## 21.8  Conversion with $a, x_k$ as Unknowns

*Differential inclusion* (Ref. Se) has the feature that the controls are eliminated from the optimization problem so that the unknown parameter vector becomes

$$X = [\, a^T \; x_2^T \; \cdots \; x_N^T \,]^T \tag{21.73}$$

and contains only $l + n(N-1)$ elements. Here, $m$ of the $n$ state equations

$$\dot{x} = g(t, x, u, a) \tag{21.74}$$

are solved for the controls which are then eliminated from the remaining equations. This process results in $n - m$ constraints of the form

$$F(t, x, \dot{x}, a) = 0. \tag{21.75}$$

In Ref. Be, these constraints are imposed at the midpoint of each interval as

$$\bar{R}_k = F(t_m, x_m, \dot{x}_m, a) = 0, \tag{21.76}$$

where

$$t_m = \frac{t_k + t_{k+1}}{2}, \quad x_m = \frac{x_k + x_{k+1}}{2}, \quad \dot{x}_m = \frac{x_{k+1} - x_k}{t_{k+1} - t_k}. \tag{21.77}$$

Upon convergence, the integration has been performed by the midpoint rule. Note that the control parameters can be eliminated from Eq.(21.65), that is, $R_k = 0$, in each interval to form Eq.(21.76). Relative to the parameter optimization problem, $F(X)$ and $D(X)$ are the same as those for the previous method, but the equality constraint function becomes

$$C(X) = \begin{bmatrix} \beta(x_N, a) \\ R_1(a, x_2) \\ R_2(a, x_2, x_3) \\ \vdots \\ R_{N-1}(a, x_{N-1}, x_N) \end{bmatrix} \tag{21.78}$$

and has $p + (n - m)(N - 1)$ elements. Once the design and state parameters are known, the control parameters are obtained from their defining equations.

## 21.9 Remarks

In recent years, several methods have been proposed for converting optimal control problems into parameter optimization problems. They have been referred to in the literature as direct shooting, direct multiple shooting, collocation or direct transcription, and differential inclusion. Because these methods have a many similarities not immediately obvious by reading the papers, the purpose of this chapter has been to summarize these methods and to categorize them in terms of the numerical integration technique (explicit or implicit) used to integrate the state differential equation, the order of the numerical integration technique, and the choice of the unknowns. Another purpose of this chapter has been to use differentials to obtain the equations of condition for numerical integration methods.

Explicit integration has been reviewed in terms of structure and order. In the conversion method called direct shooting, the values of the controls at the nodes are the unknowns. For long integration times, the accuracy can be improved by guessing the states at several intermediate nodes and restarting the integration. This method is called direct multiple shooting here.

Implicit integration has also been reviewed in terms of structure and order. The second-order midpoint rule has been used in differential inclusion; the fourth-order Simpson one-third rule has been used in collocation; and direct transcription is a name that applies to the use of any implicit method.

It is interesting to note the similarities between these conversion methods and the indirect methods proposed for solving the two point boundary value problem associated with the optimal control problem. The shooting method and quasilinearization are two of these methods. In the shooting method, values are guessed for the initial Lagrange multipliers, and the state and multiplier equations are integrated from the initial time to the final time in one pass. Because of the sensitivity of the shooting method to the initial Lagrange multiplier guesses, multiple shooting has been developed. The conversion methods involving explicit integration bear a strong resemblance to the shooting methods, but they are not as sensitive because the control history is guessed. In quasilinearization, the state and multiplier histories are the unknowns. This approach is similar to that used with implicit integration. Finally, the approach used in differential inclusion of eliminating the control is inherent in the idea of forming the two point boundary value problem.

# Appendix A

# First and Second Differentials by Taylor Series Expansion

Throughout the text, the calculus of differentials is used to derive the first and second differentials of the augmented performance index. In this appendix, the first and second differentials of the standard optimal control problem are derived using the process of Taylor series expansion. Not only can the processes be compared, but the correctness of the final results can be verified.

The standard optimal control problem is to find the control history that minimizes the performance index

$$J = \phi(t_f, x_f) + \int_{t_0}^{t_f} L(t, x, u) dt, \tag{A.1}$$

subject to the differential constraints

$$\dot{x} = f(t, x, u), \tag{A.2}$$

the prescribed initial conditions

$$t_0 = t_{0_s}, \quad x_0 = x_{0_s}, \tag{A.3}$$

and the prescribed final conditions

$$\psi(t_f, x_f) = 0. \tag{A.4}$$

Adjoining the differential constraints by variable Lagrange multipliers $\lambda(t)$ and the final conditions by constant Lagrange multipliers $\nu$, the augmented performance index is given by

$$J' = G(t_f, x_f, \nu) + \int_{t_0}^{t_f} [H(t, x, u, \lambda) - \lambda^T \dot{x}] dt, \tag{A.5}$$

where

$$G = \phi + \nu^T \psi, \quad H = L + \lambda^T f. \tag{A.6}$$

The difference of the augmented performance index evaluated on an admissible comparison path (subscript *) and that evaluated on the minimal path (no subscript) is written as

$$\begin{aligned}
\Delta J' &= G(t_{f*}, x_{f*}, \nu_*) + \int_{t_0}^{t_{f*}} [H(t, x_*, u_*, \lambda_*) - \lambda_*^T \dot{x}_*]dt \\
&\quad - G(t_f, x_f, \nu) - \int_{t_0}^{t_f} [H(t, x, u, \lambda) - \lambda^T \dot{x}]dt.
\end{aligned} \tag{A.7}$$

Isolating the integral over $t_f$, $t_{f*}$ and rearranging terms leads to

$$\begin{aligned}
\Delta J' &= G(t_{f*}, x_{f*}, \nu_*) - G(t_f, x_f, \nu) \\
&\quad + \int_{t_f}^{t_{f*}} [H(t, x_*, u_*, \lambda_*) - \lambda_*^T \dot{x}_*]dt \\
&\quad + \int_{t_0}^{t_f} [H(t, x_*, u_*, \lambda_*) - \lambda_*^T \dot{x}_* - H(t, x, u, \lambda) + \lambda^T \dot{x}]dt.
\end{aligned} \tag{A.8}$$

At this point, each starred variable is written as follows:

$$\begin{aligned}
t_{f*} &= t_f + \Delta t_f, \quad x_{*_f} = x_f + \Delta x_f, \quad \nu_* = \nu + \Delta \nu \\
x_* &= x + \tilde{\Delta} x, \quad u_* = u + \tilde{\Delta} u, \quad \lambda_* = \lambda + \tilde{\Delta} \lambda,
\end{aligned} \tag{A.9}$$

where $\Delta$ denotes a small total change (time-free) and $\tilde{\Delta}$ denotes a time-constant total change. The total change is the sum of all-order terms

$$\Delta(\ ) = d(\ ) + \frac{1}{2!}d^2(\ ) + \ldots, \quad \tilde{\Delta}(\ ) = \delta(\ ) + \frac{1}{2!}\delta^2(\ ) + \cdots \tag{A.10}$$

and must be used because the above quantities are not all independent. Next, Eqs.(A.9) are substituted into Eq.(A.8) and each term is expanded in a Taylor series. Each term is discussed separately.

The expansion of the endpoint terms gives

$$\begin{aligned}
&G(t_f + \Delta t_f, x_f + \Delta x_f, \nu_f + \Delta \nu) \\
&= G(t_f, x_f, \nu) + G_{t_f}\Delta t_f + G_{x_f}\Delta x_f + G_\nu \Delta \nu
\end{aligned}$$

$$+ \frac{1}{2!}(\Delta t_f G_{t_f t_f} \Delta t_f + \Delta t_f G_{t_f x_f} \Delta x_f + \Delta t_f G_{t_f \nu} \Delta \nu \tag{A.11}$$

$$+ \Delta x_f^T G_{x_f t_f} \Delta t_f + \Delta x_f^T G_{x_f x_f} \Delta x_f + \Delta x_f^T G_{x_f \nu} \Delta \nu$$

$$+ \Delta \nu^T G_{\nu t_f} \Delta t_f + \Delta \nu^T G_{\nu x_f} \Delta x_f + \Delta \nu^T G_{\nu \nu} \Delta \nu) + \cdots .$$

Here, $G_\nu = \psi^T = 0$, $G_{\nu t_f} = \psi_{t_f}^T$, $G_{\nu x_f} = \psi_{x_f}^T$, and $G_{\nu \nu} = 0$ because $G$ is linear in $\nu$. Also, since the admissible comparison path must satisfy the final conditions

$$\psi(t_{f*}, x_{f*}) = \psi(t_f + \Delta t_f, x_f + \Delta x_f) = 0 \tag{A.12}$$

or

$$\psi_{t_f} \Delta t_f + \psi_{x_f} \Delta x_f = O(\Delta^2). \tag{A.13}$$

Hence, all the terms involving $\Delta \nu$ either have zero coefficients or are of higher order. Thus,

$$G(t_{f*}, x_{f*}, \nu_*) - G(t_f, x_f, \nu) =$$
$$G_{t_f} \Delta t_f + G_{x_f} \Delta x_f + \frac{1}{2!}(\Delta t_f G_{t_f t_f} \Delta t_f \tag{A.14}$$
$$+ \Delta t_f G_{t_f x_f} \Delta x_f + \Delta x_f^T G_{x_f t_f} \Delta t_f + \Delta x_f^T G_{x_f x_f} \Delta x_f) + \cdots .$$

The expansion of the second line in Eq.(A.8) has been discussed adequately in Section 6.6 and is given by

$$\int_{t_f}^{t_{f*}} [H(t, x_*, u_*, \lambda_*) - \lambda_*^T \dot{x}_*] dt = \tag{A.15}$$
$$(H_* - \lambda_*^T \dot{x}_*)_f \Delta t_f + \frac{1}{2!}[\frac{d}{dt}(H_* - \lambda_*^T \dot{x}_*)]_f \Delta t_f^2.$$

The expansion required by the last line of Eq.(A.8) is written as

$$H(t, x + \tilde{\Delta} x, u + \tilde{\Delta} u, \lambda + \tilde{\Delta} \lambda) - (\lambda + \tilde{\Delta} \lambda)^T (\dot{x} + \tilde{\Delta} \dot{x}) =$$

$$H - \lambda^T \dot{x} + H_x \tilde{\Delta} x + H_u \tilde{\Delta} u + H_\lambda \tilde{\Delta} \lambda - \tilde{\Delta} \lambda^T \dot{x} - \tilde{\Delta} \lambda^T \tilde{\Delta} \dot{x}$$

$$+ \frac{1}{2!}[\tilde{\Delta} x^T H_{xx} \tilde{\Delta} x + \tilde{\Delta} x^T H_{xu} \tilde{\Delta} u + \tilde{\Delta} x^T H_{x\lambda} \tilde{\Delta} \lambda \tag{A.16}$$

$$+ \tilde{\Delta} u^T H_{ux} \tilde{\Delta} x + \tilde{\Delta} u^T H_{uu} \tilde{\Delta} u + \tilde{\Delta} u^T H_{u\lambda} \tilde{\Delta} \lambda$$

$$+ \tilde{\Delta} \lambda^T H_{\lambda x} \tilde{\Delta} x + \tilde{\Delta} \lambda^T H_{\lambda u} \tilde{\Delta} u + \tilde{\Delta} \lambda^T H_{\lambda \lambda} \tilde{\Delta} \lambda].$$

Here, $H_\lambda = f^T, H_{x\lambda} = f_x^T, H_{u\lambda} = f_u^T$, and $H_{\lambda\lambda} = 0$ because $H$ is linear in $\lambda$. Also, the admissible comparison path must satisfy the differential equation (A.2), that is,

$$\frac{d}{dt}(x + \tilde{\Delta}x) = f(t, x + \tilde{\Delta}x, u + \tilde{\Delta}u) \tag{A.17}$$

or

$$\tilde{\Delta}\dot{x} = f_x\tilde{\Delta}x + f_u\tilde{\Delta}u + O(\tilde{\Delta}^2). \tag{A.18}$$

Hence, all the terms involving $\tilde{\Delta}\lambda$ either have zero coefficients or are of higher order. Thus,

$$H_* - \lambda_*^T\dot{x}_* - H + \lambda^T\dot{x} = H_x\tilde{\Delta}x + H_u\tilde{\Delta}u - \lambda^T\tilde{\Delta}\dot{x} \tag{A.19}$$

$$+ \frac{1}{2!}(\tilde{\Delta}x^T H_{xx}\tilde{\Delta}x + \tilde{\Delta}x^T H_{xu}\tilde{\Delta}u + \tilde{\Delta}u^T H_{ux}\tilde{\Delta}x + \tilde{\Delta}u^T H_{uu}\tilde{\Delta}u) + \cdots .$$

From this relationship, it is observed that, to the order needed in Eq.(A.15),

$$H_* - \lambda_*^T\dot{x}_* = H - \lambda^T\dot{x} + H_x\tilde{\Delta}x + H_u\tilde{\Delta}u - \lambda^T\tilde{\Delta}\dot{x} + \cdots$$

$$\frac{d}{dt}(H_x - \lambda_*^T\dot{x}_*) = \frac{d}{dt}(H - \lambda^T\dot{x}) + \cdots . \tag{A.20}$$

In view of the above expansions, Eq.(A.8) becomes

$$\Delta J' = (G_{t_f} + H - \lambda^T\dot{x})\Delta t_f + G_{x_f}\Delta x_f$$

$$+ \int_{t_0}^{t_f}(H_x\tilde{\Delta}x + H_u\tilde{\Delta}u - \lambda^T\tilde{\Delta}\dot{x})dt \tag{A.21}$$

$$+ \frac{1}{2!}(\Delta t_f G_{x_f x_f}\Delta t_f + \Delta t_f G_{t_f x_f}\Delta x_f + \Delta x_f^T G_{x_f t_f}\Delta t_f + \Delta x_f^T G_{x_f x_f}\Delta x_f)$$

$$+ (H_x\tilde{\Delta}x + H_u\tilde{\Delta}u - \lambda^T\tilde{\Delta}\dot{x})_f\Delta t_f + \frac{1}{2!}[\frac{d}{dt}(H - \lambda^T\dot{x})]_f\Delta t_f^2$$

$$+ \frac{1}{2!}\int_{t_0}^{t_f}[\tilde{\Delta}x^T \tilde{\Delta}u^T]\begin{bmatrix} H_{xx} & H_{xu} \\ H_{ux} & H_{uu} \end{bmatrix}\begin{bmatrix} \tilde{\Delta}x \\ \tilde{\Delta}u \end{bmatrix} dt.$$

The next step is to integrate the term $\lambda^T\tilde{\Delta}\dot{x} = \lambda^T d(\tilde{\Delta}x)/dt$ by parts and relate first-order time-fixed changes outside the integral to time-free changes using the following expression from Section (6.4):

$$\tilde{\Delta}x = \Delta x - \dot{x}\Delta t - \tilde{\Delta}\dot{x}\Delta t - \frac{1}{2!}\ddot{x}\Delta t^2 + \cdots . \tag{A.22}$$

After some minor manipulations, Eq.(A.21) can be rewritten as

$$\Delta J' = (G_{t_f} + H_f)\Delta t_f + (G_{x_f} - \lambda_f^T)\Delta x_f$$

$$+ \int_{t_0}^{t_f} [(H_x + \dot{\lambda}^T)\tilde{\Delta} x + H_u \tilde{\Delta} u] dt \qquad (A.23)$$

$$+ \frac{1}{2!}\Delta t_f (G_{t_f t_f} + \dot{H}_f - \lambda_f^T \dot{x}_f)\Delta t_f$$

$$+ \frac{1}{2!}(\Delta t_f G_{t_f x_f}\Delta x_f + \Delta x_f^T G_{x_f t_f}\Delta t_f + \Delta x_f^T G_{x_f x_f}\Delta x_f)$$

$$+ \Delta t_f (H_x \tilde{\Delta} x + H_u \tilde{\Delta} u)_f$$

$$+ \frac{1}{2!}\int_{t_0}^{t_f} [\tilde{\Delta} x^T \tilde{\Delta} u^T] \begin{bmatrix} H_{xx} & H_{xu} \\ H_{ux} & H_{uu} \end{bmatrix} \begin{bmatrix} \tilde{\Delta} x \\ \tilde{\Delta} u \end{bmatrix} dt.$$

Note that Eq.(A.22) causes terms that are linear in $\Delta(\ )$ to make contributions to terms that are quadratic in $\Delta(\ )$.

At this point, all total changes are related to first and second-order parts as

$$\Delta(\ ) = d(\ ) + \frac{1}{2!}d^2(\ ) + \cdots, \quad \tilde{\Delta}(\ ) = \delta(\ ) + \frac{1}{2!}\delta^2(\ ) + \cdots, \qquad (A.24)$$

and the terms are collected into the first and second differentials. The first differential is then given by

$$dJ' = (G_{t_f} + H_f)dt_f + (G_{x_f} - \lambda_f^T)dx_f$$

$$+ \int_{t_0}^{t_f} [(H_x + \dot{\lambda}^T)\delta x + H_u \delta u] dt \qquad (A.25)$$

and the second differential by

$$d^2 J' = (G_{t_f} + H_f)d^2 t_f + (G_{x_f} - \lambda_f^T)d^2 x_f$$

$$+ \int_{t_0}^{t_f} [(H_x + \dot{\lambda}^T)\delta^2 x + H_u \delta^2 u] dt \qquad (A.26)$$

$$+ dt_f (G_{t_f t_f} + \dot{H}_f - \lambda_f^T \dot{x}_f)dt_f$$

$$+ dt_f G_{t_f x_f} dx_f + dx_f^T G_{x_f t_f} dt_f + dx_f^T G_{x_f x_f} dx_f$$

$$+ dt_f (2H_x \delta x + 2H_u \delta u)_f$$

$$+ \int_{t_0}^{t_f} [\delta x^T \delta u^T] \begin{bmatrix} H_{xx} & H_{xu} \\ H_{ux} & H_{uu} \end{bmatrix} \begin{bmatrix} \delta x \\ \delta u \end{bmatrix} dt.$$

In the application of the first differential, it is shown that

$$G_{t_f} + H_f = 0, \quad G_{x_f} + \lambda_f^T = 0$$

$$H_x + \dot{\lambda}^T = 0, \quad H_u = 0.$$

(A.27)

Hence, the coefficients of $d^2(\ )$ vanish as does a term in the fifth line of Eq.(A.26). Next,

$$\dot{H}_f = (H_t + H_x \dot{x} + H_u \dot{u} + H_\lambda \dot{\lambda})_f$$

$$= (H_t + H_x \dot{x} + f^T \dot{\lambda})_f$$

(A.28)

and

$$dx_f = \delta x_f + \dot{x}_f dt_f.$$

(A.29)

Then, the second differential becomes

$$d^2 J' = dt_f [G_{t_f t_f} + (H_f)_{t_f} + (H_f)_{x_f} \dot{x}_f$$

$$+ G_{t_f x_f} \dot{x}_f + \dot{x}_f^T G_{x_f t_f} + \dot{x}_f^T G_{x_f x_f} \dot{x}_f] dt_f$$

$$+ dt_f [G_{t_f x_f} + (H_f)_{x_f} + \dot{x}_f^T G_{x_f x_f}] \delta x_f$$

$$+ \delta x_f^T [G_{x_f t_f} + (H_f)_{x_f} + G_{x_f x_f} \dot{x}_f] dt_f$$

$$+ \delta x_f^T G_{x_f x_f} \delta x_f$$

$$+ \int_{t_0}^{t_f} [\delta x^T \delta u^T] \begin{bmatrix} H_{xx} & H_{xu} \\ H_{ux} & H_{uu} \end{bmatrix} \begin{bmatrix} \delta x \\ \delta u \end{bmatrix} dt$$

(A.30)

since $(H_t)_f = (H_f)_{t_f}$ and $(H_x)_f = (H_f)_{x_f}$.

For the function

$$\Omega(t_f, x_f, u_f, \nu) = G_{t_f} + L_f + G_{x_f} f_f,$$

(A.31)

it is seen that

$$
\begin{aligned}
\Omega_{t_f} &= G_{t_f t_f} + (L_f)_{t_f} + G_{x_f} f_{f t_f} + f_f^T G_{x_f t_f} \\
&= G_{t_f t_f} + (H_f)_{t_f} + f_f^T G_{x_f t_f}
\end{aligned}
\tag{A.32}
$$

and that

$$
\begin{aligned}
\Omega_{x_f} &= G_{t_f x_f} + (L_f)_{x_f} + G_{x_f} (f_f)_{x_f} + f_f^T G_{x_f x_f} \\
&= G_{t_f x_f} + (H_f)_{x_f} + f_f^T G_{x_f x_f}.
\end{aligned}
\tag{A.33}
$$

If these expressions are substituted into the definition

$$
\Omega' \triangleq \Omega_{t_f} + \Omega_{x_f} \dot{x}_f,
\tag{A.34}
$$

the following result is obtained:

$$
\begin{aligned}
\Omega' &= G_{t_f t_f} + (H_f)_{t_f} + f_f^T G_{x_f t_f} \\
&\quad + G_{t_f x_f} f_f + (H_f)_{x_f} f_f + f_f^T G_{x_f x_f} f_f.
\end{aligned}
\tag{A.35}
$$

As a consequence, the second differential (A.30) can be rewritten as

$$
\begin{aligned}
d^2 J' &= [\delta x_f^T \; dt_f]
\begin{bmatrix} G_{x_f x_f} & \Omega_{x_f}^T \\ \Omega_{\dot{x}_f} & \Omega' \end{bmatrix}
\begin{bmatrix} \delta x_f \\ dt_f \end{bmatrix} \\
&\quad + \int_{t_0}^{t_f} [\delta x^T \; \delta u^T]
\begin{bmatrix} H_{xx} & H_{xu} \\ H_{ux} & H_{uu} \end{bmatrix}
\begin{bmatrix} \delta x \\ \delta u \end{bmatrix} dt.
\end{aligned}
\tag{A.36}
$$

Eqs.(A.25) and (A.36) are the same first and second differentials derived in Chapter 13 for the free final time problem.

Note that the second differential is relative to the minimal path and not an arbitrary nominal path. For the latter, additional terms would be present. These include the first two lines of Eq.(A.26) and the terms where the expression (A.27) for $\lambda_f$ has been used.

# References

AM Ascher, U.M., Mattheij, R.M.M., and Russell, R.D., *Numerical Solution of Boundary Value Problems for Ordinary Differential Problems*, Prentice- Hall, Englewood Cliffs, NJ, 1988.

Be Betts, J.T., Using sparse nonlinear programming to compute low thrust orbit transfers, *Astronaut. Sci.*, Vol. 41, No. 3, July–September, pp. 349–371, 1993.

BH Breakwell, J.V. and Ho, Y.C., On the conjugate point condition for the control problem, *Int. J. Engng. Sci.*, Vol. 2, pp. 565–579, 1965.

BJ Bell, D.J. and Jacobson, D.H., *Singular Optimal Control Problems*, Academic, New York, 1975.

Bl Bliss, G.A., *Lectures on the Calculus of Variations*, University of Chicago Press, Chicago, 1946.

Bo Bolza, O., *Lectures on the Calculus of Variations*, Chelsea, New York, 1904.

Br Bryson, A.E., *Dynamic Optimization*, Addison-Wesley Longman, Menlo Park, 1999.

BrH Bryson, A.E. and Ho, Y.C., *Applied Optimal Control.* Blaisdell, Waltham, 1969, Hemisphere, Washington, 1975.

Bu Butcher, J.C., Implicit Runge-Kutta processes, *Math. Comput.*, Vol. 18, No. 85, January, pp. 50–64, 1964.

Ch Chen, C.-T., *Linear System Theory and Design*, Holt, Rinehart, and Winston, New York, 1984.

CA Clements, D.J., and Anderson, B.D.O., *Singular Optimal Control: The Linear-Quadratic Problem*, Springer-Verlag, New York, 1978.

CW Clohessey, W.H. and Wiltshire, R.S., Terminal guidance systems for satellite rendezvous, *Aerospace Sci.*, September, pp. 653–658 and 674, 1960.

Co Courant, R., *Differential and Integral Calculus*, Interscience, New York, 1957.

Dr  Dreyfus, S.E., *Dynamic Programming and the Calculus of Variations*, Academic, New York, 1966.

EC  Enright, P.J. and Conway, B.A., Discrete approximations to optimal trajectories using direct transcription and nonlinear programming", *Guidance, Control, Dynamics*, Vol. 15, No. 4, July–August, pp. 994-1001, 1992.

Ga  Gantmacher, F.R., *Matrix Theory, Vol. I*, Chelsea, New York, 1959.

GF  Gelfand, I.M. and Fomin, S.V., *Calculus of Variations*, Prentice-Hall, Englewood Cliffs, NJ, 1963.

Gr  Greenberg, M.D., *Foundations of Applied Mathematics*, Prentice-Hall, Englewood Cliffs, NJ, 1978.

H1  Hull, D.G., Variations, potentials, and numerical integration, AAS/AIAA Space Flight Mechanics Conference, San Antonio, TX, January, 2002.

H2  Hull, D.G., Conversion of optimal control problems into parameter optimization problems, *Guidance, Control, Dynamics*, Vol. 20, No. 1, January–February, 1997.

Ha  Hancock, H., *Theory of Maxima and Minima*, Dover, New York, 1960, also Ginn, Boston, 1917.

He  Hestenes, M.R., *Calculus of Variations and Optimal Control Theory*, Wiley, New York, 1966.

Hem  Hempel, P., Application of implicit Runge-Kutta methods, Paper No. 76-818, AIAA/AAS Astrodynamics Conference, San Diego, August, 1976.

Hi  Hildebrand, F.B., *Methods of Applied Mathematics*, Prentice-Hall, Englewood Cliffs, NJ, 1952.

Ho  Hoffman, J.D., *Numerical Methods for Engineers and Scientists*, McGraw-Hill, New York, 1992.

HP  Hargraves, C.R. and Paris, S.W., Direct trajectory optimization using nonlinear programming and collocation, *Guidance, Control, Dynamics*, Vol. 14, No. 2, March–April, pp. 431–439, 1991.

HS Hull, D.G. and Speyer, J.L., Optimal reentry and plane-change trajectories," *Astronaut. Sci.*, Vol. XXX, No. 2, April–June, pp. 117–130, 1982.

IS Ilgen, M.R., Speyer, J.L., and Leondes, C.T., Robust approximate optimal guidance strategies for aeroassisted plane change missions: A game theoretica pproach, *Control and Dynamic Systems, Vol. 52*, C.T. Leondes (Ed.), Academic, San Diego, 1992.

JL Jacobson, D.H., Lele, M.M., and Speyer, J.L., New necessary conditions of optimality for control problems with state-variable inequality constraints, J. Math. Anal. App., Vol. 35, No. 2, pp. 255–284, 1971.

Ka Kalman, R.E., The calculus of variations and optimal control theory," Chapter 16 in *Mathematical Optimization Techniques*, R. Belman (Ed.), University of California Press, Berkeley, 1963.

KK Kelly, H.J., Kopp, R.E., and Meyer, A.G., Singular extremals, Chapter 3 in *Topics in Optimization, Vol. II*, G. Leitmann (Ed.), Academic, New York, 1966.

Ki Kirk, D.E., *Optimal Control Theory, An Introduction*, Prentice-Hall, Englewood Cliffs, NJ, 1970.

Ku Kuhn, H.W., Nonlinear programming: A historical note, in *History of Mathematical Programming*, J. Lenstra, A. Rinooy Kan, and A. Schrijver (Eds.), CWI-North Holland, New York, 1991.

KT Kuhn, H. and Tucker, A.W., Nonlinear Programming, *Second Berkeley Symposium of Mathematical Statistics and Probability*, University of California Press, Berkeley, CA, 1951.

La Lanczos, C., *The Variational Principles of Mechanics*, University of Toronto Press, Canada, 1949.

LM Lee, E.B. and Marckus, L., *Foundations of Optimal Control Theory*, Krieger, Malabar, 1986.

L1 Leitmann, G. (Ed.), *Optimization Techniques*, Academic, New York, 1962.

L2 Leitmann, G., *The Calculus of Variations and Optimal Control*, Plenum, New York, 1981.

Le  Lewis, F.L., *Optimal Control*, Wiley-Interscience, New York, 1986.

Mi  Miele, A., *Theory of Optimum Aerodynamic Shapes*, Academic, New York, 1965.

PB  Pontryagin, L.S., Boltyanskii, V.G., Gamkrelidze, R.V., and Mishchenko, E.F., *The Mathematical Theory of Optimal Processes*, Interscience, New York, 1962.

Ro  Roy, A.E., *Foundations of Astrodynamics*, Macmillan, New York, 1965.

Se  Seywald, H., Trajectory optimization based on differential inclusion, *Guidance, Control, Dynamics*, Vol. 17, No. 3, May-June, pp. 480–487, 1994.

Sp  Speyer, J.L., The linear-quadratic control problem, in *Control and Dynamic Systems*, C.T. Leondes (Ed.), Vol. 23, Part 2, Academic, New York, 1986.

SB  Speyer, J.L. and Bryson, A.E., Optimal programming problems with a bounded state space," *AIAA J.*, Vol. 6, pp. 1488–1492, 1968.

Ta  Taylor, A.E., *Advanced Calculus*, Ginn, New York, 1955.

Va  Valentine, F.A., The problem of Lagrange with differential Inequalities as added side conditions, in *Contributions to the Calculus of Variations 1933-37*, University of Chicago Press, Chicago, 1937.

Wa  Walsh, G.R., *Methods of Optimization*, Wiley, New York, 1975.

WB  Wood, L.J. and Bryson, A.E., Second-order optimality conditions for variable end time terminal control problems, *AIAA J.*, Vol. 11, pp. 1241–1246, 1973.

W1  Wood, L.J. Second-order optimality conditions for the Bolza problem with both endpoints variable," *J. Aircraft*, Vol. 11, pp. 212–221, 1974.

W2  Wood, L.J., Second-order optimality conditions for the Bolza problem with path constraints, IEEE CDC Conference, San Diego, pp. 606–613, 1973.

# Index

377

# Mechanical Engineering Series

J. Angeles, **Fundamentals of Robotic Mechanical Systems: Theory, Methods, and Algorithms, 2nd ed.**

P. Basu, C. Kefa, and L. Jestin, **Boilers and Burners: Design and Theory**

J.M. Berthelot, **Composite Materials: Mechanical Behavior and Structural Analysis**

I.J. Busch-Vishniac, **Electromechanical Sensors and Actuators**

J. Chakrabarty, **Applied Plasticity**

G. Chryssolouris, **Laser Machining: Theory and Practice**

V.N. Constantinescu, **Laminar Viscous Flow**

G.A. Costello, **Theory of Wire Rope, 2nd ed.**

K. Czolczynski, **Rotordynamics of Gas-Lubricated Journal Bearing Systems**

M.S. Darlow, **Balancing of High-Speed Machinery**

J.F. Doyle, **Nonlinear Analysis of Thin-Walled Structures: Statics, Dynamics, and Stability**

J.F. Doyle, **Wave Propagation in Structures: Spectral Analysis Using Fast Discrete Fourier Transforms, 2nd ed.**

P.A. Engel, **Structural Analysis of Printed Circuit Board Systems**

A.C. Fischer-Cripps, **Introduction to Contact Mechanics**

A.C. Fischer-Cripps, **Nanoindentation**

J. García de Jalón and E. Bayo, **Kinematic and Dynamic Simulation of Multibody Systems: The Real-Time Challenge**

W.K. Gawronski, **Dynamics and Control of Structures: A Modal Approach**

K.C. Gupta, **Mechanics and Control of Robots**

D.G. Hull, **Optimal Control Theory for Applications**

J. Ida and J.P.A. Bastos, **Electromagnetics and Calculations of Fields**

M. Kaviany, **Principles of Convective Heat Transfer, 2nd ed.**

M. Kaviany, **Principles of Heat Transfer in Porous Media, 2nd ed.**

# Mechanical Engineering Series

E.N. Kuznetsov, **Underconstrained Structural Systems**

P. Ladevèze, **Nonlinear Computational Structural Mechanics:
New Approaches and Non-Incremental Methods of Calculation**

A. Lawrence, **Modern Inertial Technology: Navigation, Guidance, and
Control, 2nd ed.**

R.A. Layton, **Principles of Analytical System Dynamics**

F.F. Ling, W.M. Lai, D.A. Lucca, **Fundamentals of Surface Mechanics With
Applications, 2nd ed.**

C.V. Madhusudana, **Thermal Contact Conductance**

D.P. Miannay, **Fracture Mechanics**

D.P. Miannay, **Time-Dependent Fracture Mechanics**

D.K. Miu, **Mechatronics: Electromechanics and Contromechanics**

D. Post, B. Han, and P. Ifju, **High Sensitivity Moiré:
Experimental Analysis for Mechanics and Materials**

F.P. Rimrott, **Introductory Attitude Dynamics**

S.S. Sadhal, P.S. Ayyaswamy, and J.N. Chung, **Transport Phenomena
with Drops and Bubbles**

A.A. Shabana, **Theory of Vibration: An Introduction, 2nd ed.**

A.A. Shabana, **Theory of Vibration: Discrete and Continuous Systems,
2nd ed.**